INDUSTRIAL SAFETY

AND HEALTH MANAGEMENT

PRENTICE HALL INTERNATIONAL SERIES
IN INDUSTRIAL AND SYSTEMS ENGINEERING
W. J. Fabrycky and J. H. Mize, Editors

ALEXANDER *The Practice and Management of Industrial Ergonomics*
AMOS AND SARCHET *Management for Engineers*
ASFAHL, *Industrial Safety and Health Management*, 2/E
BANKS AND CARSON *Discrete-Event System Simulation*
BEIGHTLER, PHILLIPS, AND WILDE *Foundations of Optimization*, 2/E
BANKS AND FABRYCKY *Procurement and Inventory Systems Analysis*
BLANCHARD *Logistics Engineering and Management*, 3/E
BLANCHARD AND FABRYCKY *Systems Engineering and Analysis*, 2/E
BROWN *Systems Analysis and Design for Safety*
BUSSEY *The Economic Analysis of Industrial Projects*
CANADA/SULLIVAN, *Economic and Multi-Attribute Evaluation of Advanced Manufacturing Systems*
CHANG AND WYSK *An Introduction to Automated Process Planning Systems*
CLYMER, *Systems Analysis Using Simulation and Markov Models*
ELSAYED AND BOUCHER *Analysis and Control of Production Systems*
FABRYCKY, GHARE, AND TORGERSEN *Applied Operations Research and Management Science*
FRANCES AND WHITE *Facility Layout and Location: An Analytical Approach*
GIBSON, *Modern Management of the High-Technology Enterprise*
HAMMER *Occupational Safety Management and Engineering*, 4/E
HUTCHINSON *An Integrated Approach to Logistics Management*
IGNIZIO *Linear Programming in Single- and Multiple-Objective Systems*
KUSIAK, *Intelligent Manufacturing Systems*
MONDEL *Improving Productivity and Effectiveness*
MUNDEL *Motion and Time Study: Improving Productivity*, 6/E
OSTWALD *Cost Estimating*, 2/E
PHILLIPS AND GARCIA-DIAZ *Fundamentals of Network Analysis*
SANDQUIST *Introduction to System Science*
SMALLEY *Hospital Management Engineering*
TAHA *Simulation Modeling and SIMNET*
THUESEN AND FABRYCKY *Engineering Economy*, 7/E
TURNER, MIZE, AND CASE *Introduction to Industrial and Systems Engineering*, 2/E
WHITEHOUSE *Systems Analysis and Design Using Network Techniques*
WOLFF *Stochastic Modeling and the Theory of Queues*

second edition

INDUSTRIAL SAFETY
AND HEALTH MANAGEMENT

C. RAY ASFAHL

University of Arkansas

Prentice Hall, Englewood Cliffs, New Jersey 07632

Library of Congress Cataloging-in-Publication Data

Asfahl, C. Ray [date]
    Industrial safety and health management / C. Ray Asfahl. — 2nd
ed.
      p. cm.
    Bibliography: p. 392
    Includes index.
    ISBN 0-13-463258-3
    1. Industrial safety. 2. Industrial hygiene. I. Title.
T55.A83 1990   89–35700
613.6′2—dc20   CIP

Editorial/production supervision and
  interior design: Fred Dahl
Cover design: Wanda Lubelska
Manufacturing buyer: Denise Duggan

© 1990, 1984 by Prentice-Hall, Inc.
A Division of Simon & Schuster
Englewood Cliffs, New Jersey 07632

Printed in the United States of America
10 9 8 7 6 5 4 3 2 1

ISBN 0-13-463258-3

Prentice-Hall International (UK) Limited, *London*
Prentice-Hall of Australia Pty. Limited, *Sydney*
Prentice-Hall Canada Inc., *Toronto*
Prentice-Hall Hispanoamericana, S.A., *Mexico*
Prentice-Hall of India Private Limited, *New Delhi*
Prentice-Hall of Japan, Inc., *Tokyo*
Simon & Schuster Asia Pte. Ltd., *Singapore*
Editora Prentice-Hall do Brasil, Ltda., *Rio de Janeiro*

To Vicki and to Eloise Crowley
of Republic Safety Institute

Contents

PREFACE, xiii

1 THE SAFETY AND HEALTH MANAGER, 1

A Reasonable Objective, 3
Safety versus Health, 4
Role in the Corporate Structure, 5
Resources at Hand, 6
Summary, 10
Exercises and Study Questions, 10

2 DEVELOPMENT OF THE SAFETY AND HEALTH FUNCTION, 12

Workers' Compensation, 13
Recordkeeping, 15
Accident Cause Analysis, 27
Organization of Committees, 28
Safety and Health Economics, 28
Training, 35
Summary, 37
Exercises and Study Questions, 38

3 CONCEPTS OF HAZARD AVOIDANCE, 42

The Enforcement Approach, 43
The Psychological Approach, 45
The Engineering Approach, 46
The Analytical Approach, 52
Hazards Classification Scale, 62
Summary, 66
Exercises and Study Questions, 66

4 IMPACT OF FEDERAL REGULATION, 69

Standards, 69
NIOSH, 74
Enforcement, 74
Public Uproar, 79
Role of the States, 80
Future Trends, 81
Exercises and Study Questions, 83

5 INFORMATION SYSTEMS, 85

Hazard Communication, 86
Environmental Protection Agency, 91
Computer Information Systems, 96
Summary, 98
Exercises and Study Questions, 99

6 BUILDINGS AND FACILITIES, 100

Walking and Working Surfaces, 101
Exits, 112
Illumination, 113
Miscellaneous Facilities, 113
Sanitation, 116
Summary, 117
Exercises and Study Questions, 117

7 HEALTH AND ENVIRONMENTAL CONTROL, 119

Baseline Examinations, 120
Toxic Substances, 120
Measures of Exposure, 129

Standards Completion Project, 134
Detecting Contaminants, 136
Ventilation, 141
Industrial Noise, 151
Radiation, 167
Computer Terminals, 168
Summary, 168
Exercises and Study Questions, 169

8   HAZARDOUS MATERIAL, 175

Flammable Liquids, 175
Sources of Ignition, 180
Spray Finishing, 184
Dip Tanks, 186
Explosives, 188
Liquefied Petroleum Gas, 188
Conclusion, 190
Exercises and Study Questions, 190

9   PERSONAL PROTECTION AND FIRST AID, 192

Hearing Protection, 193
Eye and Face Protection, 195
Respiratory Protection, 196
Head Protection, 207
Miscellaneous Personal Protective Equipment, 208
First Aid, 210
Conclusion, 211
Exercises and Study Questions, 212

10   FIRE PROTECTION, 213

Industrial Fires, 214
Fire Prevention, 215
Emergency Evacuation, 215
Fire Brigades, 216
Fire Extinguishers, 217
Standpipe and Hose Systems, 219
Automatic Sprinkler Systems, 221
Fixed Extinguishing Systems, 221
Summary, 223
Exercises and Study Questions, 223

11 MATERIALS HANDLING AND STORAGE, 225

Materials Storage, 226
Industrial Trucks, 227
Cranes, 233
Slings, 248
Conveyors, 252
Lifting, 254
Summary, 255
Exercises and Study Questions, 256

12 MACHINE GUARDING, 259

General Machine Guarding, 259
Safeguarding the Point of Operation, 267
Power Presses, 276
Grinding Machines, 298
Saws, 301
Belts and Pulleys, 306
Summary, 310
Exercises and Study Questions, 311

13 WELDING, 313

Process Terminology, 313
Gas Welding Hazards, 317
Arc Welding Hazards, 324
Resistance Welding Hazards, 326
Fires and Explosions, 327
Eye Protection, 328
Protective Clothing, 328
Gases and Fumes, 329
Summary, 332
Exercises and Study Questions, 333

14 ELECTRICAL HAZARDS, 335

Electrocution Hazards, 335
Fire Hazards, 348
Test Equipment, 354
Frequent Violations, 357
Summary, 358
Exercises and Study Questions, 359

15 CONSTRUCTION, 362

General Facilities, 362
Personal Protective Equipment, 363
Fire Protection, 366
Tools, 366
Electrical, 368
Ladders and Scaffolds, 370
Floors and Stairways, 373
Cranes and Hoists, 373
Heavy Vehicles and Equipment, 378
Trenching and Excavations, 381
Concrete Work, 385
Steel Erection, 386
Demolition, 387
Explosive Blasting, 388
Electric Utilities, 389
Summary, 390
Exercises and Study Questions, 391

BIBLIOGRAPHY, 392

APPENDICES

A OSHA Permissible Exposure Limits, 397

B Medical Treatment, 433

C First Aid Treatment, 434

D Classification of Medical Treatment, 435

E EPA List of Extremely Hazardous Substances, 437

F Standard Industrial Classification (SIC) Code, 446

G States Having Federally Approved State Plans, 447

GLOSSARY, 448

INDEX, 452

Preface

Significant changes have taken place recently to reshape the role of the Safety and Health Manager for the 1990s. Most visible among these changes is hazard communication and the increased emphasis on safety and health information systems. To an increasing extent, the computer has become a tool of the Safety and Health Manager.

Chapter 5, a new chapter in this text, is devoted to hazard communication and computer information systems. In addition, every chapter of this book has been revised in some way. Case studies have been added, and additional exercises and study questions have been supplied to facilitate use of this text as a teaching tool. Many of the exercises are designed to assist the reader who is preparing to take the Certified Safety Profession (CSP) examination.

To those doubters who said, "The end of OSHA is near!" when Ronald Reagan was elected President of the United States in 1980, be informed that in 1990, a decade later, OSHA is alive and well. And it is writing citations for record-high penalties for those firms that fail to take a proactive stance in combatting "egregious violations" (OSHA's new label for flagrant violations) of serious hazards and who fail to abate conditions found in repeated inspections. A more sophisticated and mature OSHA now recognizes that complex safety and health problems, such as carpal tunnel syndrome, cannot be solved by a straightforward specification standard. Sometimes a redesign of the workplace is necessary, using analyses applying the principles of ergonomics. Performance standards have become more important as OSHA turns to educated professionals in the field both to identify firms that fail to comply as well as to assist and bring recognition to those firms whose managements are attempting to solve complex problems. The Safety and Health Manager is likewise advancing professionally and in technical skills. Examination and certification as either a Safety Professional (CSP) or Industrial Hygienist (CIH) is becoming more important to the Safety and Health Manager in the 1990s.

Somewhat a surprise to the profession was OSHA's dramatic change in strategy to revise the entire list of Permissible Exposure Limits (PELs) for hazardous concentrations of airborne contaminants, proposed in 1988 and made final early in 1989. The revision was the first general revision of the PELs since publication of the first enforceable table in the early 1970s and was heralded as a much-needed modernization of the regulations to reflect current Threshold Limit Values (TLVs) established by the American Conference of Governmental Industrial Hygienists' (ACGIH) Committee on Industrial Ventilation and research findings by NIOSH. Prior to this general revision, OSHA had made painstaking attempts in the "standards completion project" to write standards one-at-a-time for prominent health hazards such as asbestos, vinyl chloride, lead, and benzene. It became obvious in the 1980s that the "standards completion project" would never be completed, so OSHA decided to change PELs on an across-the-board basis. This revised edition includes the recent PEL revision in the Appendix, and case studies, exercises, and study questions have been revised to include the latest available PEL data where appropriate.

I would like to acknowledge the contributions of the following OSHA personnel to my knowledge: Gilbert J. Saulter, Administrator, Region VI, Glen R. Williamson, former Area Director and current Deputy Regional Administrator, and Compliance Safety and Health Officers Jim Rogers, Dick Zier, Howard Watkins, Jerry Loux, Linda Holt Sullivan, Bill McConnell, Jim Males, Carroll Jones, and Mike Talmont.

The professional consultants of the Arkansas Department of Labor have been very helpful, especially Jim Bailey, Ron Burnett, Richard Chitwood, Claud Edwards, Clark Thomas, Jerry Hendrix, Deborah Moore, Mary Frances Branton, George McAllister, Bud Daven, Ed Baker, Robert Childers, Bernie Fullaway, John Jarratt, Glenn Purifoy, Clif Shepard, Norman Brooks, Gary Parish, Tom Rimmer, Reiman Diles, and Richard Steward.

Colleagues at the University of Arkansas who have contributed are Sandra Parker, Susan Johnsen, David Boyster, Darlene Butler, Jerri Wilkerson, Jim Moorhead, Bob Belcher, and Nancy Sloan.

Richard Perry of Aetna Life and Casualty and Ron Cross of McClinton-Anchor Company have added industrial insights, as have Thomas W. Lawrence of the Monsanto Company, David Nigus of Wausau Insurance Companies, Kim Anderson of A.O. Smith, Inc., and Rebecca Anderson of Marquette University.

Special thanks go to Dr. Jim Orr who reviewed the manuscript and provided valuable suggestions.

RAY ASFAHL

INDUSTRIAL SAFETY

AND HEALTH MANAGEMENT

1

The Safety and Health Manager

Everyone wants a safe and healthful workplace, but what each person is willing to do to achieve this worthwhile objective can vary a great deal. As a result, the management of each firm must decide at what level, along a broad spectrum, the safety and health effort will be aimed. Some managers deny this responsibility and attempt to leave the decision to employees. This strategy seems to square with hallowed principles of personal freedom and individual responsibility. But such a denial of responsibility by management results in a decision by default, and usually the result is a relatively low level of safety and health in the workplace.

Is the foregoing an indictment on the judgment of the individual worker? Not really, because without a commitment on the part of management, the worker usually is unable singlehandedly to build safety into his or her job station. It has been proven that the attitude of the worker is the most important determinant for his or her safety, but attitude alone cannot make a dangerous job safe. Furthermore, even if a given worker has a strong inclination to be careful and to guard his or her health, there are plenty of production incentives and other quite natural incentives to undermine safe attitudes when management has not made a commitment to safety and health.

One person, usually designated as "Safety Director" or "Industrial Hygienist," sets the tone of the safety and health program within a firm. In fact, right at the start it says something about the commitment of management when a firm decides to designate a person by title to the responsibility of safety and health. But naming someone "Safety Director" or "Manager of Safety and Health" is just a beginning step. Many such persons have little authority and have been largely ignored by management and worker alike, especially in the past. It was not unusual for a Safety Director's work to be typified by public relations activities, such as posting motivational signs and compiling statistics. These are still important functions, but much more responsibility for this function is now recognized.

Something happened in the 1970s to change dramatically the role of the typical Safety Director in industrial firms throughout the country. The passage of the Occupational Safety and Health Act of 1970 created the Occupational Safety and Health Administration (OSHA), a federal agency whose regulations would have a large impact on the role of the typical Safety Director. Chapter 4 discusses this impact in detail, but the balance of this chapter discusses the enlarged role of the person charged with industrial safety and health.

There is little doubt that OSHA enhanced the authority of the Safety Manager in the typical industrial plant in the United States. Prior to OSHA, few Safety Managers dared to interfere with production schedules to alleviate a safety or health problem. But prominent OSHA cases in the news media have brought to the attention of top management personnel the dire consequences that can ensue when serious safety or health problems are left untended.

The field of occupational health has probably benefitted even more from OSHA than has the field of occupational safety. Prior to OSHA, occupational health seemed to be a matter too remote to really concern anyone, except perhaps the plant nurse. And the plant nurse had little authority to influence policy or even to take action to prevent hazards. Prior to OSHA, the plant nurse was chiefly concerned with first aid, after the fact, and physical examinations, not with hazard abatement and prevention.

In describing the functions of today's executive charged with the safety and health reponsibility, this text will use the designation *Safety and Health Manager,* recognizing the dual nature of the job. Also, the term *manager* envisions the enlarged scope of responsibility, which includes analysis of hazards, compliance with standards, and capital investment planning, in addition to the conventional functions described earlier. The purpose of this text is to provide tools and guidelines to Safety and Health Managers to help them execute their enlarged duties.

Dealing with applicable standards is one of the greatest challenges facing today's Safety and Health Manager, and to meet this challenge is a primary purpose of this book. Since only 10% of the standards generate 90% of the activity, Safety and Health Managers need guides to the important parts of the standards. Frequently cited standards should receive prime attention because they indicate areas in which industries are having difficulty complying or areas in which enforcement agencies are giving a great deal of attention. In either event, Safety and Health Managers have a need to know these frequently cited standards so that they can bring their facilities within compliance. Besides the frequency of citation, Safety and Health Managers need to know the "why" behind the standards. Until the Safety and Health Manager learns what hazards a particular standard is intended to prevent, he or she will have a difficult time persuading either management or employees that a given situation needs correction.

A REASONABLE OBJECTIVE

Top management sometimes turns a deaf ear to the pleas of the Safety and Health Manager for plant improvements. But the Safety and Health Manager is sometimes a crusader with a one-track mind. Any Safety and Health Manager who feels that elimination of workplace hazards is an indisputable goal is naive. In the real world we must choose between:

1. Hazards that are physically infeasible to correct
2. Hazards that are physically feasible, but are economically infeasible, to correct
3. Hazards that are economically feasible to correct

Until the Safety and Health Manager comes to grips with this reality, he or she cannot expect to enjoy the approval of top management. Some Safety and Health Managers have faced this reality on the surface, but within their hearts they resent the attitudes of top executives who are unwilling to support their efforts to eliminate all workplace hazards. But this resentment is unjustified because it is an unrealistic and naive strategy to attempt to eliminate all hazards.

It may be surprising to some readers to discover that this book, which is supposed to be about safety and health, does not really advocate elimination of all workplace hazards. Such a goal is unattainable, and to reach for it is poor strategy because it ignores the need for discriminating among hazards to be corrected. To see how such a naive strategy is not even in the interest of safety or health, consider the following case study.

Case Study 1.1   A Safety and Health Manager receives three suggestions from three different operating personnel as follows:

1. Install a drain to remove water that occasionally collects around the die-casting area.
2. Post a warning sign that advises forklift truck drivers to slow down.
3. Improve sanitation by cleaning the rest rooms more frequently.

There is a safety and/or health rationale for the correction of all three of these problems. Should they be corrected?

Some managers would accept the safety and health rationale as *all they need* to begin action to correct the problems listed in Case Study 1.1. But this would be a naive response. More data are needed to decide what to do. While busying the plant maintenance department to correct the foregoing three problems, which may or may not be consequential, a serious electrocution or respiratory hazard may be going unchecked or maybe even unnoticed. By reacting to every hazard that happens to show up, the Safety and Health Manager may be missing opportunities to have a really significant impact on worker safety and health. At the same time such overreaction may also be deteriorating the Safety and Health Manager's credibility with top management. Even the law does not call for the elimination of all hazards, just the "recognized" ones. Therefore, let it be clearly understood that our objective is to eliminate unreasonable risks, not all risks, in the workplace. The goal of this book, then, is to assist the Safety and Health Manager in detecting hazards and in deciding which ones are worth correcting. The goal is an ambitious one, and this book certainly claims no breakthroughs for solving this difficult problem. But any light that can be shed on the mystery of which hazards are more significant and which standards, among the thousands, are most important is sorely needed by Safety and Health Managers everywhere.

SAFETY VERSUS HEALTH

This chapter has already implied that early "safety directors" did not emphasize health problems. It is essential that today's Safety and Health Manager give sufficient attention not only to safety hazards but also to health hazards, which are steadily gaining in importance as new data about industrial disease are being uncovered.

What really is the difference between safety and health? The words are so common that almost everyone has a firm image of the concept of safety versus the concept of health. There is no question that machine guarding is a safety consideration, and airborne asbestos is a health hazard, but some hazards, such as those associated with paint spray areas and welding operations, are not so easy to classify. Some situations may be both a health and a safety hazard. This text will draw the following line between safety and health:

Safety deals with acute hazards, whereas health deals with chronic hazards.

An acute effect is a sudden reaction to a severe condition; a chronic effect is a long-term deterioration due to a prolonged exposure to a milder adverse condition. Everyday concepts of health and safety fit this definition, which separates the two. Industrial noise, for instance, is usually a health hazard because it is usually the long-term exposure to noise levels in the range 90 to 100 decibels that does the permanent damage. But noise can also be a safety hazard because a sudden acute exposure to impact noise can *injure* the hearing system. Many chemical exposures have both an acute and a chronic effect and are thus both safety and health hazards.

Industrial hygienists, those who concentrate on health hazards, are known by their sophisticated instruments and scientific expertise. These tools are necessary to the industrial hygienist because of the tiny effects they must measure in order to determine whether a chronic hazard exists. By contrast, the safety specialist, instead of being an expert with precise scientific instruments, usually has more industrial process experience and practical on-the-job knowledge. This difference in backgrounds generates some confrontation between safety and health professionals, and though they should be partners, they often compete.

The bases of competition between safety and health professionals are classics: young versus old, new versus old, and education versus experience. Safety professionals are usually older and have more industrial experience; their career field is more traditional and more entrenched in industrial organizations. Health professionals are typically younger, have more college education, and occupy newer job positions.

Degree of hazard is another point of contention between safety and health: both sides think their hazards are more grave. Safety professionals can point to fatalities on the job and feel an urgency in protecting the worker from imminent danger from accidents. To them the industrial hygienists with their meters, pumps, and test tubes are checking for microscopic hazards which are not really pressing. But to the industrial hygienist, the safety professional's work sometimes seems superficial. The industrial hygienist is fighting those insidious unseen hazards on the job which can be just as lethal as a falling crane. There are probably more occupational health fatalities than safety fatalities, but the statistics will not reflect this difference because the health fatalities are delayed and are often never diagnosed.

Because of age and experience, a safety professional will often have managerial control over health functions, and this can rankle the industrial hygienist. The Safety and Health Manager should be aware of this sensitivity and be careful to recognize the importance of both fields of interest. A growing number of well-educated industrial hygienists are being given overall responsibility for both safety and health.

ROLE IN THE CORPORATE STRUCTURE

Most Safety and Health Managers wear several hats, especially in smaller firms. Often they are responsible also for security, although the security function has little in common with workplace safety and health. Some Safety and Health Managers are also personnel managers, and even more frequently they report to the personnel manager. This is a fairly natural arrangement in that it emphasizes the importance of worker training, statistics, job placement, and the industrial relations aspect of safety and health. The growing importance of engineering to workplace safety and health, however, strains the placement of the Safety and Health Manager within the personnel department, which traditionally has little interaction with engineering.

The Safety and Health Manager is virtually never associated with the purchasing function, but one of the first goals of the Safety and Health Manager should be to obtain some input to the purchasing process. Used-equipment dealers, even new-equipment dealers, often have bargain-priced machines, compressors, tractors, forklifts, and other pieces of equipment that fail in some way or another to meet standards. The purchasing agent is usually not knowledgeable in safety and health standards and is easy prey for these dealers because the price is right. What is needed is a knowledgeable person to check specifications and prevent the costly purchasing error of buying equipment that does not meet current safety and health standards.

When standards change, another category of equipment sometimes becomes obsolete, and the Safety and Health Manager should warn the purchasing department when these changes occur.

A recent concept of the Safety and Health Manager is as a liaison with government agencies, a condition brought about by the arrival of OSHA. Some Safety and Health Managers have a dual responsibility for environmental protection activities. Sometimes the Safety and Health Manager is considered a staff member of the legal department. This arrangement stresses the adversarial role and is not recommended because it tends to detract both from workplace safety and health and from constructive relations with enforcement agencies.

A related field is consumer product safety. The Consumer Product Safety Commission (CPSC) is a federal agency whose organization is obviously patterned after OSHA's. The Product Safety and Liability Act was passed the year after the passage of the OSHA Act, and the wording of the two laws is remarkably similar. Although both fields consider the safety of machines and equipment, the CPSC concentrates on the responsibility of the manufacturers of the machines and equipment, whereas OSHA concentrates on the responsibility of the employer who places the equipment into use in the workplace.

The decade of the 1980s saw renewed interest in environmental protection and especially in the prevention and emergency clean-up of disastrous accidental releases of toxic substances. In 1984, in a shocking disaster in Bhopal, India, at least 2,500 civilians were killed in a single industrial accidental release of deadly methyl isocyanate gas, and without doubt this incident had its impact upon public policy in the United States. Because of the close relationship to worker safety and health, the responsibility for compliance with Environmental Protection Agency (EPA) requirements is often made a part of the duties of the Safety and Health Manager. Chapter 5 of this book reveals more of this relationship between safety inside and safety outside the plant.

RESOURCES AT HAND

The demands for training materials, the need for ideas for hazard correction, and the importance of the latest interpretations of standards guarantee that the successful Safety and Health Manager will not try to work alone, insulated from the career field. A variety of resources have arisen to meet these needs of the Safety and Health Professional.

Professional Certification

Safety and Health Managers can establish themselves with their peers as well as with their employers by achieving professional certification. This requires relevant work experience, letters of recommendation, and an examination. Education is also required, but there is partial trade-off between education and experience. Safety professionals should apply to

Board of Certified Safety Professionals of America
208 Burwash
Savoy, IL 61874

Health professionals should apply to

American Board of Industrial Hygiene
4600 W. Saginaw, Suite 101
Lansing, MI 48917

The examinations are quite difficult, and the background screening process is stringent. Few individuals qualify for certification in both fields.

Professional Societies

Two professional societies are foremost in the career field of occupational safety and health:

American Society of Safety Engineers (ASSE)
1800 E. Oakton St.
Des Plaines, IL 60016

American Industrial Hygiene Association (AIHA)
475 Wolf Ledges Pkwy.
Akron, OH 44311

A somewhat narrower but influential organization is

American Conference of Government Industrial Hygienists (ACGIH)
Bldg. D-7, 6500 Glenway Ave.
Cincinnati, OH 45211

An important committee of this organization is

Committee on Industrial Ventilation
PO Box 16153
Lansing, MI 48901

This committee's publication *Industrial Ventilation* (ref. 45) is probably the best recognized manual of recommended practice in the field of ventilation.

National Safety Council

Although not a professional society in the strict sense of the term, the National Safety Council (NSC) is the most important organization to the field. The Council is headquartered at

National Safety Council
444 North Michigan Avenue
Chicago, IL 60611

The NSC is broad in scope and encompasses all kinds of safety, not just occupational safety. The membership of the NSC consists principally of organizations and businesses. It has been in existence for over 70 years.

Corporate membership in the NSC affords many benefits for the Safety and Health Manager. The Council is the principal focal point for information about safety hazards. The library at the Council's national headquarters contains a wealth of information, and although it is open to the public, companies who are members of the Council have special privileges for copying services and research. Each year the Council publishes comprehensive summaries of accident statistics in its book *Accident Facts* (ref. 1). The NSC's *Accident Prevention Manual for Industrial Operations* (ref. 52), now in its eighth edition and in two volumes, is the most authoritative and comprehensive reference in the field.

The NSC makes use of its current statistical data to continually recognize and award member firms with excellent-safety awards. The annual edition of *Work Injury and Illness Rates* (ref. 92) lists company names by Standard Industrial Classification (SIC) Code that have the "best records known in industry."

Standards Institutes

The age of OSHA enforcement has brought increased recognition of the national standards-producing organizations. The following are the most prominent among these:

American National Standards Institute (ANSI)
1430 Broadway
New York, NY 10018

National Fire Protection Association (NFPA)
Batterymarch Park
Quincy, MA 02269

American Society of Mechanical Engineers (ASME)
345 East 47th Street
New York, NY 10017

American Society for Testing and Materials (ASTM)
1916 Race Street
Philadelphia, PA 19103

These organizations have standing committees that invite public comment and prepare voluntary standards for occupational safety and health. At the outset, OSHA issued many existing standards produced by these organizations, designating such existing standards as representing the "national consensus." Most of OSHA's national consensus standards are derived from either ANSI or NFPA standards.

Trade Associations

If a problem pertains to a specific industry or type of equipment, an association of manufacturers may be called upon to furnish safety and health data. Some people decry such use of trade association data as biased, but many careful studies of safety

Amer Councal Govern I H
ACGIH

and health problems have been performed by trade associations. The following are useful associations in this regard:

1. American Foundrymen's Society (AFS)
2. American Iron and Steel Institute (AISI)
3. American Metal Stamping Association (AMSA)
4. American Petroleum Institute (API)
5. American Welding Society (AWS)
6. Associated General Contractors of America (AGCA)
7. Compressed Gas Association (CGA)
8. Industrial Safety Equipment Association (ISEA)
9. Institute of Makers of Explosives (IME)
10. National Electrical Manufacturers Association (NEMA)
11. National LP-Gas Association (NLPGA)
12. National Machine Tool Builders Association (NMTBA)
13. Scaffolding, Shoring, and Forming Institute (SSFI)

National trade associations are especially useful in providing audiovisual training materials if one is willing to accept the fact that such materials also serve to promote the industry's products.

Government Agencies

A program of free consultation is usually provided by state agencies and in some states by private consulting firms. There is an understandable reluctance to call a government agency for help with a safety standards problem, but the purpose of these consultation agencies is to assist, not to write citations. In most states the consultation function is performed by a completely separate agency from the enforcement agency. In states that have comprehensive state plans, with one agency responsible for both enforcement and consultation, care is taken to maintain confidentiality of records.

The National Institue for Occupational Safety and Health (NIOSH) has a wealth of research data on the hazards of specific materials and processes. NIOSH uses these data to write criteria for recommended new standards. In addition to its research function, NIOSH acts as a source of technical information for questions about safety and health. Vivian Morgan, acting chief of the technical information branch of the agency, was quoted in 1987 as saying, "We get 37,000 letters from the public each year and 4,000 additional ones from the private sector and state and federal government agencies requesting technical information. We receive about 40 phone calls per day; 10 to 15 of these are for technical information." NIOSH has a hot line for this purpose: 1-800-35-NIOSH.

OSHA itself can be of value to the Safety and Health Manager who seeks information. Some Safety and Health Managers would never consider calling OSHA to discuss a problem for fear of precipitating an inspection. However, problems can be posed in a hypothetical setting, and OSHA personnel can understand the need for keeping such questions hypothetical. Most OSHA personnel will be glad to furnish whatever answers to questions are available in the interest of encouraging employers to keep their facilities safe and healthful. OSHA has also opened the doors to its national training institute in Des Plaines, Illinois, for training the general public in "voluntary compliance."

SUMMARY

Safety and Health Manager is a career field that has been both enhanced and made more challenging by the institution of the federal OSHA agency. This chapter has provided a brief introduction to the career field and to some of the organizations and boards that have given the field identity and assistance in fulfilling its mission.

Chapter 2 proceeds to describe how Safety and Health Managers can go about performing their responsibilities within their organizations. Chapter 3 addresses the principal goal of the Safety and Health Manager—the reduction of workplace hazards—describing four basic approaches to this problem. OSHA has had such an impact on the field that it deserves a chapter of its own; Chapter 4 describes OSHA in general, including both the positive and the negative aspects of this controversial agency. The remaining chapters address specific hazard categories, advising Safety and Health Managers on what to do to eliminate hazards while complying with established standards. Any factual data or approach strategies that can assist the Manager in understanding the hazard mechanisms and in tackling the most significant problems first will be helpful, even if many questions are left unanswered.

EXERCISES AND STUDY QUESTIONS

1.1. Why are some safety and health standards cited more frequently than others?

1.2. What else, besides frequency of citation, does a Safety and Health Manager need to know about safety and health standards?

1.3. Should Safety and Health Managers attempt to eliminate all workplace hazards? Why or why not?

1.4. Identify three categories of hazards with respect to feasibility of correction.

1.5. Describe at least two disadvantages of overreacting to minor hazards in the workplace.

1.6. How does a safety hazard differ from a health hazard?

1.7. Name three safety hazards and three health hazards.

1.8. Name some chemicals that are both health and safety hazards.

1.9. Name some physical agents that can be both health and safety hazards.

1.10. Why does an industrial hygienist need more scientific instruments to evaluate hazards than does a safety specialist?

1.11. Which type of hazard appears to be more grave: safety or health?

1.12. What aspects of the Safety and Health Manager's job are related to the personnel department?

1.13. What disadvantage is associated with placing the Safety and Health Manager within the personnel department?

1.14. What disadvantage can be seen in placing the Safety and Health Manager within the legal department of a firm?

1.15. Compare the missions of the two federal agencies, OSHA and CPSC.

1.16. What national organization is of the greatest importance to the Safety and Health Manager?

1.17. What is ANSI, and what relationship does it have to the field of safety and health?

1.18. Compare the missions of OSHA and EPA. Why might the same individual within a given industrial plant have responsibility for dealing with both agencies?

2

Development of the Safety and Health Function

The safety and health function has both line and staff characteristics, and the Safety and Health Manager needs to recognize which elements of the function belong to which. The physical accomplishment of workplace safety and health is a line function. For example, operator work practices are the responsibility of the workers themselves as directed by their line supervisor. In industries where maintenance departments are recognized as another line function, the correction of facilities problems is again the direct responsibility of the maintenance operators and their line supervisors. The Safety and Health Manager then performs a staff function by acting as a "facilitator" in assisting, motivating, and advising the line function in achieving worker safety and health.

The interest of line personnel in receiving this advice and assistance from the Safety and Health Manager is going to depend on how important the goal of safety and health is to top management. The successful Safety and Health Manager will be keenly aware of this need for top management support. Respect and approval of top management must be won by responsible decisions and actions by the Safety and Health Manager. A necessary ingredient of these decisions and actions is a recognition of the important principle stated in Chapter 1, that the goal is to eliminate unreasonable hazards, not all hazards. The respect and approval of top management is difficult to win because Safety and Health Managers are too often such emotional crusaders for the cause that they lose their credibility and with it lose their eligibility to be considered "managers." On the other hand, federal regulation has added a measure of urgency and credibility to efforts to make industries safe and healthful, and this has substantially strengthened the position of the Safety and Health Manager in the management hierarchy.

12

Similarities can be drawn between the safety function and other staff functions, such as quality control and production control. Like safety and health, quality and production goals must be achieved by line personnel, facilitated by the staff function. This principle is recognized in such clichés as "You can't inspect quality into a product" and "Safety is everybody's business."

Once the approval of top management has been won, the Safety and Health Manager is advised to document this approval in a written safety and health policy statement, issued by top management. This written policy becomes the documented authority to line personnel that top management does have safety and health goals and wants these goals met. The everyday actions of management then reinforce the written policy, as discussed in Chapter 1. If, however, top management fails to practice what it preaches in the policy statement, it is the duty of the Safety and Health Manager to go back to management and redetermine its level of commitment to safety and health.

Having established by word and deed management's commitment to safety and health, the Safety and Health Manager is ready to proceed with the staff function of facilitating the safety and health program throughout the plant. To do this, plant operating personnel will have needs that the Safety and Health Manager can satisfy. To make workers aware of hazards, supervisors and the workers themselves need regular training in hazard recognition and correction. Statistics and accident records are needed to keep management and operating personnel advised of how well the company and its departments are doing in achieving their safety and health goals. Sometimes the Safety and Health Manager can provide a plantwide stimulus for worker safety and health by means of contests and awards for safety performance. Finally, of increasing importance is the Safety and Health Manager's role in dealing with safety and health standards and in assisting operating personnel in achieving compliance with these standards. The remainder of this chapter itemizes staff functions of the Safety and Health Manager's office with guidance for the development of each.

WORKERS' COMPENSATION

"Workers' Compensation" laws, formerly known as "Workmen's Compensation" laws, provided an initial structure to industrial safety. The first such laws were introduced in state legislatures in 1909, and now all states have comparable legislation. It is invariably the duty of the Safety and Health Manager to implement the Workers' Compensation system within the plant, so it is considered here as the first staff function to be discussed.

Workers' Compensation legislation has the ostensible purpose of protecting the worker by providing statutory compensation levels to be paid by the employer for various injuries that may be incurred by the worker. There is an ulterior feature, however, which provokes labor to be dissatisfied with the Workers' Compensation system. This feature is the immunity from additional liability that the Workers' Compensation system grants to the employer, except in cases in which "gross negligence" can be proved.

Table 2.1 lists examples of statutory compensation levels for various permanent injury types. To most people the various levels seem too low to compensate adequately for the permanent injury to the worker. Historical evolution of the rates is slow, and it can be seen that public sensitivity to worker injuries has increased over the years. This sensitivity has resulted in outcries for reform of the Workers' Compensation system. On the other side of the issue is management's position that industry can never fully compensate monetarily for everything that may happen to workers in the course of their duties. Since some risk is inescapable in any line of work, management's general position is that in consideration of the salary and wages that workers receive, part of the normal risk of injury must be borne by the worker.

Typically, the firm does not pay the Workers' Compensation payments directly; rather, it carries insurance against compensation claims. The insurance company is vitally interested in the safety and health within the plant, and this provides a ma-

TABLE 2.1   Sample Statutory Compensation Levels for Permanent Injuries (Compensation at 66⅔% of Average Weekly Pay)

Type of Permanent Injury	Compensation Level[a] (weeks)
Arm amputated	
At or above the elbow	210
Below the elbow	158
Leg amputated	
At or above the knee	184
Below the knee	131
Hand amputated	158
Thumb amputated	63
Finger(s) amputated	
First	37
Second	32
Third	21
Fourth	16
Foot amputated	131
Toe amputated	
Great toe	32
Other toes, per toe	11
Loss of sight in one eye	105
Loss of hearing in one ear	42
Loss of hearing in both ears	158
Loss of testicle	53
Loss of both testicles	158

[a]These compensation levels are in addition to any compensation paid for the healing period. The sample compensation levels were obtained from workers' compensation levels for Arkansas and are intended only as an approximate guide. Exceptions and special cases exist, and tables vary somewhat from state to state.

Source: Arkansas Workers' Compensation Commission (ref. 3).

jor impetus to the development of the safety and health program. The accident experience of the firm is reflected in the levels of the insurance premium, which can be adjusted up or down depending on plant safety experience. The insurance industry applies an "experience rating," expressed as a decimal fraction to be multiplied by the standard premium rate. The experience rating is based upon a three-year average of the firm's actual claims experience and can be less than or greater than 1.00. An experience rating of 1.00 would represent no modification at all and would be applied to a firm that is estimated by the insurance company to have a standard, typical risk. A large company with an experience rating of perhaps 0.80 can save thousands of dollars in annual Workers' Compensation insurance premiums. Even in the first year of coverage under Workers' Compensation insurance, a company can profit from an effective safety and health program, because the insurance underwriter depends upon prior loss data and an initial assessment of the firm's hazards before setting the initial annual premium.

A good insurance company will make regular inspections of facilities to be sure that installations and practices are safe. This is a direct and measurable monetary stimulus to the safety program.

Some companies choose to self-insure against Workers' Compensation claims. This may make economic sense when claims experience and premium levels are compared. But the intangible benefits provided by the insurance company must also be considered in order to make a rational decision. Besides the regular inspections mentioned earlier, the insurance companies are valuable sources of technical advice to their clients. Many insurance companies provide training films and other valuable aids to the conduct of the safety and health program. If the Safety and Health Manager is not receiving these aids and services from the insurance company, he or she should request them and also perhaps consider alternative vendors when policy renewal time presents itself.

The number of companies that have elected to self-insure has led to a new type of consultant called a "loss control representative." This consultant's objective is to keep Workers' Compensation claims low by supplying the type of services normally provided by the insurance carrier. A significant part of these services is maintaining close relationships with employees who do file claims. This serves the purpose of showing an interest in the well-being and encouragement of the honest claimant and at the same time uncovering evidence of fraudulent claims in the case of the dishonest worker who is either malingering or truly injured but whose injury occurred off the job.

RECORDKEEPING

The National Safety Council established the first national system of industrial safety recordkeeping. This system was standardized and designated the Z16.1 system by the American National Standards Institute. In the 1970s, the federal OSHA agency set mandatory recordkeeping requirements very similar to the Z16.1 system, which was voluntary. There were some differences, however, which make year-by-year comparisons of safety and health records infeasible when one year is based on traditional Z16.1 records and another on the federal system. This is particularly unfortunate in that it confounds attempts to determine from statistical records whether the federal agency has had any beneficial impact on worker safety and health. Some specific industries and hazard categories, such as trenching and excavation cave-ins in construction, have shown visible improvements since OSHA's inception, but other gains have been obscured by the change in the statistical records system. Other variations in conditions, such as employment levels and recession cycles, have also acted to blur statistical comparisons.

The recording of worker fatalities is more consistent than that of injuries and illnesses; thus fatality statistics can be used to observe trends both before and after federal regulation. Figure 2.1 shows that the long-range trend of industrial fatalities is downward. Note that the implementation of the OSHA law in the early 1970s has had very little visible impact on this trend when all industries are considered together.

Traditional Indexes

Familiar statistical measures are *frequency* and *severity,* which were defined by the old Z16.1 system. Frequency measured the numbers of cases per standard quantity of workhours, and severity measured the total impact of these cases in terms of "lost workdays" per standard quantity of workhours.

Some injuries, such as amputations, are quite severe but might result in few or no lost workdays. To avoid a distortion in severity rates in such cases, standard lost-workday charges were arbitrarily set for permanent injuries such as amputations or loss of eyesight. The greatest need for arbitrary severity charges was for fatalities, because when you think about it, a fatality is not really a lost-workday case in the literal sense of the term; neither is a permanent total disability, because the worker never works again.

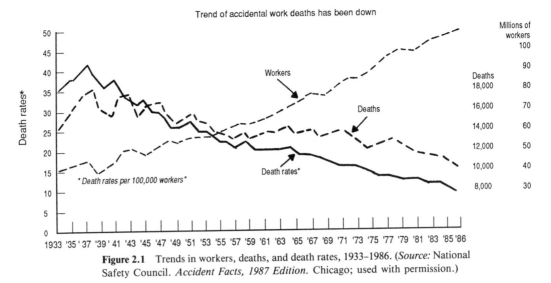

Figure 2.1   Trends in workers, deaths, and death rates, 1933–1986. (*Source:* National Safety Council. *Accident Facts, 1987 Edition.* Chicago; used with permission.)

Another obsolete term is *seriousness,* which was the ratio of severity to frequency. This produced a measure of the relative average importance of injuries and illnesses without regard to the number of hours worked during the study period.

Incidence Rates

The current system of recordkeeping represents an enlargement of the old Z16.1 system. The total injury/illness treatment incidence rate includes all injuries or illnesses which require medical treatment, plus fatalities. Compare this with the traditional "frequency" rate, which included only those cases in which the worker missed at least a day of work.[1] "Medical treatment" does not include simple first aid, preventive medicine (such as tetanus shots), or medical diagnostic procedures with negative results. First aid is described as "one-time treatment and subsequent observation of minor scatches, cuts, burns, splinters, and so forth, which do not ordinarily require medical care" and is not considered medical treatment even if it is administered by a physician or registered professional personnel. Regardless of treatment, if an injury involves loss of consciousness, restriction of work or motion, or transfer to another job, the injury is required to be recorded. Certainly regulating agencies, by their recordkeeping criteria, would not want to discourage medical treatment for an injury that should receive attention, so the United States Bureau of Labor Statistics (ref. 74) has listed sample types of medical treatment, as shown in Appendix B. Any injury that receives one or more of these types of treatment or should have received such treatment is almost always considered recordable. Appendix C gives examples of first aid given for injuries that are not normally recordable unless they qualify for recording for another reason, such as loss of consciousness or transfer to another job.

To compute the incidence rate, the number of injuries is divided by the number of hours worked during the period covered by the study and then is multiplied by a standard factor to make the rate more understandable. Specifically,

$$\text{total injury/illness incident rate} = \frac{\text{number of injuries and illnesses including fatalities} \times 200{,}000}{\text{total hours worked by all employees during the period covered}} \qquad (2.1)$$

# of hours works spent by 100 works

Without the factor of 200,000, the incidence rate would be a very small fraction indeed, as it should be. One should expect a very small number of recordable injuries and illnesses per single hour worked! The choice of the number 200,000 is not entirely arbitrary. A full-time worker typically works approximately 50 weeks per year at 40 hours per week. Thus the number of hours worked per year per worker is approximately

40 hours/week × 50 weeks/year = 2,000 hours/year

50 40 =

---

[1]The ANSI Z16.1 system labeled such lost-workday cases as "disabling injuries" whether the disability was temporary or permanent.

So 200,000 hours represents the number of workhours spent by 100 workers in a year:

$$100 \text{ workers} \times 2{,}000 \text{ hours/year/worker} = 200{,}000 \text{ hours/year}$$

Thus the total injury/illness incidence rate represents the number of injuries expected by a 100-employee firm in a full year, if injuries and illnesses during the year follow the same frequency as observed during the study period. Note from equation (2.1) that the actual period for gathering the incidence rate data need not be a year or any other specific time period. A fairly long period is needed, however, to obtain a representative number of cases, especially when the incidence is low. A typical data collection period is one year.

Sometimes the Safety and Health Manager will want to relate current total injury/illness incidence rates to the traditional "frequency" rate. The old "frequency" rate used a factor of 1,000,000 hours instead of 200,000. Thus rates were standard as "per million manhours," as they were called in those days. Note that such a factor related to a standard year for a firm employing 500 employees, not 100 employees. Thus the old "frequency" rates should be higher than the current total injury/illness incidence rates—but they generally are not because one must remember that the current rate includes all cases involving medical treatment, not just lost-workday cases. Also, days in which the worker was still on the job but was unable to perform his or her regular job due to an injury or illness are now taken into account. These days are called "restricted work activity" days and may be lumped together with lost workdays or considered separately, depending on the statistic desired. Unless specifically stated otherwise, today's interpretation of "lost workday" includes days in restricted work activity as well as days away from work.

The term *incidence rate* is really a general term and in addition to the total injury/illness incidence rate includes the following:

1. Injury incidence rate
2. Illness incidence rate
3. Fatality incidence rate
4. Lost-workday-cases incidence rate
5. Number-of-lost-workdays rate
6. Specific-hazard incidence rate
7. Lost-workday-injuries rate (LWDI)

All of the foregoing rates use the standard 200,000 factor. Note the difference between rates numbered 4 and 5 in the foregoing list. Rate 4 counts *cases* in which one or more workdays were lost or in which the worker was transferred to another job. Rate 5 counts the total *number* of workdays lost or in which the worker was transferred to another job. Rate 4, the lost-workday-cases incidence rate, is of special significance in that it corresponds most closely to the old "frequency rate." One must remember, however, that the newer lost-workday-cases incidence rate uses a factor of 200,000 employee-hours worked instead of the 1,000,000 employee-hour factor used in the older frequency rates.

In counting the number of lost workdays, the date of the injury or onset of illness should not be counted, even though the employee may leave work for most of that day. Thus if the employee returns to his or her *regular* job and is able to perform all regular duties full time on the day after the injury or illness, no lost workdays are counted. Also, when counting lost workdays, weekends or other normal days off should not be counted if the worker would not have worked those days anyway. The number-of-lost-workdays rate compares to the old severity rate except that no arbitrary charges are assessed for permanent partial disabilities, and except for the 200,000 factor.

The specific hazard incidence rate is useful in observing a narrow slice of the total hazards picture. For specific hazards, injury incidence, illness incidence, fatality incidence, and all of the other rates can be computed. Care must be taken in selecting the corresponding total hours worked to be used in the denominator in calculating specific hazard incidence rates. Since specific hazards are more narrow and fewer workers are exposed, data should be collected over several years to achieve meaningful results for specific hazards incidence rates.

To the Safety and Health Manager, the most important incidence rate of all is the "lost-workday-injuries rate," or LWDI, as it is called by federal OSHA enforcement officials. The LWDI is the criterion used by OSHA in deciding whether to conduct a full inspection of a facility. Thus, on a general-schedule inspection, in an initial review of records at a facility, if OSHA finds that the firm has a lower-than-average LWDI, the firm might not undergo a full inspection. Recent program directives specify conditions for partial inspections for firms with low LWDIs, and these conditions will be covered in detail in Chapter 4. On the other hand, if an initial OSHA review of the firm's records reveals an LWDI higher than the national average, the firm will likely receive a full inspection.

A peculiarity of the LWDI is that it considers only injuries, not illnesses. OSHA directives specify that the enforcement officer "make note of any significant recorded illnesses and submit a health referral if appropriate," but illnesses are not used in the calculation of the LWDI. Also, according to current policy, fatalities are not included in the LWDI calculation.

The US Bureau of Labor Statistics (BLS) has computed incidence rates for industries based upon nationwide surveys and has reported these rates by Standard Industrial Classification (SIC) code. A chart comparing the lost-workday-cases incidence rate, the cases-without-lost-workdays incidence rate, and the total incidence rate for various industry general classifications is shown in Figure 2.2.

Recordkeeping Forms

The format for keeping injury and illness records has been standardized. The basic form is the "Log of Occupational Injuries and Illnesses," displayed in Figure 2.3 (see pp. 22–23). The right half of the form is used as a summary to post annually, for

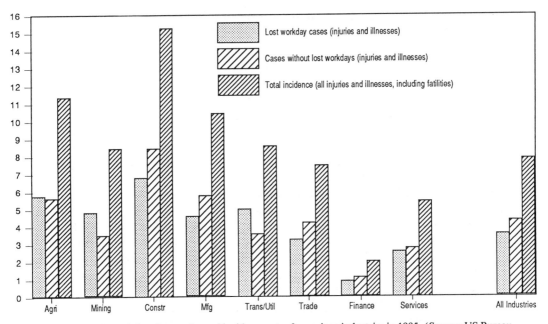

Figure 2.2. Comparison of incidence rates for various industries in 1985. (*Source:* US Bureau of Labor Statistics, ref. 95. Total incidence rates may not exactly represent the sum of the lost-workday-cases rate plus the cases-without-lost-workdays rate because of rounding and because the total incidence rate includes fatalities, whereas neither of the other two rates includes fatalities.)

employees to see what injuries and illnesses have been recorded for the year. The summary is required to be posted in a prominent position in the workplace on February 1 each year and to remain posted for 30 days. It is the employer's responsibility to enter data correctly into the log/summary. General records are required to be saved for a period of at least five years.

The person responsible for completing the log/summary may need some guidance in distinguishing between occupational injuries and illnesses. Examples of occupational injuries include lacerations, fractures, sprains, and amputations that result from a work accident or from an exposure involving a single incident in the work environment. Animal bites, such as insect or snake bites, are considered injuries. Even chemical exposures can be considered injuries if they result from a one-time exposure.

An illness is any abnormal condition or disorder, not classified as an injury, caused by exposure to environmental factors associated with employment. Illnesses are usually associated with chronic exposures, but some acute exposures can be considered illness if the exposure is the result of more than a single incident or accident. More detailed classification in the log/summary is required for illnesses than for injuries, as can be seen by examining column 7 of the log/summary shown in Figure 2.3. Some guidance in classifying illnesses among the subcategories in column 7 can be found in Appendix D.

Besides the log/summary, there is a "Supplementary Record of Occupational Injuries and Illnesses" (see Fig. 2.4 on p. 24). Each page of the supplementary record corresponds to a single-line entry in the log.

To illustrate the calculation of the various incidence rates and to demonstrate the use of the standard forms, a case study will now be analyzed.

Case Study 2.1 A metal products fabrication and assembly plant employs 650 workers and has the following injury/illness experience for the year (workers are employed on a regular 40-hour-workweek basis):

Files 1, 9, 14	Workers sprain ankles in slips and falls; treatment received; workers remain on job
File 2	Worker cuts hand on metal strapping in receiving department; first aid received; no treatment; worker stays on job
Files 3, 11	Workers' skin irritated by continuous, daily exposure to solvents; treatment received; workers remain on job
File 4	Worker falls from platform and breaks leg; misses 1 week of work; spends 6 additional weeks on a secondary activity in the plant because of inability to walk
File 5	Worker electrocuted by portable electric drill on February 6, 1989
File 6	Lifting injury to back; 1 week off work; 2 weeks in a different job not requiring lifting
File 7	Severe dermatitis from continuous, daily exposure to solvent; 2 days away from work; 3 days of restricted work in areas not using solvents
File 8	Carbon tetrachloride poisoning effects observed from long-term exposure; worker became ill; 4 weeks in restricted activity
File 10	Fingertip amputated on box stitching machine; worker loses 2 days' work
File 12	Worker strikes head on overhead conveyor structure; 20 stitches taken; 1 workday lost
File 13	Worker burns hand (second-degree burn) on hot casting; treated, released; no lost workdays
File 15	Wrist sprain; worked 2 days as maintenance dispatcher before returning to regular job
File 16	Worker injured in weekend water-skiing accident; loses 1 week of work

Bureau of Labor Statistics
Log and Summary of Occupational
Injuries and Illnesses

| NOTE: | This form is required by Public Law 91-596 and must be kept in the establishment for 5 years. Failure to maintain and post can result in the issuance of citations and assessment of penalties. (See posting requirements on the other side of form.) | | | RECORDABLE CASES: You are required to record information about every occupational death; every nonfatal occupational illness; and those nonfatal occupational injuries which involve one or more of the following: loss of consciousness, restriction of work or motion, transfer to another job, or medical treatment (other than first aid). (See definitions on the other side of form.) | |

Case or File Number	Date of Injury or Onset of Illness	Employee's Name	Occupation	Department	Description of Injury or Illness
Enter a nonduplicating number which will facilitate comparisons with supplementary records.	Enter Mo./day.	Enter first name or initial, middle initial, last name.	Enter regular job title, not activity employee was performing when injured or at onset of illness. In the absence of a formal title, enter a brief description of the employee's duties.	Enter department in which the employee is regularly employed or a description of normal workplace to which employee is assigned, even though temporarily working in another department at the time of injury or illness.	Enter a brief description of the injury or illness and indicate the part or parts of body affected.

Typical entries for this column might be: Amputation of 1st joint right forefinger; Strain of lower back; Contact dermatitis on both hands; Electrocution—body. |
(A)	(B)	(C)	(D)	(E)	(F)
					PREVIOUS PAGE TOTALS →
					TOTALS (Instructions on other side of form.) →

OSHA No. 200

Figure 2.3. OSHA Form No. 200: Log of Occupational Injuries and Illnesses.

U.S. Department of Labor

For Calendar Year 19 _____   Page ___ of ___

Company Name

Establishment Name

Establishment Address

Form Approved
O.M.B. No. 1220-0029

Extent of and Outcome of INJURY						Type, Extent of, and Outcome of ILLNESS												
Fatalities	Nonfatal Injuries					Type of Illness							Fatalities	Nonfatal Illnesses				
Injury Related	Injuries With Lost Workdays				Injuries Without Lost Workdays	CHECK Only One Column for Each Illness (See other side of form for terminations or permanent transfers.)							Illness Related	Illnesses With Lost Workdays				Illnesses Without Lost Workdays
Enter DATE of death. Mo./day/yr.	Enter a CHECK if injury involves days away from work, or days of restricted work activity, or both.	Enter a CHECK if injury involves days away from work.	Enter number of DAYS away from work.	Enter number of DAYS of restricted work activity.	Enter a CHECK if no entry was made in columns 1 or 2 but the injury is recordable as defined above.	Occupational skin diseases or disorders	Dust diseases of the lungs	Respiratory conditions due to toxic agents	Poisoning (systemic effects of toxic materials)	Disorders due to physical agents	Disorders associated with repeated trauma	All other occupational illnesses	Enter DATE of death. Mo./day/yr.	Enter a CHECK if illness involves days away from work, or days of restricted work activity, or both.	Enter a CHECK if illness involves days away from work.	Enter number of DAYS away from work.	Enter number of DAYS of restricted work activity.	Enter a CHECK if no entry was made in columns 8 or 9.
(1)	(2)	(3)	(4)	(5)	(6)	(a)	(b)	(c)	(d)	(e)	(f)	(g)	(8)	(9)	(10)	(11)	(12)	(13)
						(7)												

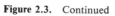

INJURIES   ILLNESSES

Certification of Annual Summary Totals By _____ Title _____ Date _____

OSHA No. 200   **POST ONLY THIS PORTION OF THE LAST PAGE NO LATER THAN FEBRUARY 1.**

FOLD

Figure 2.3. Continued

OSHA No. 101
Case or File No. _____

Form approved
OMB No. 44R 1453

Supplementary Record of Occupational Injuries and Illnesses

EMPLOYER

1. Name _____

2. Mail address _____
   (No. and street)　　　　　　　(City or town)　　　　　(State)

3. Location, if different from mail address _____

INJURED OR ILL EMPLOYEE

4. Name _____ Social Security No. _____
   (First name)　　　(Middle name)　　　(Last name)

5. Home address _____
   (No. and street)　　　　(City or town)　　　(State)

6. Age _____　7. Sex: Male_____ Female_____ (Check one)

8. Occupation _____
   (Enter regular job title, *not* the specific activity he was performing at time of injury.)

9. Department _____
   (Enter name of department or division in which the injured person is regularly employed, even though he may have been temporarily working in another department at the time of injury.)

THE ACCIDENT OR EXPOSURE TO OCCUPATIONAL ILLNESS

10. Place of accident or exposure _____
    (No. and street)　　　　(City or town)　　　　(State)
    If accident or exposure occurred on employer's premises, give address of plant or establishment in which it occurred. Do not indicate department or division within the plant or establishment. If accident occurred outside employer's premises at an identifiable address, give that address. If it occurred on a public highway or at any other place which cannot be identified by number and street, please provide place references locating the place of injury as accurately as possible.

11. Was place of accident or exposure on employer's premises? _____ (Yes or No)

12. What was the employee doing when injured? _____
    (Be specific. If he was using tools or equipment or handling material,
    _____
    name them and tell what he was doing with them.)
    _____

13. How did the accident occur? _____
    (Describe fully the events which resulted in the injury or occupational illness. Tell what
    _____
    happened and how it happened. Name any objects or substances involved and tell how they were involved. Give
    _____
    full details on all factors which led or contributed to the accident. Use separate sheet for additional space.)

OCCUPATIONAL INJURY OR OCCUPATIONAL ILLNESS

14. Describe the injury or illness in detail and indicate the part of body affected. _____
    (e.g.: amputation of right index finger
    _____
    at second joint; fracture of ribs; lead poisoning; dermatitis of left hand, etc.)

15. Name the object or substance which directly injured the employee. (For example, the machine or thing he struck against or which struck him; the vapor or poison he inhaled or swallowed; the chemical or radiation which irritated his skin; or in cases of strains, hernias, etc., the thing he was lifting, pulling, etc.)
    _____
    _____

16. Date of injury or initial diagnosis of occupational illness _____
    (Date)

17. Did employee die? _____ (Yes or No)

OTHER

18. Name and address of physician _____

19. If hospitalized, name and address of hospital _____
    _____

Date of report _____ Prepared by _____
Official position _____

☆ U.S. Government Printing Office 1978—671-610/216

Figure 2.4. OSHA Form No. 101: Supplementary Record of Occupational Injuries and Illnesses.

Analysis. Figure 2.5 displays entries in appropriate numbered columns of the Log of Occupational Injuries and Illnesses.

Calculation of incidence rate

Injury and illness incidence rate $= \dfrac{(5 + 4 + 2 + 2) \times 200{,}000}{650 \times 2{,}000} = 2.00$

Total incidence rate (including fatalities)
$$= \dfrac{1 + 5 + 4 + 2 + 2) \times 200{,}000}{650 \times 2{,}000} = 2.15$$

Injury incidence rate $= \dfrac{(5 + 4) \times 200{,}000}{650 \times 2{,}000} = 1.38$

Illness incidence rate $= \dfrac{(2 + 2) \times 200{,}000}{650 \times 2{,}000} = 0.62$

Fatality incidence rate $= \dfrac{1 \times 200{,}000}{650 \times 2{,}000} = 0.15$

Lost-workday-cases incidence rate $= \dfrac{(5 + 2) \times 200{,}000}{650 \times 2{,}000} = 1.08$

Number-of-lost-workdays rate $= \dfrac{(13 + 42 + 2 + 4) \times 200{,}000}{650 \times 2{,}000} = 9.38$

Specific-hazard incidence rate (dermatitis and skin disease) $= \dfrac{3 \times 200{,}000}{650 \times 2{,}000} = 0.46$

The 650-employee firm in Case Study 2.1 provides ample data to show meaningful calculations for the various incidence rates. But most firms are much smaller. For very small firms the calculations are obviously inappropriate. It is not uncommon for small businesses to operate for several years without a single injury or illness. Recognizing that the general injury/illness recordkeeping system was designed for larger firms, Congress exempted small firms with ten or fewer employees from general recordkeeping requirements.

The current general recordkeeping system is based upon federal standards and has remained relatively static since the early 1970s. As was stated earlier in this chapter, the required retention period for these general records is five years. However, in the early 1980s, special recordkeeping requirements were established for toxic chemicals, in the movement that became known as "right-to-know." It will be seen in Chapter 5 that the recordkeeping requirements for toxic chemicals are much more comprehensive and have led to the development of computer information systems for safety

U.S. Department of Labor

For Calendar Year 19 _____   Page ___ of ___

Form Approved
O.M.B. No. 1220-0029

Company Name

Establishment Name

Establishment Address

Extent of and Outcome of INJURY						Type, Extent of, and Outcome of ILLNESS														
Fatalities	Nonfatal Injuries					Type of Illness								Fatalities	Nonfatal Illnesses					
Injury Related	Injuries With Lost Workdays				Injuries Without Lost Workdays	CHECK Only One Column for Each Illness (See other side of form for terminations or permanent transfers.)								Illness Related	Illnesses With Lost Workdays				Illnesses Without Lost Workdays
Enter DATE of death. Mo./day/yr. (1)	Enter a CHECK if injury involves days away from work, or days of restricted work activity, or both. (2)	Enter a CHECK if injury involves days away from work. (3)	Enter number of DAYS away from work. (4)	Enter number of DAYS of restricted work activity. (5)	Enter a CHECK if no entry was made in columns 1 or 2 but the injury is recordable as defined above. (6)	Occupational skin diseases or disorders (a)	Dust diseases of the lungs (b)	Respiratory conditions due to toxic agents (c)	Poisoning (systemic effects of toxic materials) (d)	Disorders due to physical agents (e)	Disorders associated with repeated trauma (f)	All other occupational illnesses (g)	(7)	Enter DATE of death. Mo./day/yr. (8)	Enter a CHECK if illness involves days away from work, or days of restricted work activity, or both. (9)	Enter a CHECK if illness involves days away from work. (10)	Enter number of DAYS away from work. (11)	Enter number of DAYS of restricted work activity. (12)	Enter a CHECK if no entry was made in columns 8 or 9. (13)
file 1					✓														
" 2	NOT RECORDABLE																		
" 3						✓													✓
" 4	✓	✓	5	30															
" 5	2/6/89																		
" 6	✓	✓	5	10															
" 7						✓									✓		2	3	
" 8									✓						✓			1	
" 9					✓														
" 10	✓	✓	2																
" 11						✓													✓
" 12	✓	✓	1																
" 13					✓														
" 14					✓														
" 15	✓			2															
" 16	NOT RECORDABLE																		
Totals	1	5	4	13	42	4	3		1						2	1	2	4	2

Certification of Annual Summary Totals By _____ Title _____ Date _____

OSHA No. 20C

POST ONLY THIS PORTION OF THE LAST PAGE NO LATER THAN FEBRUARY 1.

Figure 2.5. Log Summary for Case Study 2.1.

and health. The required retention period for hazardous chemical exposure records and medical records under the "right-to-know" standards is thirty years instead of five years.

ACCIDENT CAUSE ANALYSIS

So far this chapter has discussed the more visible busy functions of the job of the Safety and Health Manager, many of which are required by state or federal agencies. But even more important to the health and safety of workers are some of the jobs that the Safety and Health Manager is not required to, but should, do. One of these voluntary but important tasks is a thorough analysis of the potential causes of injuries and illnesses that have already occurred in the plant. Even accidents or incidents that may not actually have caused injuries or illnesses, but which could have, should be studied to prevent their recurrence. Any occurrence of an unplanned, unwanted event is a piece of data to consider in the prevention of future illnesses and injuries. Accident cause analysis and subsequent dissemination of this information to personnel who will be exposed to the hazards in the future is believed to be the most effective way of preventing injuries and illnesses. The literature of injury case histories is filled with accounts of cases in which workers are killed by conditions that had previously caused accidents or injuries to others. One case will be used to illustrate this point.

Case Study 2.2   A worker was struck in the head and killed by the sudden movement of a large wrench for releasing gates on the bottom of railroad hopper cars. The worker used a powerful, 3- to 4-foot-long wrench to trip a mechanical latch on the bottom gate of the car. The wrench was supposed to be of a ratchet type so that when the gate was tripped, the tremendous weight of the bulk material in the hopper car would not suddenly force the wrench back on the worker. But the ratchet wrench for some reason was not available, and workers had been using an ordinary rigid wrench to release the latch. Only a week before the fatality occurred another worker had narrowly escaped the same injury when he lost control of the same wrench in the same operation.

Sometimes the accident analysis leads to a design change in a product or process. In other cases, work procedures are changed to prevent future occurrences, or to minimize the adverse effects of these occurrences. Even when nothing can be changed to prevent a future occurrence, at least workers can be informed of what happened, what caused the accident, under what conditions the accident might occur again, and how to protect themselves in such an event. Informing workers of the facts and causes of accidents that have already happened to their co-workers is the single most effective method of training workers to avoid injury and illness. Thus accident cause analysis is the foundation upon which safety and health engineering, capital investment planning, training, motivation, and other functions are constructed. There are other types of accident analysis; statistical frequency analysis was discussed earlier, and cost analysis will be discussed later in this chapter. But no type of analysis is as important as the determination of causes of accidents that have already occurred and might occur again.

ORGANIZATION OF COMMITTEES

The value of using safety and health committees has long been recognized. Committees are appointed from the ranks of the operating personnel of the regular line organization. The appointments are temporary, so that workers throughout the organization rotate on and off a committee periodically. The committees then make facilities inspections, evaluate safety and health suggestions, analyze accident causes, and make recommendations.

Several natural advantages of the committee approach make it a winning strategy. In general, operating personnel know a lot more about their processes and machines than does the Safety and Health Manager. Many valuable and practical ideas can come from operating personnel if staff persons will listen. Also, operating personnel may more readily accept new policies and procedures if these procedures arise from other operating personnel like themselves. Then there is the advantage of exposure. Sooner or later, nearly everyone has his or her turn on a safety committee, which means that the direct activity of the safety and health program is a product of plantwide participation. Some workers have no appreciation for or sensitivity to safety and health hazards until they take their turns on the committee. Indirectly, then, the committee becomes a vehicle for safety and health training.

Despite its advantages, there are pitfalls to the committee approach. The Safety and Health Manager should provide resources and guidance to the committee so that it will have the necessary tools and knowledge to function effectively. Otherwise, the committee may make ridiculous suggestions and be disappointed when management does not approve or does not follow through with capital support. Also, committees must be conditioned not to expect miracles. Some orientation or training is necessary so that committee members will comprehend the goal of eliminating recognized and *unreasonable* hazards, but not all hazards. Finally, committees should not be allowed to degenerate into spy parties with the objective of discrediting the processes or procedures of other departments.

SAFETY AND HEALTH ECONOMICS

Safety and Health Managers are sometimes dismayed to discover that top management bases safety and health decisions on dollars and cents. But the cold reality is that business exists to make profits, and everything a business does is either directly or indirectly related to economics. Safety and Health Managers who are naive enough to think that the humanitarian objective of worker safety and health transcends the more crude issues of profit and loss should ask themselves the following question: How *much* safety and health staff activity is justified by the humanitarian objective?

The prevention of employee injuries and illnesses can be formulated as an economic objective; such a formulation is more meaningful to management than vague humanitarian aspirations. Accidents, injuries, and illnesses have undeniable costs that contribute nothing to the value of products manufactured by the firm or services performed. Occupational injuries alone have been estimated to total over $20 billion annually. The annual cost of injuries and illnesses in many industries dwarfs the total profits picture. This is a reality that almost any top manager will want to consider. While it is true that many of these costs are subtle and difficult to estimate, the existence of these costs is in no way diminished by this fact.

One obvious and direct category of costs from injuries and illnesses is the payment of Workers' Compensation insurance premiums, which are based on the firm's injury and illness experience. Self-insured firms have the actual claims data on which to calculate these direct costs. In addition to these claims are medical costs that may be covered by insurance. Since these costs are directly identified to injuries and illnesses in accounting records, they are sometimes called "direct costs" of injuries and illnesses. Workers' Compensation costs typically range from 1% to 2% of the total payroll of the firm. Certain hazardous industries may have Workers' Compensation costs of around 3% of payroll, according to a 1980 survey by the United States Chamber of Commerce.

The figure 1% or 2% of payroll, which may or may not seem significant to the reader, has been referred to by some analysts as the "tip of the iceberg." The intangible costs of accidents, although hidden, appear to be much greater than the so-called "direct costs." It is the job of the Safety and Health Manager to attempt to estimate these costs and to keep management apprised so that rational investment decisions can be made.

The National Safety Council, in its *Accident Prevention Manual for Industrial Operations*,[2] lists the following categories of hidden costs of accidents:

1. **Cost of wages paid for time lost by workers who were not injured.** These are employees who stopped work to watch or assist after the accident or to talk about it, or who lost time because they needed equipment damaged in the accident or because they needed the output or the aid of the injured worker.

2. **Cost of damage to material or equipment.** The validity of property damage as a cost can scarcely be questioned. Occasionally, there is no property damage, but a substantial cost is incurred in putting back in order material or equipment which has been thrown into a state of disorder. The charge should, however, be confined to the net cost of repairing or putting in order material or equipment that has been damaged or displaced, or to the current worth of the equipment less salvage value if it is damaged beyond repair.

---

[2]*Accident Prevention Manual for Industrial Operations: Administration and Programs Volume*, 8th ed. Chicago: National Safety Council, 1981, pp. 214–215 (used with permission).

An estimate of property damage should have the approval of the cost accountant, particularly if the current worth of the damaged property used in the cost estimate differs from the depreciated value established by the accounting department.

3. **Cost of wages paid for time lost by the injured worker,** other than workers' compensation payments. Payments made under workers' compensation laws for time lost after the waiting period are not included in this element of cost.

4. **Extra cost of overtime work necessitated by the accident.** The charge against an accident for overtime work necessitated by the accident is the difference between normal wages and overtime wages for the time needed to make up lost production, and the cost of extra supervision, heat, light, cleaning, and other extra services.

5. **Cost of wages paid supervisors for time required for activities necessitated by the accident.** The most satisfactory way of estimating this cost is to charge the wages paid to the foreman for the time spent away from normal activities as a result of the accident.

6. **Wage cost caused by decreased output of injured worker after return to work.** If the injured worker's previous wage payments are continued despite a 40% reduction in his output, the accident should be charged with 40% of his wages during the period of such low output.

7. **Cost of learning period of new worker.** If a replacement worker produces only half as much in his first two weeks as the injured worker would have produced for the same pay, then half of the new worker's wages for the two weeks' period should be considered part of the cost of the accident that made it necessary to hire him.

   A wage cost for time spent by supervisors or others in training the new worker also should be attributed to the accident.

8. **Uninsured medical cost borne by the company.** This cost is usually that of medical services provided at the plant dispensary. There is no great difficulty in estimating an average cost per visit for this medical attention.

   The question may be raised, however, whether this expense may properly be considered a variable cost. That is, would a reduction in accidents result in lower expenses for operating the dispensary?

9. **Cost of time spent by higher supervision and clerical workers** on investigations or in the processing of compensation application forms. Time spent by supervision (other than the foreman or supervisor covered in Item 5) and by clerical employees in investigating an accident, or settling claims arising from it, is chargeable to the accident.

10. **Miscellaneous usual costs.** This category includes the less typical costs, the validity of which must be clearly shown by the investigator on individual accident reports. Among such possible costs are public liability claims, cost of renting equipment, loss of profit on contracts canceled or orders lost if the accident causes a net long-run reduction in total sales, loss of bonuses by the company,

cost of hiring new employees if the additional hiring expense is significant, cost of *excess* spoilage (above normal) by new employees, and demurrage. These cost factors and any others not suggested above would need to be well substantiated.

Every firm is different, and if time and staff resources permit, the best way to estimate hidden costs of accidents is to survey and analyze the individual company's recent accident data. When performing such an analysis, it must be remembered that noninjury accidents can also be costly and are generally caused by the same types of conditions and practices that result in injury accidents. Therefore, noninjury accidents should also be included when one attempts to assess the total costs of accidents.

Most firms cannot afford the luxury of a comprehensive, statistically reliable, in-house study of hidden accident costs. An alternative is to turn to national studies of average costs of various accident categories, and apply these survey averages as estimates of in-house costs. Two well-known studies of uninsured costs of accidents were reported by Grimaldi and Simonds (ref. 34) and Imre (ref. 42) in 1975. Although the data were gathered over a span of several years, when the dollar figures were adjusted for inflation to a common representative year, the results of the two studies were shown to corroborate each other, recognizing that rough approximations are all that can be hoped for in such studies.

To make use of the Grimaldi and Simonds and Imre data, the Safety and Health Manager must determine or estimate the numbers of accident cases that fall into the following four general categories, as standardized by the National Safety Council:[3]

- *Class 1:* Cases involving lost workdays, if records are kept under OSH Act, or permanent partial disabilities and temporary total disabilities, if records are based on ANSI Z16.1.
- *Class 2:* Medical treatment cases requiring the attention of a physician outside the plant.
- *Class 3:* Medical treatment cases requiring only first aid or local dispensary treatment and resulting in property damage of less than $20.00 or loss of less than 8 hours' working time.
- *Class 4:* Accidents which either cause no injury or cause minor injury not requiring the attention of a physician, and which result in property damage of $20.00 or more, or loss of 8 or more employee-hours.

Class 1 and 2 cases are taken almost directly from required records kept in-house. Class 3 and 4 cases require additional recordkeeping. Class 3 cases are minor accidents, and although their individual costs are usually small, the total cost generated by this category is significant because of the large numbers of accidents in this category.

[3]*Accident Prevention Manual for Industrial Operations: Administration and Programs Volume,* 8th ed. Chicago: National Safety Council, 1981, p. 209 (used with permission).

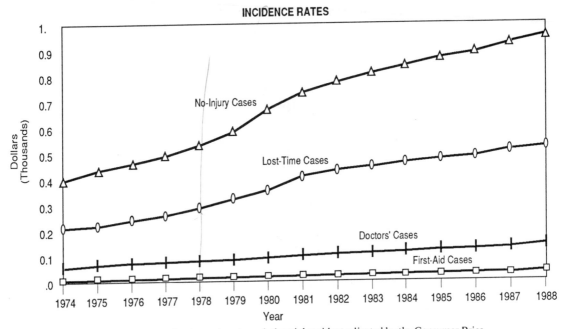

Figure 2.6.   Estimate of uninsured costs per industrial accident adjusted by the Consumer Price Index.

The total number of cases in Class 3 is typically larger than for any of the other categories, and if such is not found to be true of any given study, the analyst should probe deeper for more Class 3 cases that may have been overlooked. Class 4 cases may be even more difficult to track. Class 3 cases at least could be counted by examining infirmary records. Since Class 4 cases are difficult to recognize, the more major cases are the ones that are counted most often. Since this was no doubt also true of the cost studies that generated survey averages, the average cost figures for Class 4 cases turn out to be surprisingly large.

The following figures are suggested for general use for estimates of uninsured costs, as recommended by Grimaldi and Simonds[4] in terms of 1974 dollars:

Lost-time cases	$220	First-aid cases	12
Doctors' cases	55	No-injury cases	400

[4]John V. Grimaldi, and Rollin H. Simonds, *Safety Management*, 3rd ed. Homewood, IL: Richard D. Irwin, Inc., © copyright 1975, p. 614 (used with permission).

The National Safety Council suggests the Consumer Price Index for adjusting accident cost estimates for inflation. Figure 2.6 uses this method to adjust the survey data through 1987.

Having gathered the numbers of accident cases in each of the four classes, the Safety and Health Manager can then compute total uninsured costs by multiplying by the cost factors obtained from Figure 2.6.

Costs of Workers' Compensation insurance and other direct costs should be added to the uninsured cost estimates. Cost estimates of potential fines and legal costs can also be added, although these costs are usually dwarfed by other costs that are sure to be incurred whether or not an enforcement inspection takes place.

Case Study 2.3 Estimate the total annual uninsured cost of accidents in a firm that has records of the following numbers of accidents in the first six months of 1987.

Lost-time cases   12
Doctors' cases   10
First-aid cases   46
No-injury cases   18

Calculation

Estimate of Uninsured Costs per Accident by Type, Adjusted by Consumer Price Index (see Figure 2.6).

	1974	CPI Ratio	1987
Lost-time cases	220	2.49	548
Doctors' cases	55	2.49	137
First-aid cases	12	2.49	30
No-injury cases	400	2.49	996

The annual costs[5] are therefore:

Lost-time:	2(12 cases × $ 548/case) =	$13,152/year
Doctors':	2(10 cases × $ 137/case) =	$ 2,740/year
First-aid:	2(46 cases × $ 30/case) =	$ 2,760/year
No-injury:	2(18 cases × $ 996/case) =	$35,856/year
Total		$54,508/year

---

[5]Note that a factor of 2 is used in the calculation to transform the raw data, based upon a six-month period, to an annual rate.

Case Study 2.4   A Safety and Health Manager presents a status report in a staff meeting and displays a table as follows:

Plant Accident Summary
1991

Lost time accidents. 14
Doctors' cases. 18
First-aid cases. 56
No-injury accidents. 9

   The plant manager asks which of these categories contributes the most to overall cost. If you were the Safety and Health Manager, how would you respond?

Analysis.   The Safety and Health Manager could explain that costs fall into two categories: insured costs and uninsured costs. The insurance company carrying the industrial loss-control policy for the company could provide estimates of the contribution to premium costs for each of the categories of risk. This would answer the plant manager's question with regard to insured costs.

   The larger category of cost would be the uninsured costs, and these are also more difficult to estimate. Using the Grimaldi and Simonds and Imre data (ref. 34, 42), the unadjusted cost estimates for each category could be used to determine estimates for the contribution to cost for each of the accident categories in the company in question. Note that in such a calculation it is unnecessary to adjust the original cost figures for inflation using the Consumer Price Index, because the plant manager is interested in learning which categories contribute the most cost, not in an estimate of the actual dollar cost of each category in today's dollars. Therefore it can be estimated that the "no-injury" accident category contributes the most uninsured cost to the accident problem.

Calculation

Accident Cause	Cost Estimate (1974 $)	Percent of Total
Lost-time accidents	14 × 220 = 3,080	37
Doctors' cases	18 × 55 = 990	12
First-aid cases	56 × 12 = 672	8
No-injury accidents	9 × 400 = 3,600	43
Total	8,342	100

TRAINING

Training or training support may be the most important staff function to be performed by the Safety and Health Manager. Despite a recent trend toward concentration on unsafe conditions, experts still attribute most worker injuries and illnesses to unsafe acts. Unsafe work habits are deeply rooted, even in new, young workers. Our society and its standards of status, as influenced by the media, especially television, places a premium on high-risk activity. From an early age children learn that heroes are people who are daring, lucky, and risk their lives, especially in their life work. In some jobs, such as space exploration, the military, law enforcement, and firefighting, it is occasionally both necessary and rational to take big risks. And those who take these risks indeed deserve to be called heroes. Unfortunately, though, the desire for recognition, status, and the esteem of their peers causes people to take unnecessary risks in activities that do not warrant such risk. A good example of this phenomenon is exhibited in the automobile driving habits of people of all ages. Deep-rooted unsafe habits and lack of knowledge about specific job hazards are major barriers to worker safety and health. It is upon these two problems that the training program should be focused, and training is perhaps the most challenging and important function of the Safety and Health Manager.

One of the biggest mistakes a Safety and Health Manager can make is to assume that he or she is the principal trainer in safety and health. The principal trainers in safety or health or in any other aspect of the job are the first-line supervisors. Their direct contact with the workers will determine how the job will be done. A corollary to this principle is that most training in safety and health is informal and is conducted on the job. In fact, training by example is a very important delivery mode, and new workers are more influenced by what the supervisor and experienced workers *do* than by what they *say*.

Recognizing that most training takes place between supervisor and worker, there is still a need for classroom training in safety and health principles, standards, and hazards recognition, especially for supervisors. The Safety and Health Manager can provide this training directly or can act as a facilitator in bringing useful information and training aids into the plant in subjects needed.

A word of caution is in order, however: Safety and Health Managers should not "try to reinvent the wheel" in their development of training materials. Audiovisual packages and outlines are available, and when the Safety and Health Manager's time, including overhead, is considered, it is usually much more reasonable to purchase or rent training materials than to attempt to create original material in-house. Chapter 1 enumerated some sources to assist Safety and Health Managers in developing this aspect of their function. Chapter 3 will also address the subject of training and how it becomes a part of the overall objective of hazard avoidance.

Drug and Alcohol Abuse

Safety and Health Managers are taking a more assertive posture to control the effects of drug and alcohol abuse in the workplace. Drug and alcohol abuse has been shown to be a greater problem than was once thought. Consider the experience of the Aluminum Company of America's Vancouver, Washington, plant (ref. 94). Following the lead of a sister ALCOA plant in Davenport, Iowa, the company decided to try a preemployment drug screening test for all applicants. To the surprise of management, half of the 750 applicants in three months of testing failed the test. The test was a urinalysis designed to indicate whether drugs had been used in the preceding two or three days and was conducted by a hospital laboratory service. The test results indicated that the use of marijuana was the most prevalent. ALCOA hired 130 of the applicants who passed the test and, according to the personnel manager, found the people hired to be better workers than the company had had before the drug screening program had been added to the hiring process.

It is not difficult to justify a carefully planned and executed drug and alcohol abuse plan for any company, and the Safety and Health Manager should take the lead in establishing one. Indeed there is no choice in certain sectors of the transportation industry subject to mandatory testing for marijuana, cocaine, opiates, amphetamines, and PCP, under rules issued by the US Department of Transportation. The program prescribes random, preemployment, periodic, reasonable cause, and postaccident testing (ref. 21). Drug and alcohol testing may make even more sense in other industries. And when a treatment program becomes necessary, few parties have as much influence over an employee's decision to enter treatment as does his or her employer, whether the treatment is for drug abuse or alcohol abuse. As former Congressman Wilbur Mills once said, "An alcoholic may be insane, but he's not crazy. He'll go to a facility any time rather than lose his job. He has to have money to finance his habit" (ref. 56).

A key question to ask management is whether they can imagine a situation in which the firm might some day need to terminate an employee because excessive drug or alcohol abuse has affected his or her job. If the answer is yes, the firm is exposed to litigation risks if a policy on drug and alcohol abuse is not in place. If your company has a rule against drug abuse, employees should be informed, and new employees should be required to consent in writing that they are working under the company's drug abuse policy as a condition of employment (ref. 89). By making preemployment drug screening a requisite for employment, safety and health hazards can be prevented, but to be consistent, the company should apply the same rules against drug abuse to existing employees. Otherwise the firm may face a discrimination charge from an applicant who has been refused employment.

Besides screening programs to detect the problem, both in the cases of new applicants and existing employees, many employers are instituting "employee-assistance programs" to deal with the difficulties of employees who have a recognized drug or alcohol abuse problem. The rationale is that some well-trained and competent employees are too valuable to lose because of a drug or alcohol problem. Therefore, instead of viewing the problem as a matter of discipline, the condition is viewed as a sickness requiring treatment and therapy to restore the worker to full usefulness. Such programs have the intangible benefit of conveying to workers in general that the company cares about the well-being of its employees and would rather see them cured than terminated.

As pervasive as the drug and alcohol abuse problem is, there is no doubt that Safety and Health Managers will continue to encounter increased responsibilities to establish and maintain programs to control the hazards these problems present.

AIDS

The discovery and disturbing rise in incidence of the viral disease acquired immune deficiency syndrome (AIDS) in the 1980s has created two problems for the Safety and Health Manager:

1. Protecting workers in selected occupations in which there is a potential exposure to the AIDS virus while the worker is on the job.
2. Through worker education, preventing hysteria over the disease from blocking productivity and the flow of business when risks of contracting AIDS are not present in the workplace.

As of August 1987, only six years after the discovery of the disease, the Centers for Disease Control (CDC) had received reports of more than 40,000 cases of AIDS, including 23,000 deaths. Approximately 1.5 million more people were estimated to be carriers without symptoms of the disease. The seriousness of the problem has attracted the attention of OSHA, with the result that information brochures have been distributed by the agency recommending practices for protection against occupational exposure to AIDS (ref. 91). Various OSHA standards, such as those for personal protective equipment, sanitation, and waste disposal, can be brought to bear against industries in which workers are subject to the hazards of AIDS. In addition, OSHA can rely upon the General Duty Clause of the OSH Act that applies to general hazards. The General Duty Clause is covered in detail in Chapter 4.

SUMMARY

Safety and health on the job, like production quality or any other desirable factory characteristic, is achieved by the workers themselves. Thus the actual achievement of safety and health is a line function. The Safety and Health Manager, then, has a staff function in facilitating the line organization, especially first-line supervisors, in achieving the goal of safety and health.

It cannot be assumed that safety and health is really a "goal" of line management. Everyone wants safety and health in the workplace, more or less, but the degree of management's commitment to this goal must be assessed and documented by the Safety and Health Manager.

Once management's commitment to the goal of safety and health is ascertained, the Safety and Health Manager can get to the important functions of dealing with Workers' Compensation, collecting and analyzing statistical records, economic analysis, safety and health training, and dealing with both hazards and violations of safety and health standards.

EXERCISES AND STUDY QUESTIONS

2.1. Is worker safety a line or staff function?

2.2. What is the Safety and Health Manager's relationship to the line organization?

2.3. Why is it often difficult for Safety and Health Managers to win the respect and approval of top management?

2.4. What principle is embodied in the statement "Safety is everybody's business"?

2.5. What should the Safety and Health Manager do if top management does not "practice what it preaches" in safety and health policies?

2.6. What is the difference between the terms *Workmen's Compensation* and *Workers' Compensation*?

2.7. What are the ostensible and ulterior purposes of the Workers' Compensation system?

2.8. Should employers bear all the risk of injury to their workers?

2.9. Who pays the Workers' Compensation claims of injured workers?

2.10. What is a "loss-control representative"?

2.11. What is the Z16.1 system?

2.12. Why is it difficult to prove by statistics whether OSHA has been of any benefit?

2.13. What is the disadvantage of using "lost workdays" as a measure of injury or illness severity?

2.14. What constitutes a "recordable injury or illness"?

2.15. A firm employing 300 workers has 25 recordable injuries/illnesses in a year. What is its total injury/illness incidence rate?

2.16. How does the injury/illness incidence rate compare with the traditional "frequency" rate?

2.17. Compare the terms *frequency, severity,* and *seriousness.*

2.18. What is important about February 1 in the life of a Safety and Health Manager?

2.19. How long are files of records required to be kept?

2.20. Are there advantages and pitfalls to the use of safety and health committees?

2.21. Compare the importance of direct and hidden costs of accidents.

2.22. Name some categories of hidden costs of accidents.

2.23. How can records of noninjury accidents be important to the Safety and Health Manager?

2.24. Who are the principal trainers in safety and health in industry?

2.25. During a six-month period, a firm employing 50 employees has 18 injuries and illnesses requiring medical treatment, in 4 of which cases the employee lost at least 1 day from work.
  (a) What is the general injury/illness incidence rate?
  (b) What is the traditional "frequency" rate?
  (c) Is this a very dangerous industry?

2.26. For the year 1990, a firm with 25 employees has two medical-treatment injuries plus one injury in which the worker lost three days of work. Calculate the injury incidence rate and the LWDI.

2.27. A firm has 62 employees. During the year there are seven first-aid cases, three medical treatment injuries, an accident in which an injured employee was required to work one week in restricted work activity, a work-related illness in which the employee lost one week of work, a work-related illness in which the employee lost six weeks of work, and a fatality resulting from an electrocution. Calculate the total incidence rate, the lost-work-day-cases incidence rate, the number-of-lost-workdays rate, and the LWDI.

2.28. Top management has set a safety and health objective for the year for a plant employing 135 employees. The objective is to reduce the overall injury incidence rate for the firm to a level that would exempt the firm from general schedule inspections by federal enforcement authorities. The target injury incidence frequency rate is 4.7. By May 1 the safety and health manager has logged twelve first-aid cases, three recordable injuries, and two illnesses, both of which resulted in hospitalization. Based upon these preliminary results, does it appear that the firm will meet top management's objective for the year? Show calculations to justify your conclusion.

2.29. A chemical plant employing 900 employees (employees work a regular 40-hour workweek) has the following safety/health record for the year 1987:

File 1:	Forklift truck drops pallet load of packaged raw material; no injuries; some material wasted; pallet destroyed; extensive cleanup required
File 2:	Worker suffers heat cramps (illness) from continuous exposure to hot process; admitted to hospital for treatment; two weeks off
File 3:	Worker burns hand on steam pipe; first aid received and worker returns to workstation
File 4:	Worker suffers dermatitis from repeated contact with solvent; 1 week of work lost; another 4 weeks of work restricted to an assembly job
File 5:	Worker fractures finger in packaging machine; worker sent to hospital for treatment; back on the job the next day
File 6:	Maintenance worker lacerates hand when screwdriver slips; five sutures taken; worker back on the job the next day
File 7:	Pressure vessel explodes; extensive damage to processing area; miraculously, no one is injured
File 8:	Worker gets poison ivy from exposure a week earlier while removing weeds around the plant perimeter fence; worker receives doctor's treatment but no workdays are lost
File 9:	Worker becomes ill from continuous exposure to hydrogen sulfide leaks from furnace area; misses 2 weeks' work; leaks are repaired
File 10:	Worker gets severe poison ivy from weekend outing with Boy Scout troop; misses 2 days of work
File 11:	Maintenance worker falls from fractionating tower and is killed
File 12:	Worker fractures arm in power transmission system powering pulverizer mill; loses 3 days of work and an additional 6 weeks of work is in the production scheduling office before returning to regular job

(a) Calculate the following incidence rates:
1. Total injury/illness incidence rate
2. Injury incidence rate
3. Illness incidence rate
4. Fatality incidence rate
5. Lost-workday-cases incidence rate
6. Number-of-lost-workdays rate

7. Specific-hazard incidence rate (fractures)
8. LWDI

(b) Estimate the hidden costs of accidents using survey average cost data.

(c) Does the use of survey average cost data seem appropriate in this case? Why or why not?

(d) How does the safety/health record of this firm compare with that of other manufacturing companies and with industries in general?

2.30. On February 1, a 50-employee firm posts its annual OSHA log for the previous year, as shown in Figure 2.7. Complete the table and calculate the following:

(a) Injury incidence rate

(b) Illness incidence rate

(c) Number-of-lost-workdays rate

(d) Incidence rate used by OSHA as the criterion for determining whether to conduct an inspection

2.31. A certain large firm with 1,400 employees in 1988 pays an annual Workers' Compensation insurance premium of $120,000. The experience modifier for this firm for 1988 was 1.05. By the year 1991, the firm has shown a dramatic improvement in worker safety and health and accordingly the insurance rating bureau revises the experience modifier for this firm to 0.80. What is the actual and percent savings in Workers' Compensation premium for this firm in 1991 as compared with the premium in 1988?

U.S. Department of Labor

For Calendar Year 19 _____   Page ____ of____

Company Name

Establishment Name

Establishment Address

Form Approved
O.M.B. No. 1220-0029

Extent of and Outcome of INJURY						Type, Extent of, and Outcome of ILLNESS												
Fatalities	Nonfatal Injuries					Type of Illness							Fatalities	Nonfatal Illnesses				
Injury Related	Injuries With Lost Workdays				Injuries Without Lost Workdays	CHECK Only One Column for Each Illness *(See other side of form for terminations or permanent transfers.)*							Illness Related	Illnesses With Lost Workdays				Illnesses Without Lost Workdays
Enter DATE of death. Mo./day/yr. (1)	Enter a CHECK if injury involves days away from work, or days of restricted work activity, or both. (2)	Enter a CHECK if injury involves days away from work. (3)	Enter number of DAYS away from work. (4)	Enter number of DAYS of restricted work activity. (5)	Enter a CHECK if no entry was made in columns 1 or 2 but the injury is recordable as defined above. (6)	Occupational skin diseases or disorders (a)	Dust diseases of the lungs (b)	Respiratory conditions due to toxic agents (c)	Poisoning (systemic effects of toxic materials) (d)	Disorders due to physical agents (e)	Disorders associated with repeated trauma (f)	All other occupational illnesses (g)	Enter DATE of death. Mo./day/yr. (8)	Enter a CHECK if illness involves days away from work, or days of restricted work activity, or both. (9)	Enter a CHECK if illness involves days away from work. (10)	Enter number of DAYS away from work. (11)	Enter number of DAYS of restricted work activity. (12)	Enter a CHECK if no entry was made in columns 8 or 9. (13)
	✓			8														
											✓			✓			40	
								✓						✓	✓	10		
								✓					10/10/89					
	✓	✓	3															
					✓													
					✓													
					✓													
	✓	✓	5	10														
					✓													
						✓								✓	✓	2	10	
												✓						✓
					✓													
	✓	✓	2	6														
								✓										✓
					✓													
						✓								✓			5	
						✓												✓

Certification of Annual Summary Totals By _____ Title _____ Date _____

OSHA No. 200   **POST ONLY THIS PORTION OF THE LAST PAGE NO LATER THAN FEBRUARY 1.**

Figure 2.7. OSHA Annual Summary for Exercise 2.30.

3

Concepts of Hazard Avoidance

Hazards involve risk or chance, and these words deal with the unknown. As soon as the unknown element is eliminated, the problem is no longer one of safety or health. For example, everyone knows what would happen if he or she jumped off a ten-story building. Immediate death would be virtually a certainty, and such an act is not properly described as unsafe; it is described as suicidal. But to work on the roof of a ten-story building with no intention of falling off becomes a matter of safety. Workers without fall protection on the unguarded rooftop of a building are exposed to a recognized hazard. This is not to say that such workers will be killed or even to say that they will come to any harm whatsoever, but there is that chance, that unknown element.

Dealing with the unknown makes the job of the Safety and Health Manager a difficult one. If the Safety and Health Manager pushes for a capital investment to enhance safety or health, who is able to *prove* later that the investment was worthwhile? Improved injury and illness statistics help and may look impressive, but they do not actually prove that the capital investment was worthwhile because no one really knows what the statistics would have shown had the investment not been made. It is in the realm of the unknown.

Since safety and health deal with the unknown, there is no step-by-step recipe for eliminating hazards within the workplace. Instead, there are merely concepts or approaches to take to whittle away at the problem. All the approaches have merit, but none is a panacea. Drawing on their own strengths, different Safety and Health Managers tend to concentrate on certain favorite approaches familiar to them. The objective of this chapter is to present various approaches so that the Safety and Health Manager will have a variety of tools, not just one or two, to deal with the unknown elements of worker safety and health. Both the good and the bad will be discussed for each approach. The good is often obvious or taken for granted. But the drawbacks of each approach must be squarely faced, too, so that Safety and Health Managers can see the limitations and draw on the strengths of each approach to accomplish their missions.

THE ENFORCEMENT APPROACH

This is the approach initially taken by OSHA, but OSHA was certainly not the first to employ it. Safety rules with penalties for breaking them have existed almost since people first began to deal with risks. The pure enforcement approach says that since people do not properly assess hazards and take prudent precautions, they should be told rules to follow and be subject to penalties for breaking these rules.

The enforcement approach is simple and direct; there is no question that it has an impact. The enforcement must be swift and sure and the penalties sufficiently severe, but if these conditions are met, people will follow rules to some extent. Using the enforcement approach, OSHA has without doubt forced thousands of industries to comply with regulations that have changed workplaces and made millions of jobs safer and more healthful. The preceding statement sounds like a glowing success story for OSHA, but the reader knows that the enforcement approach has failed to do the whole job. It is difficult to see any general improvement in injury and illness statistics as a result of enforcement, although some categories, such as trenching and excavation cave-ins, have shown marked improvement. Despite its advantages, there are some basic weaknesses in the enforcement approach, as the statistics suggest. These weaknesses will now be examined.

At the foundation of any enforcement approach lies a set of mandatory standards. Mandatory standards must be worded in absolutes, such as "always do this" or "never do that." The wording of complicated exceptions can alleviate the problem somewhat but requires the anticipation of every circumstance to be encountered. Within the framework of the stated scope of the standard, recognizing all exempt situations, each rule must be absolutely mandatory to be enforceable. But mandatory language employing the words *always* and *never* is really inappropriate when dealing with the uncertainties of safety and health hazards. To see how this is true, consider the following case study.

CASE STUDY 3.1   Suppose that a properly grounded electrical appliance for the resuscitation of injured employees is equipped with a three-prong plug. But in the midst of an emergency it is discovered that the wall receptacle is the old, ungrounded, two-hole variety. With no adapter in sight and an employee in desperate need of the appliance, who would not bend back or cut off the grounding plug and proceed to save the employee's life?

Of course this example states an extreme case, and we must be "reasonable" and use our "professional judgment," but in the arena of enforcement and mandatory standards, who is going to say what is "reasonable"? Everyone knows what is reasonable in such an extreme case, but countless borderline cases occur every day in which it is not certain whether the proper course of action is to violate or not to violate the rule. Consider the following actual case.

CASE STUDY 3.2 A dangerous fire was in progress as flammable liquids were burning in tanks. To shut off the source of fuel, a thinking employee quickly turned off the adjacent tank valves to avert a more dangerous fire which could have cost many lives, not to speak of property damage. Did the employee receive a medal for his meritorious act? No, the company received an OSHA citation because the employee was not wearing gloves! The valves were hot, and because the employee went ahead and closed the valves, burning his hands, the company was issued a citation.

If a government agency will issue a citation for failure to wear gloves while closing a valve in an emergency, who will have the courage to "be reasonable" and to go ahead and act even if a violation is the consequence? In the face of a pure enforcement approach, industry employees and employers alike will gradually retreat to a defensive position, failing to produce, and blaming their lack of productivity on the government.

As stated earlier, OSHA did not invent the enforcement approach for dealing with hazards. Other mandatory safety rules and laws are familiar to everyone. Sometimes overzealous and oppressive rules can destroy themselves by alienating the very persons they are intended to protect. A notorious example is the mandatory helmet law for motorcycle riders. The helmet manufacturers can present impressive statistics that show that helmets save lives, at least in some accidents. Such statistics are a strong motivation to motorcyclists to wear helmets. But in certain situations, the use of a helmet has disadvantages that cause motorcyclists to hate the law that requires them *always* to wear a helmet. Thus it is illegal to give a friend a trial ride around the block without obtaining an extra helmet for the passenger for that one ride. If a temporary skin rash or scalp treatment prevents an operator from wearing the helmet for a day or two, he or she must not ride, even if the motorcycle is the sole means of transportation. Where to put the helmet during a brief stop can also become extremely awkward in some situations. One motorcyclist in Houston became so frustrated that he complied with the letter of the law while he defied its spirit by riding his motorcycle around and about Houston wearing his helmet on his elbow!

THE PSYCHOLOGICAL APPROACH

Contrasted with the enforcement approach is an approach that attempts to reward safe attitudes. This is the approach employed by many Safety and Health Managers and may be identified as the psychological approach. The familiar elements of the psychological approach are posters and signs reminding employees to work safely. A large sign may be at the front gate of the plant displaying the number of days since a lost-time injury. Safety meetings, departmental awards, drawings, prizes, and picnics can be used to recognize and reward safe attitudes.

Religion versus Science

The psychological approach emphasizes the religion of safety and health versus the science. Safety meetings at which the psychological approach is used are typified by attempts at persuasion, sometimes called "pep talks." The idea is that employees can be rewarded into wanting to have safe work habits. Peer pressure can be brought to bear on the individual when the whole department may suffer when one person has an injury or illness.

Top Management Support

The psychological approach is very sensitive to the support of top management, and if such support is absent, the approach is very vulnerable. Recognition pins, certificates, and even monetary prizes are small reward if workers feel that to win these rewards they are not pursuing the real goals of top management.

Workers are able to sense the extent of management's commitment to safety by the day-to-day decisions it makes, not by written proclamations that everyone should "be safe." A rule requiring safety glasses to be worn in the production area is undermined when top management does not wear safety glasses when visiting the production floor. If safe practices are ordered to be cast aside when production must be expedited to fill an order on time, workers find out just how much worker safety and health mean to top management. Most Safety and Health Managers will want to get a written endorsement of the plant safety program from top management, but unless top management really understands and believes in the safety and health program, the written endorsement is not very valuable. The true orientation of top management will soon show through. Safety and Health Managers should beware of this pitfall when seeking such a written endorsement.

Young Workers

New workers, especially new young workers, are particularly influenced by the psychological approach to safety and health. Workers in their late teens or early twenties enter the workplace having recently emerged from a social structure that places a great deal of importance on daring and risk taking. These new workers are watching supervisors and more experienced co-workers to determine what kind of behavior or work habits earn respect in the industrial setting. If their older, more experienced peers wear respirators or ear protection, the young workers may also adopt safe habits. If highly respected co-workers laugh at or ignore safety principles, young workers may get off to a very bad start, never taking safety and health seriously.

Accident reports confirm that a large percentage of injuries are caused by unsafe acts by workers. This fact underlines the importance of the psychological approach in developing good worker attitudes toward safety and health. This approach can be bolstered by training in the hazards of specific operations. Once the subtle hazards are made known to workers who would not know about these hazards from their general experience, the development of safe attitudes becomes less difficult.

THE ENGINEERING APPROACH

For decades, safety engineers have attributed most workplace injuries to unsafe worker acts, not unsafe conditions. The origin of this thinking has been traced to the great pioneering work in the field by the late H. W. Heinrich (ref. 39), the first safety engineer so recognized. Heinrich's studies resulted in the widely known ratio 88:10:2:

Unsafe acts	88%
Unsafe conditions	10%
Unsafe causes	2%
Total causes of workplace accidents	100%

Recently, Heinrich's ratios have been questioned, and efforts to recover Heinrich's original research data have produced sketchy results. The current trend is to give increasing emphasis to the workplace machinery, environment, guards, protective systems (i.e., the *conditions* of the workplace). Accident analyses are probing more deeply to determine whether incidents which at first appear to be caused by "worker carelessness" could have been prevented by a process redesign. This development has greatly enhanced the importance of the "engineering approach" to dealing with workplace hazards.

Three Lines of Defense

From the profession has emerged a definite preference for the engineering approach when dealing with health hazards. When the process is noisy or presents airborne exposures to toxic materials, the firm should first try to redesign or revise the process to "engineer out" the hazard. Thus engineering controls receive first preference in what might be called "three lines of defense" against health hazards. These three lines of defense are identified as follows:

1. Engineering controls
2. Administrative or work practice controls
3. Personal protective equipment

The advantages of the engineering approach are obvious. Engineering controls deal directly with the hazard by removing it, ventilating it, suppressing it, or otherwise rendering the workplace safe and healthful. This removes the necessity of living with the hazard and minimizing its effects as contrasted with the strategies of administrative controls and the use of personal protective equipment. This preference of strategies will be considered again in Chapters 7 and 9.

Safety Factors

Engineers have long recognized the chance element in safety and know that margins for error must be provided. This basic principle of engineering design appears at various places in safety standards. For example, the safety factor for the design of scaffold components is 4:1. For overhead crane hoists the factor is 5:1, and for scaffold *ropes*, the factor is 6:1 (i.e., scaffold ropes are designed to withstand six times the intended load).

The selection of safety factors is an important responsibility. It would be nice if all safety factors could be 10:1, but there are trade-offs which make such large safety factors unreasonable, even infeasible, in some situations. Cost is the obvious, but not the only trade-off. Weight, supporting structure, speed, horsepower, and size are all factors that may be affected by selection of too large a safety factor. The drawbacks of large safety factors must be weighed against the consequences of system failure, in order to arrive at a rational decision. There are many degrees of difference between situations when evaluating the consequences of system failures. Compare the importance of safety factor in the Kansas City hotel disaster[1] of 1981 to a failure in which the only loss is some material or damaged equipment. Obviously, the former situation should employ a larger safety factor than the latter. Selection of safety factors depends on the evaluation or classification of degree of hazard, a subject that will be treated in depth later in this chapter.

Fail-Safe Principles

Besides the engineering principle of safety factors, there are additional principles of engineering design that consider the consequences of component failure within the system. These principles are labeled here as "fail-safe principles," and three are identified as follows:

1. General fail-safe principle
2. Fail-safe principle of redundancy
3. Principle of worst case

Each of these principles will be considered here, and their applications will appear again and again in subsequent chapters dealing with specific hazards.

---

[1]Two skywalks collapsed in the crowded multistory lobby of the Hyatt Regency Hotel in Kansas City, Missouri, on July 17, 1981, killing 113 persons.

The resulting status of a system, in event of failure of one of its components, shall be in a safe mode.

Systems or subsystems generally have two modes: active and inert. With most machines the inert mode is the safer of the two. Thus product safety engineering is usually quite simple: If you "pull the plug" on the machine, it cannot hurt you. But the inert mode is not always the safer mode. Suppose that the system is a complicated one, with subsystems built in to protect the operator and others in the area in the event of a failure within the system. In this case, "pulling the plug" to disconnect the machine might deactivate the safety subsystems so essential to protect the operator and others in the area. In the case of such a system, disconnecting the power might render the system more unsafe than it was when the power was on. Design engineers need to consider the general fail-safe principle so as to ensure that a failure within the system will result in a safe mode. Thus it may be necessary to provide backup power for the proper functioning of safety subsystems. The following examples will illustrate the concept.

CASE STUDY 3.3 An electric drill has a trigger switch which might be continuously depressed to operate the drill. The trigger switch is loaded with a spring so that if some failure (on the part of the operator in this case) results in the release of the trigger, the machine will return to a safe mode (off, in this case). Such a switch is often called a "deadman control." This example illustrates the common situation in which the inert state of the system is the safer one.

CASE STUDY 3.4 This example illustrates the more uncommon situation in which the inert state of the system is the more dangerous. Consider an automobile with power steering and power brakes. When the engine dies, both steering and braking become very difficult; so at least as far as these subsystems are concerned, the inert state is more dangerous than the active one.

As will be seen in later chapters, industrial systems have characteristics similar to those described in the foregoing examples. Sometimes safety standards specify that system failures be accommodated in such a way that subsystems for safety will continue to be operative.

The general fail-safe principle is the one that embodies the literal meaning of the words *fail-safe*. However, industry and technology often associate another concept with the term *fail-safe*, and that is the concept of *redundancy*, stated as follows:

Fail-Safe Principle of Redundancy   *back-up systems*

A critically important function of a system, subsystem, or components can be preserved by alternate parallel or standby units.

The design principle of redundancy has been widely used in the aerospace industry. When systems are so complicated and of such critical importance as a large aircraft or a space vehicle, the function is too important to allow the failure of a tiny component to bring down the entire system. Therefore, engineers back up primary subsystems with standby units. Sometimes, dual units can be specified right down at the component level. For extremely critical functions, three or four backup systems can be specified. In the field of occupational safety and health, some systems are seen as so vital as to require design redundancy. Mechanical power presses are an example.

Still another fail-safe design principle is the principle of "worst case":

Principle of Worst Case

The design of a system should consider the worst situation to which it may be subjected in use.

This principle is really a recognition of "Murphy's law," which states that "if anything can go wrong, it will." Murphy's law is no joke; it is a simple observation of the result of chance occurrences over a long period of time. Random events that have a constant hazard of occurrence are called "Poisson processes." The design of a system must consider the possibility of the occurrence of some chance event that can have an adverse effect on safety and health.

An application of the principle of worst case is seen in the specification of explosion-proof motors in ventilation systems for rooms in which flammable liquids are handled. Explosion-proof motors are much more expensive than ordinary motors, and industries may resist the requirement to install explosion-proof motors, especially in those processes in which the vapor levels of the substances mixed never even get close to the explosive range. But consider the scenario presented by a hot summer day in which a spill happens to occur. The hot weather raises the vapor level of the flammable liquid being handled. A spill at such an unfortunate time dramatically increases the liquid surface exposure, which makes the problem many times worse. At no other time would the ventilation system be more important. But if the motor is not explosion-proof and is exposed to the critical concentration of vapors, a catastrophic explosion would occur as soon as the ventilation system switches on.

The concept of "defensive driving" is well known to all drivers and serves to explain the principle of worst case. Defensive drivers control their vehicles to the extent that they are prepared for the worst random event they can reasonably imagine.

Engineering Pitfalls

It is easy to get caught up in the idea that technology can solve our problems, including elimination of workplace hazards. Certainly, an inventor of a new gadget to prevent injuries or illnesses can quickly become enamored with it and present a convincing argument that the new invention should be installed in workplaces everywhere. When standards writers are persuaded by these arguments, they often require all appropriate industries to install the new device. Several things can go wrong, however.

Recalling the case against the enforcement approach, certain unusual circumstances can make the engineering solution inappropriate or even unsafe. A good example is the use of spring-loaded cutoff valves in air lines for compressed air tools. The purpose of the cutoff valve is to prevent hose whip action by stopping the flow of air if the tool accidentally separates from the hose. The sudden flow of air overcomes the spring-loaded valve and closes it, stopping the flow. The problem comes when several tools are operated from the same main hose and flow reaches maximum even during normal use. The cutoff then becomes a nuisance and impedes production.

A second problem with the engineering approach is related to the first: Workers remove or defeat the purpose of engineering controls or safety devices. The most obvious example is the removal of guards from machines. Before faulting the worker for such behavior, take a close look at the guard design. Some guards are so awkward that they make the work nearly impossible. Some machine guards are so impractical that they conjure up doubts about the motives of the equipment manufacturer. There is a legal motivation to install an impractical guard on a new machine so that users will take the guard off before putting the machine in service. When the user modifies a machine by removing a guard, the manufacturer is absolved of guilt for any accident that later occurs but that might reasonably have been prevented by the guard.

An irony of the engineering approach is that if the engineered system does not do the job for which it was intended, it can do more harm than good by engendering a false sense of security. A printing press operator once lost the end of his finger when he attempted to reach into the danger zone, which was protected by a photoelectric sensing device. Believing the system to be infallible, he reached as fast as he could, thinking that the control system would stop the machine in time; but it did not.

Such false sense of security can even lead to new operator procedures which depend on the safety device to control the operation so that the work can be speeded up. The best example that comes to mind is the hoist limit switch on an overhead crane. The switch is tripped, shutting off the hoist motor, if the hoist load block approaches too close to the bridge. The idea sounds good, but the operator can take advantage of the device by *depending* on the switch to stop the load during normal operation. The hoist limit switch is not intended as an operating control, but workers can and do use it that way. The only defense against such use appears to be proper training and safe attitudes on the part of the operator—that is, the psychological approach.

Finally, the engineered system can sometimes cause a hazard, as illustrated in the following example in which a pneumatic press ram pinned an operator's hand on the *upstroke* (see Figure 3.1). The press was equipped with a two-hand control that was designed for safety's sake not to allow the press to be actuated except by both hands of the operator. Ironically, the two-hand control created a hazard. This press was later redesigned to place a shield in front of the ram so that the operator would be unable to reach into the area above the ram.

As another example, robots are being used to work in hot, noisy environments, to lift heavy objects, and to otherwise serve where humans might be injured or suffer health hazards. Most industrial robots are simply mechanical arms computer programmed to feed material to machines or to do welding. But the mindless swinging of these mechanical arms can cause injury to workers who get in the path of the robot.

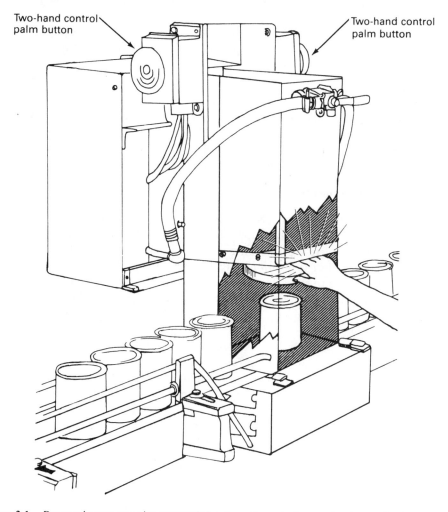

Figure 3.1. Pneumatic press ram pins operator's hands on the upstroke; two-hand control safety device prevents the operator from reactivating the press to release her hand.

The irony is that a hazard is created by the robot, the very purpose of which was to reduce hazards. One solution is to make the robot more sophisticated, giving it sensors to detect when a foreign object or person gets into its path. Another solution is simply to install guardrails around the robot or otherwise keep personnel out of the danger zone.

In summary, the engineering approach is a good one and deserves the emphasis it is receiving in the current decade. However, there are pitfalls, and the Safety and Health Manager needs a certain sophistication to see both the advantages and disadvantages in the proposed capital equipment investments in safety and health systems. Upon review of the foregoing examples of engineering pitfalls, it can be seen that almost every problem can be dealt with if some additional thought is given to the design of the equipment or its intended operation. The conclusion to be reached is that engineering can solve safety and health problems, but the Safety and Health Manager should not naively assume that the solutions will be simple.

THE ANALYTICAL APPROACH

The analytical approach deals with hazards by studying their mechanisms, analyzing statistical histories, computing probabilities of accidents, conducting epidemiological and toxicological studies, and weighing costs and benefits of hazards elimination. Many, but not all, analytical methods involve computations.

Accident Analysis

Accident and incident (near miss) analysis is so important that it has already been discussed extensively in Chapter 2. No safety and health program within an industrial plant is complete without some form of review of mishaps that have actually occurred. The subject is mentioned again here only to classify it as within the analytical approach and to show its relationship to other methods of hazard avoidance. Its only drawback is that after an accident has occurred it is too late to prevent the consequences of that accident. But the value of the analysis for future accident prevention is critical.

Accident analysis is not used nearly enough to assist in the other approaches to hazard avoidance. The enforcement approach would be much more palatable to the public if the enforcing agency would spend more time analyzing accident histories so that citations are written only for the most important violations. The psychological approach could also be strengthened a great deal by substantiating persuasive appeals with actual results of accidents. The engineering approach needs accident analysis to know where the problems are and to design the solution to deal with all the accident mechanisms.

Failure Modes and Effects Analysis

Sometimes a hazard has several origins, and a detailed analysis of potential causes must be made. Reliability engineers use a method called *failure modes and effects analysis* (FMEA) to trace the effects of individual component failures on the overall or "catastrophic" failure of equipment. Such an analysis is equipment oriented instead of hazard oriented. In their own interest, manufacturers of equipment sometimes perform an FMEA before a new product is released. Users of these products can sometimes benefit from an examination of the manufacturer's FMEA in determining what caused a particular piece of equipment to fail in use.

The FMEA can become important to the Safety and Health Manager when the failure of a piece of equipment can result in an industrial injury or illness. If a piece of equipment is critical to the health or safety of employees, the Safety and Health Manager may desire to request a report of an FMEA performed by the manufacturer or potential bidder for the equipment. In practice, however, FMEA is usually ignored by Safety and Health Managers until *after* an accident has occurred. Certainly, Safety and Health Managers should at least know what the initials FMEA represent so that they are not dazzled by the term in the courtroom should equipment manufacturers use it to defend the safety of their products.

One beneficial way of using FMEA *before* an accident occurs is in preventive maintenance. Every component of equipment has some feasible mechanism for eventual failure. To simply use equipment until it eventually fails can sometimes have tragic consequences. Consider, for example, the wire rope on a crane, or the chain links in a sling, or the brakes on a forklift truck. The FMEA can direct attention to critical components that should be set up on a preventive maintenance policy which permits parts to be inspected and replaced *before* failure.

Fault Tree Analysis

A very similar but more general method of analysis than FMEA is "fault tree analysis." Whereas FMEA focuses upon component reliability, fault tree analysis concentrates on the end result, which is usually an accident or some other adverse consequence. Accidents are caused at least as often by procedural errors as by equipment failures, and fault tree analysis considers all causes—procedural and/or equipment. The method was developed by Bell Laboratories under contract with the US Air Force in the early 1960s. The objective was to avoid a potential missile system disaster.

The term *fault tree* arises from the appearance of the logic diagram that is used to analyze the probabilities associated with the various causes and their effects. The leaves and branches of the fault tree are the myriad individual circumstances or events that can contribute to an accident. The base or trunk of the tree is the catastrophic accident or other undesirable result being studied. Figure 3.2 shows a sample fault tree diagram of the network of causal relationships that contribute to electrocution of a worker using a portable electric drill.

The diagram in Figure 3.2 reveals the use of two symbols in coding causal relationships. Figure 3.3 deciphers this code. It is essential that the analyst be able to distinguish between the "AND" versus the "OR" relationship for event conditions. All of the causal event conditions are required to be present to cause a result to occur when these conditions are connected by an AND gate. However, a single condition is sufficient to cause a result to occur when conditions are connected by an OR gate.

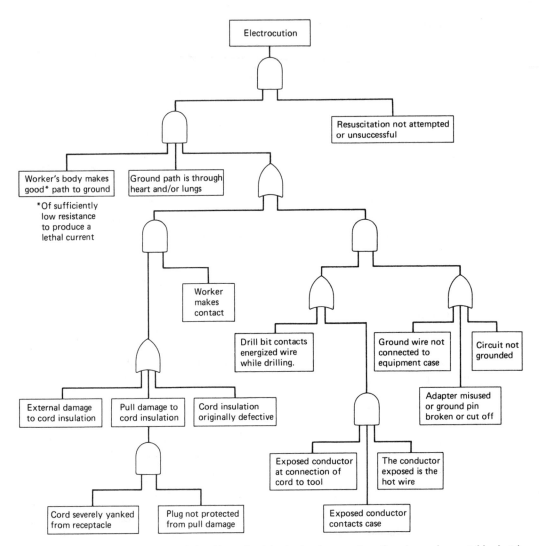

Figure 3.2. Fault tree analysis of hazard origins in the electrocution of workers using portable electric drills (not double-insulated).

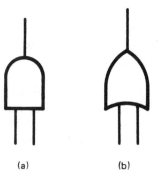

(a)   (b)

Figure 3.3. Logic code for fault tree diagrams: (a) AND-gate symbol; (b) OR-gate symbol.

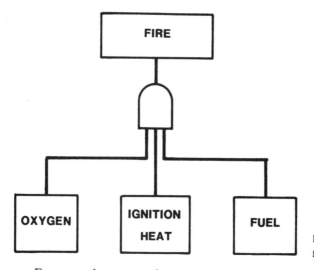

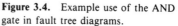

Figure 3.4. Example use of the AND gate in fault tree diagrams.

For example, oxygen, heat, and fuel are all required to produce fire, so they are connected by an AND gate, as shown in Figure 3.4. But *either* an open flame or a static spark may be sufficient to produce ignition heat for a given substance, so these conditions are connected by an OR gate, as shown in Figure 3.5. Note that Figures 3.4 and 3.5 could be combined to start the buildup of a branch of a fault tree.

One difficulty with fault tree analysis is that it requires each condition to be stated in absolute "yes/no" or "go/nogo" language. The analysis breaks down if a condition as stated may or may not cause a specified result. When the analyst is confronted with a "maybe" situation, it usually means that the cause has not been sufficiently analyzed to determine what additional conditions are also necessary to effect the result. Therefore, the difficulty of dealing with a "maybe" situation forces the analyst to look more deeply into the fault relationships, so the "difficulty" may be a benefit after all.

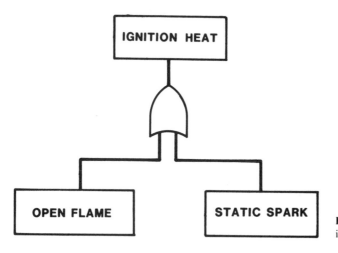

Figure 3.5. Sample use of the OR gate in fault tree analysis.

Fault tree analysis permits the analyst to compute quantitative measures of the probabilities of accident occurrence. The computation is tricky, however, and at best is only as good as the estimates of the probabilities of occurrence of the causal conditions. It seems intuitive that one should add the probabilities of events leading into an OR gate and multiply the probabilities of events leading to an AND gate. This intuition is wrong in both cases and will be shown for the OR gate as follows. Suppose in the example shown in Figure 3.5 that the probabilities of occurrence of the two event causes were as follows:

Event cause	Probability of occurrence (%)
Open flame	50
Static spark	50

To add these two probabilities would result in a 100% chance of reaching "ignition heat," an obviously erroneous result. A 50–50 chance of either of two possible causes is insufficient basis to find a 100% certainty of any resulting event. To make the logic even more convincing, the reader should rework the example using the following probabilities:

Event cause	Probability of occurrence (%)
Open flame	60
Static spark	70

Obviously, no result can have a probability of 130%! The correct probability computation is to subtract the "intersection," or probability that both independent causes would occur as follows:

$$P[\text{ignition heat}] = P[\text{open flame}] + P[\text{static spark}]$$
$$- P[\text{open flame}] \times P[\text{static spark}]$$
$$= 0.6 + 0.7 - 0.6 \times 0.7$$
$$= 1.3 - 0.42 = 0.88 = 88\%$$

For OR gates in which there are several event causes, the computation becomes very complex. Adding to the complexity is the question of whether the event causes are "independent"; that is, the occurrence of one event does not affect the likelihood of the occurrence of the other event cause(s). A special case of "dependent" events are events which are "mutually exclusive"; that is, the occurrence of one event *precludes* the occurrence of the other(s). The "mutually exclusive" condition simplifies the computation, but event causes in fault tree diagrams are typically not mutually exclusive.

Fully resolving the problem of fault tree computations would require a treatise on probability theory, which is beyond the scope of this book. Suffice it to say here that fault tree probability computations are more complex than most people think and are usually done incorrectly.

Despite the computational problems associated with fault tree computations, the fault tree diagram itself is a useful analytical tool. The diagramming process itself forces the analyst to think about various event causes and their relationship to the overall problem. The completed diagram permits certain logical conclusions to be reached without computation. For instance, in Figure 3.2 the event "worker makes contact" is key because from the diagram it can be seen that prevention of this event will preclude any of the five events to the lower left in the diagram from causing electrocution. Even more important is the event "worker's body makes good path to ground." Prevention of this one single event is sufficient to prevent electrocution, according to the diagram. The reader can no doubt draw other interesting conclusions from the diagram in Figure 3.2. These conclusions result in a more sophisticated understanding of the hazard and in turn may lead to revisions to make the diagram more realistic. Such a developmental process leads to the overall goal of hazard avoidance.

Loss Incident Causation Models

Closely related to both fault tree analysis and failure modes and effects analysis is a model that emphasizes the causes of "loss incidents," whether or not the incident results in personal injury, reported by Robert E. McClay (ref. 51). McClay's model attempts to take a universal perspective in which the entire causal system is examined, including primary causes, labeled *proximal* causes, and secondary causes, labeled *distal* causes. A proximal cause would be a direct hazard in the conventional sense of the word—for example, a missing guard on a punch press. By contrast, an example of a distal cause would include a management attitude or policy that is deficient in allocating resources and attention to the elimination of hazards. Distal causes are as important as proximal causes, because although the effect of distal causes is less direct and immediate, it is the distal causes that create and shape the proximal causes.

A critical point in the progression of the loss incident causation model is the "point of irreversibility." McClay identifies this as the point at which the various interacting proximal causes will result in a loss incident. Despite the number and variety of proximal causes, only a few select cases will result in a sequence of events in which the point of irreversibility is reached. Once this point is reached, a loss incident is unavoidable. This is still not to say that an injury will occur. A loss incident can

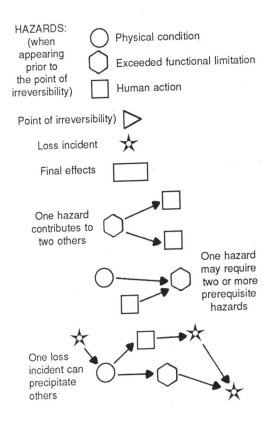

Figure 3.6. Symbols used in diagramming loss incidents. (*Source:* Reprinted by permission of Professional Safety, Des Plaines, IL.)

occur, and still no personnel might be exposed, or perhaps whatever exposure does occur is not injurious. Factors such as personnel exposure in the event of the loss incident affect the severity of the effects of the incident after the point of irreversibility has been exceeded. Such factors affecting the outcome can be either negative or positive; that is, they can be "aggravating factors," which make the outcome more severe, or they can be "mitigating factors," which make the outcome less severe.

A diagramming convention has been proposed by McClay to assist the analyst in visualizing the universal model of causation. Figure 3.6 defines the symbols to be used in depicting the model. Note in Figure 3.6 that proximal causes are represented by three different symbols that represent respectively three categories of hazards: physical conditions, exceeded functional limitations, and human actions. Each of these three types of hazards can have a causal relationship with either of the other two, and even a loss incident can in turn have a causal relationship upon other proximal causes or other loss incidents, as is illustrated in the examples in the diagram. The diagram in Figure 3.7 summarizes the universal model of loss incident causation, revealing the relationship between distal and proximal causes and representing the region prior to the point of irreversibility as the "sphere of control." The value in visualizing a loss incident causation system in a diagram such as that

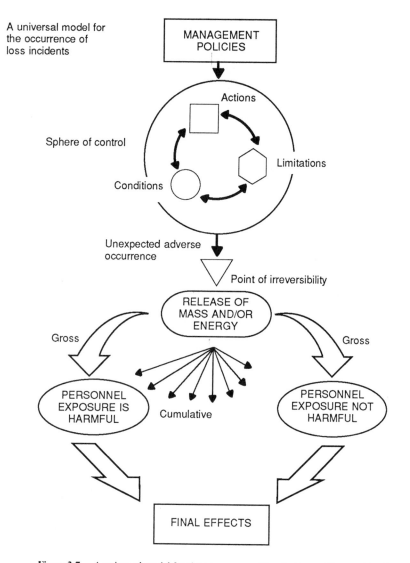

A universal model for
the occurrence of
loss incidents

MANAGEMENT
POLICIES

Actions

Sphere of control

Limitations

Conditions

Unexpected adverse
occurrence

Point of irreversibility

RELEASE OF
MASS AND/OR
ENERGY

Gross

Gross

PERSONNEL
EXPOSURE IS
HARMFUL

Cumulative

PERSONNEL
EXPOSURE NOT
HARMFUL

FINAL EFFECTS

Figure 3.7. A universal model for the occurrence of loss incidents. (*Source:* Reprinted by permission of Professional Safety, Des Plaines, IL.)

in Figure 3.7 is in permitting the analyst or the Safety and Health Manager to distinguish the factors and conditions that can be controlled and to perceive the consequences, good and bad, to be derived from applying resources to eliminate hazards or to mitigate their effects. As such, the analytical approach can be useful in assisting the Safety and Health Manager to define and accomplish "reasonable objectives" as suggested in Chapter 1.

Toxicology

Toxicology is the study of the nature and effects of poisons. Industrial toxicology is especially concerned with identifying what industrial materials or contaminants can harm workers and what should be done to control these materials. This is really a broad statement because virtually every material is harmful to living organisms if the exposure rate or quantity is great enough.

Many toxicological studies are performed on animals to provide the basis for conclusion about the hazards to humans. These animal studies are essential because most toxicological experiments would cause death or serious harm to human subjects. The disadvantage is that animal defenses to various toxic substances vary between species. The more the field of toxicology advances, however, the more various toxic materials can be classified and their effects predicted even before experiment. Rabbit-man, monkey-man, mouse-man, and guinea pig-man comparisons of species defenses to various agents are becoming fairly well known.

Epidemiological Studies

Epidemiology is contrasted with toxicology in that epidemiology studies are strictly of people, not animals. The word obviously derives from the word *epidemic*, and in the literal sense epidemiology is the study of epidemics. The epidemiological approach is to examine populations of people to associate various patterns of possible disease causes with the occurrence of the disease. It draws heavily on the analytical tools of mathematical statistics.

A classic epidemiological study was the association of the disease rubella (German measles) in pregnant women to birth defects in the infants born of those pregnancies. The study began with a curiosity observed by N. McAlister Gregg, an Australian ophthalmologist, in 1941. He observed eye cataracts among infants born of mothers who had had German measles during pregnancy in 1939 and 1940. The phenomenon might have been unnoticed except for the Australian epidemic of that disease during the World War II buildup. Years of statistical epidemiological studies later confirmed a strong relationship between rubella during pregnancy and a wide variety of birth defects in the infants born of those pregnancies.

Epidemics are usually thought of as attacking a general population at a specific time in a specific geographical area. Examples are the bubonic plague in Europe in the mid-fourteenth century and the rubella epidemic in Australia in 1939-1940. But more subtle epidemics attack a specific subgroup of people who may be spread out over time and place. In other words, the victims of the epidemic may not live in one place or at one time but have some other common characteristic, such as what they *do*. This aspect of epidemiology is what makes it important to occupational safety and health. Thus lung fibrosis may not be a very common disease in any location or at any time, but when the population of only those persons who have worked with asbestos is examined, it can be seen that after a long latency period, lung fibrosis can be considered an epidemic. Epidemiological studies linking lung fibrosis to asbestos have led to the identification of a type of lung fibrosis known as "asbestosis." Other epidemiological links are "brown lung" to textile workers, "black lung" to coal miners, and angiosarcoma to vinyl chloride workers.

Both epidemiology and toxicology are important elements in the analytical approach to avoiding hazards, but the Safety and Health Manager does not typically perform such studies. The studies provide the basis for mandatory standards which are subsequently used in the enforcement approach. Safety and Health Managers may also use the results of such analytical studies to substantiate the psychological approach or as a justification for an engineering approach to a health problem.

Cost–Benefit Analysis

Chapter 1 set the record straight on the importance of hazard costs. Like it or not, people do make cost judgments on occupational safety and health—not just management, but workers, too. In the world of reality, funds do have limitations, and cost–benefit analyses must be used to compare capital investment alternatives. Safe-

---

CASE STUDY 3.5 COST–BENEFIT ANALYSIS OF INSTALLING A PARTICULAR MACHINE GUARD

Cost

Amortization of initial investment:

Initial cost	$4,000
Expected useful life	8 years
Salvage value	0
Interest cost on invested capital	20%
Annual cost = $4,000 × (20% interest factor for 8 years)	
= $4,000 × 0.26061 =	$1,042
Expected cost of annual maintenance (if any)	0
Expected annual cost due to reduction in production rate (if any)	800
Total expected annual cost	**$1,842**

Benefit

Estimated tangible costs per injury of this type	$ 350
Estimated intangible costs per injury of this type	2,400
Total costs per injury	$2,750
Average number of injuries per year on this machine due to this hazard	1.2
Expected number of injuries of this type after guarding	0.1
Expected reduction in injuries per year	1.1
Expected benefit = $2,750 × 1.1 =	**$3,025**

Since the total expected benefit of $3,025 is more than the total expected cost of $1,842, the conclusion is that it would be worthwhile to install this machine guard.

ty and Health Managers who feel that they can justify at any cost capital investment proposals that can be shown to have the possibility of preventing injuries and illnesses can easily appear naive. There is always more than one opportunity to improve safety and health, and cost–benefit analyses provide the basis for deciding which ones to undertake first.

The biggest difficulty with cost–benefit analysis is the estimation of the benefit side of the picture. Benefits to safety and health consist of hazard reduction, and to make the cost-benefit analysis computation, some quantitative assessment of hazard must be made. Such probabilities of injury or illness are very difficult to determine. Statistical data are being compiled at the state level, as discussed in Chapter 2. But these data are still usually of insufficient detail to permit a quantitative determination of existing risk. In addition to this determination, an estimate of expected risk *after* the improvement must also be made, since it cannot be assumed in general that every safety and health improvement will completely eliminate hazards. Case Study 3.5 illustrates the method of cost–benefit analysis.

In Case Study 3.5, the reader will perceive the large degree of speculation in the estimates of hazard costs. Such speculation casts doubt upon the entire analysis, but it is better to speculate and calculate than to completely ignore the cost–benefit trade-off. This is an opportune point to lead into the next section of this chapter— hazards classification. There is a great deal to be gained from a subjective analysis of hazard costs, without resorting to overly sophisticated quantitative analysis, the credibility of which "hard" data may not support.

HAZARDS CLASSIFICATION SCALE

The absence of hard data to support quantitative cost–benefit analyses leaves a void of tools or benchmarks for use by the Safety and Health Manager, safety committee, or other party on whom the decision responsibility for safety and health improvement may be thrust. Some sort of ranking or scale is needed to distinguish between serious hazards and minor ones so that rational decisions can be made to eliminate hazards on a worst-first basis.

OSHA does recognize four categories of hazards or standards violations as follows:

- Imminent danger
- Serious violations
- Nonserious violations
- De minimus violations

Chapter 4 explains these categories in more detail. However, the categories are rather loosely defined and are distinguished chiefly by the extent of the penalty authorized for each type. The "imminent danger" category qualifies OSHA to seek a US district court injunction to force the employer to remove the hazard or face a court-ordered shutdown of the operation. The "de minimus" violations, by contrast, are merely technical violations that bear little relationship to safety or health; they typically do not carry a monetary penalty. But the designations of which violations will be recognized as falling into which category of hazard is especially unclear.

It is perhaps impossible to define clear-cut categories in every instance, but much is to be gained by some type of subjective ranking of workplace hazards. The thesis of this book is that a scale of 1 to 10 should be attempted, crude as that scale might be. Until people start talking about degrees of hazards on such a quantitative scale, little progress can be made toward establishing an effective and orderly strategy for hazard elimination. On a 10-point scale, a "10" is characterized as the worst hazard imaginable and a "1" the least significant or mildest of hazards.

The 10-point scale is recommended because such a scale has become very popular in everyday speech. Facilitated by the media, especially television, the public has come to understand a statement such as "on a scale of 1 to 10, that item (tennis match, ski slope, kiss, etc.) was at least a 9." The familiarity of this popular jargon can be employed to characterize workplace hazards.

Table 3.1 is a first attempt to describe subjectively each of ten levels of hazards. The definitions address basically four types of hazards: fatalities, health hazards, industrial noise hazards, and safety (injury) hazards. A clear-cut delineation is obviously difficult, and some readers will no doubt disagree with the wording of the definitions. Criticisms of the scale will reflect both the shortcomings of the definitions and the biases of the critics themselves. Acoustical experts, for instance, may want to place a high degree of emphasis on excessive noise hazards. Other specialists will want to emphasize other areas.

One critical test is met by the proposed scale. Within each hazard type, each successive level of the scale describes a progressively more severe hazard. Individual industries may devise more suitable definitions, but the idea is to start talking and thinking about hazards in terms of a 10-point scale.

One criterion omitted from the scale definitions is cost of compliance or cost of correcting a given hazard. Cost is a completely different criterion and is almost independent of level of hazard. That is, it can easily cost just as much or more to correct a Category 2 hazard as it does to correct a Category 9 hazard. Cost is an important criterion in the decision-making algorithm but is omitted from the scale definitions to permit a clear ranking of hazard priority first. Once the hazards have been sorted out, costs of hazard correction can be estimated and capital allocated for correction according to a rational capital investment policy.

Another missing criterion, this one perhaps conspicuous, is any mention of OSHA legal definitions of "imminent danger," "serious violation," and so on. To place these legal designations into fixed positions in the hazard scale would undermine the objectivity of the classification. Many persons have a preconceived notion of what legal penalty should be imposed or not imposed for a given hazardous situation. This bias is found in OSHA officers and their industrial counterparts as well. For instance, if a plant Safety and Health Manager should happen to believe that a given situation is not serious enough to warrant a plant shutdown, he or she would

TABLE 3.1 Category Descriptions for a 10-Point Scale for Workplace Hazards

1	2	3	4	5	6	7	8	9	10
"Technical" violations; OSHA standards may be violated but no real occupational health or safety hazard exists	No real fatality hazard	Fatality hazard and not of real concern	Fatality hazard either remote or nonexistent	Fatality hazard either remote or not applicable	Fatality hazard unlikely	Fatality not very likely, but still a consideration	Fatality possible; this operation has never produced a fatality, but a fatality could easily occur at any time	Fatality likely; similar conditions have produced fatalities in the past; conditions too risky for normal operation; rescue operations are undertaken for injured workers with rescuers using personal protective equipment	Fatality imminent; risks are grave; some employees earlier in the day have died or are dying; conditions are too risky even for daring rescue operations except perhaps with exotic rescue protection
	Health hazards minor or unverified	Health hazards have exceeded designated action levels	Health hazards characterized by illnesses that are usually temporary; controls or personal protective equipment may or may not be required	Long-range health *may* be at risk; controls or personal protective equipment *advisable* or *required* by OSHA	Long-range health definitely at risk; controls or personal protective equipment *required* by OSHA	Serious long-range health hazards are proven; controls or personal protective equipment essential to prevent *serious* occupational illness	Severe long-range health hazards are *obvious*; controls or personal protective equipment essential to prevent *fatal* occupational illness		
		or	*or*	*or*	*or*	*or*	*or*		
		Sound exposure action levels exceeded (e.g., continuous exposure in the range 85–90 dBA)	Temporary hearing damage usually will result without controls or protection	Hearing damage may be permanent without controls or protection (e.g., continuous 8-hr exposure in the range 95–100 dBA)	Hearing damage likely to be permanent without controls or protection (e.g., continuous 8-hr exposure in the range 100–105 dBA)	Hearing damage obviously would be *severe* and permanent without protection (e.g., continuous 8-hr exposure in excess of 105 dBA)	Major injury is likely; amputations or other major injuries have *already* occurred in this operation in the past		
	Even minor injuries unlikely	*or*	*or*	*or*	*or*	*or*			
		Minor injury risk exists but major injury hazard is very unlikely	Minor injuries, such as cuts or abrasions, but major injury risk is low	Major injuries such as amputation not very likely	Major injury such as amputation not very likely but definitely *could* occur	Major injury such as amputation could easily occur			

64

tend to prohibit selection of any designation that would bear the legal OSHA designation "imminent danger." This is despite the fact that reason would place the Category 10 definition into the imminent danger category.

Three credible profiles of the OSHA legal classification superimposed on the proposed 10-point scale are displayed in Figure 3.8. Profile A has an industry flavor and represents the viewpoints of some business executives. Profile A cannot be considered an extreme position because thousands of American businesses are headed

HAZARD CLASSIFICATION SCALE

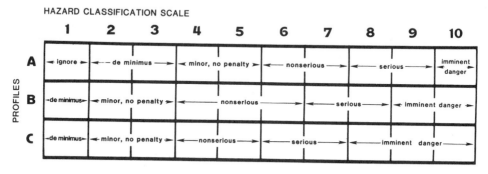

Figure 3.8. Three profiles of legal designations for hazards superimposed on the 10-point hazard classification scale.

by persons who believe that no government official should have the right to step in and close their businesses with or without a court order, regardless of hazards. Therefore, some business executives do not recognize the legal category "imminent danger." Profile A does at least recognize the imminent danger category, although the profile is skewed to the right. Profile C shows a contrasting position, being skewed to the left. Some OSHA officers have demonstrated that their positions are very close to profile C. Profile C likewise is not the ultimate in left extreme because some people believe that any fatality or amputation hazard should be classified as serious regardless of the remoteness of the hazard. Profile B represents a sort of middle-of-the-road position.

A final consideration in the classification of hazards would be the industry environment in which the hazard is portrayed. What seems hazardous to a structural steelworker working 100 feet above the ground on a narrow beam is in an entirely different category from what seems hazardous to an accountant. A similar comparison could be drawn between a coal miner and a computer analyst. It is only hoped that as a first step at least, coal miners should know what other coal miners mean when they talk about Category 4 hazards. Similarly, accountants should be uniform in their conception of Category 9 hazards, for example.

Except for some very broad categories, such as "serious" and "nonserious," federal law has little to say about degree of hazard. No consistent criterion is available to Safety and Health Managers, who must decide which problems must be tackled first. The 10-point scale recommended by this book for ranking safety and health hazards offers an opportunity to bring some order to this perplexing problem. The scale is a vehicle for encouraging all parties to focus on the degree of each hazard so that rational decisions can be made to correct those problems that are indeed significant.

SUMMARY

The sophisticated Safety and Health Manager is not content with one approach to dealing with workplace hazards. There is too much uncertainty to solve the enormous problems neatly with a simple approach such as "awards for no lost-time accidents" or "fines for anyone who breaks the rules." These two approaches, together with all the others, have their place, but an integrated program using the strengths of all approaches has the greatest potential for success.

While becoming more sophisticated, Safety and Health Managers may fall into the trap of becoming more enamored with their impressive-looking analyses, scientific formulas, and statistics. Some of the best analyses may be subjective, not quantitative. The 10-point hazard classification scale suggested in this chapter represents an opportunity for Safety and Health Managers to talk and think about workplace hazards in terms of *degree*. Corporate top management has been looking for years for that breed of Safety and Health Manager to emerge—the kind who can discern between the really significant problems, the ordinary problems, and the trivial ones.

EXERCISES AND STUDY QUESTIONS

3.1. Name the four categories of seriousness of hazards as recognized by OSHA.

3.2. What is the OSHA designation for minor standards violations that bear little or no relationship to safety and health?

3.3. Consider the following hazards and rank each on a scale of 1 to 10 (10 is worst). Also classify each into the four OSHA categories according to your opinion:
  (a) Ground plug (third prong) is cut off on a power cord for an office computer.
  (b) Ground plug is cut off on a power cord for a shop wet-vac vacuum cleaner.
  (c) An electric drill with faulty wiring causes an employee to receive a severe shock and to refuse to use it. Another employee scoffs at the hazard, claims that he is "too tough for 110 volts" and picks up the tool to continue the job.

3.4. A well-known safety rule is not to pull an electrical plug out of a wall socket by means of the cord. Have you broken this rule? Do you regularly do so? What is the reason for this rule? Do you think the rule is an effective one?

3.5. Do you believe that safety or health rules are "made to be broken?" Why or why not?

3.6. Name four basic approaches to hazard avoidance.

3.7. Recall your first supervisor in your first full-time job. Did he or she even mention safety or health on the job? Was your supervisor's influence on your safety habits positive, negative, or neutral?

3.8. What is the Heinrich ratio?

3.9. What are the three lines of defense against health hazards?

3.10. What is the standard safety factor for crane hoists? for scaffolds? for scaffold ropes?

3.11. Name three general fail-safe principles. Can you think of a real-world example of any of them?

3.12. What has Murphy's law to do with occupational safety and health?

3.13. What is the purpose of FMEA?

3.14. Construct a fault tree diagram describing potential causes of fire in paint spray areas. Compare your analysis with others in the class.

3.15. Construct a fault tree diagram that describes the ways a pair of dice can be thrown to "roll 7." Calculate the "risk" or chances of rolling 7.

3.16. Explain the difference between toxicology and epidemiology.

3.17. Alter the fault tree diagram of Figure 3.2 to consider the possibility that the portable electric drill might be "double-insulated" (i.e., sheathed in an approved plastic case to prevent electrical contact with the metal tool case).

3.18. In the universal loss incident causation model, what is the difference between proximal factors and distal factors? To which category does "management policy" belong?

3.19. Explain the concept of "point of irreversibility." Does this point guarantee that a personal injury will occur? What are the roles of "aggravating" and "mitigating" factors?

3.20. Accident causes A, B, and C each have a probability of occurrence of about 1 in 1,000, but the causes are mutually exclusive. Suppose Cause B does in fact occur in a given situation. What are the chances that Cause A will occur in this situation?

3.21. Using a pair of dice for a simulation device, suppose an outcome of 11 represents an industrial accident.
   (a) Draw a fault tree diagram to illustrate the ways this "accident" could happen.
   (b) Calculate the probability that this accident will happen in a given throw of the dice.
   (c) Are the "causes" of this accident mutually exclusive?

3.22. Following are some causes of slips and falls:
   (a) Oil leaks from fork lift trucks.
   (b) Water or wax on floor during cleaning operations.
   (c) Ice on walkways or loading docks during winter weather.
   (d) Slippery heels on footwear.
   From the perspective of a total plant safety program, are these causes mutually exclusive? Why or why not? For a given single accident, are these causes mutually exclusive? Why or why not?

3.23. Following are three among many causes of injury to an operator of a punch press:
   (a) Barrier guard of adequate size but placed too high—worker can reach under the guard.
   (b) Barrier guard of adequate size but placed too low—worker can reach over the guard.
   (c) Openings in the barrier guard are too large—worker can reach through the guard.
   Which of these causes are mutually exclusive?

3.24. A certain type of injury has tangible costs of $15,000 per occurrence and intangible costs estimated to be $250,000 per occurrence. Injury frequency is 0.01 per year but would be reduced by one half with the installation of a new engineering control system. What annual benefit would the new system provide?

3.25. A certain ventilation system would cost a firm approximately $60,000, which can be amortized over its useful life at a cost of $15,000 per year. Annual maintenance costs are expected to be $600 per year and monthly operating costs (utilities) at $150 per month.

The system is expected to facilitate production by reducing the amount of machine clean-ing for an expected annual savings of $1,200 per year. The primary benefit of the pro-posed ventilation system is expected to be the elimination of the need for respirators, which cost the company $4,000 per year in equipment, maintenance, employee train-ing, and respiratory system management. The ventilation system is expected to reduce short-term illness complaints by an average of 6 per year and long-term illnesses by an average of 0.2 per year. The short-term illnesses result in a total cost of $600 per occur-rence, including intangibles. The long-term illnesses are expected to result in a total cost of $30,000 per occurrence, including intangibles. Use a cost–benefit analysis to deter-mine whether the ventilation system should be installed. What is the primary benefit of the ventilation system?

3.26. On a scale of 1 to 10 (10 is worst), how would you rate each of the following hazards?
  (a) Ten-foot balcony not provided with a guardrail. Workers regularly work close to the edge every day without fall protection.
  (b) Same as part (a) except that the working surface is outdoors and is very slippery in rainy weather.
  (c) No guardrail on a 10-foot balcony accessed only twice per year by a maintenance worker to service an air conditioner.
  (d) An unguarded flat rooftop accessed only by air conditioner maintenance person-nel. The closest necessary approach to the edge of the roof is 25 feet.
  (e) Broken ladder rung (middle rung of a 12-foot ladder).
  (f) Overfilled waste receptacles in lunchroom.
  (g) Two-ton hoist rope with dangerously frayed and broken wires in several strands.
  (h) Mushroomed head on a cold chisel.

3.27. Consider the following accident case history:
On July 4, 1980, three workers, ages 14, 16, and 17, were installing a sign at a bait shop along a state highway. They were using a truck-mounted extensible metal ladder to un-load and position an upright steel support for the sign. Two of the workers were hold-ing and guiding the steel support while the third stood on the flatbed truck operating the extensible ladder controls. The metal ladder came into contact with a 13,200-volt overhead power line. The two workers guiding the steel support were standing on the ground and were subjected to immediate electrocution. The third worker attempted to break contact with the power line by operating the controls, but the controls had be-come useless, probably because the control wiring had become burned by the high-voltage contact. The worker then leaped from the truck and ran around to the front to try to drive the truck away to break contact. As he grasped for the door handle to the truck cab, he was still standing on the ground, which provided a path for the current through his body. The utility company had equipped the line with a "recloser" which normally would have broken the circuit under these conditions, but for a variety of reasons failed to do so in this case. Therefore, the power remained on for a rather lengthy period. The high voltage and burning current finally broke down the dielectric strength of the rubber tires and the tires blew out. This shifted the truck's position and contact with the power line was broken, but not before one worker had been burned in half, and another's legs were both burned off; all three workers died.
How could future accidents of this type be prevented? Compare the four basic approaches to hazard avoidance in preventing accidents of this type.

4

Impact of Federal Regulation

The field of industrial safety and health took a giant leap in the early 1970s. There is some question, however, whether the leap was forward or backward. On December 29, 1970, Congress passed the Williams-Steiger Occupational Safety and Health Act of 1970, creating OSHA, the Occupational Safety and Health Administration of the US Department of Labor. OSHA got off to a bad start and immediately became the target of sharp public criticism. But at the same time the agency created an immediate focus of attention on the field of industrial safety and health. Changes in federal administrations have changed OSHA's methods, and these methods need to be examined here, along with the basic approaches upon which OSHA was founded. Regardless of how OSHA may change, its impact on the field is a permanent one, and this book would not be complete without examining this impact.

Most people think of OSHA when the subject of federal safety and health regulations is mentioned. But there is also MSHA, the Mine Safety and Health Administration; TOSCA, the Toxic Substances Control Act; and CPSC, the Consumer Product Safety Commission. Most of these other agencies and legislation governing safety and health are patterned after the OSHA agency and legislation. Accordingly, this chapter examines the basic approach of OSHA and the impact it has had on the field.

STANDARDS

The most significant change that OSHA brought to industry was a book of federal standards. Never before had virtually all general industries been subjected to a book of federally prescribed, mandatory rules for worker safety and health. This set of rules formed the basis for inspection, citations, penalties, and virtually every activity in which OSHA is engaged. One rule, however, was not in the book; it is the most important rule of all and will be discussed first.

General Duty Clause

Congress decided to set up one general rule for all to follow, and this rule was included in its entirety right in the text of the law that created OSHA. This rule is called the "General Duty Clause," and might be called the First Commandment of OSHA. It is quoted as follows:

> Public Law 91-596
>
> Section 5(a)   Each employer. . .
>
> (1) shall furnish to each of his employees employment and a place of employment which are free from recognized hazards that are causing or are likely to cause death or serious physical harm to his employees. . . .

The General Duty Clause is cited by OSHA whenever a serious safety or health violation is alleged for which no specific rule seems to apply. This has drawn some criticism from industry because after a serious accident has happened, it is easy to conclude that a situation was unsafe. However, before the accident occurred, it might not be clear what should have been done, especially in the absence of any specific rule to provide guidance. Sometimes a specific rule will also apply somewhat, and OSHA has cited both the General Duty Clause and the specific standard. When a situation is clearly in violation of a specific standard, OSHA usually does not bother to cite the General Duty Clause.

If the General Duty Clause is the First Commandment of OSHA, the second is like unto it, and is quoted as follows:

> Public Law 91-596
>
> Section 5(b)   Each employee shall comply with occupational safety and health standards and all rules, regulations, and orders issued pursuant to this Act which are applicable to his own actions and conduct.

Note that Section 5(a)(1) describes a responsibility for employ*ers*, whereas Section 5(b) pertains to employ*ees*. Penalties are prescribed for employer violations, but no penalty is prescribed for employee violations. Section 5(a)(1) has been cited quite frequently, but as far as this author knows, there has never been an OSHA citation of Section 5(b).

Promulgation

In addition to the General Duty Clause, the law that created OSHA also set up machinery for OSHA to issue new standards too technical and detailed to be included one by one in the text of the law passed by Congress. The law provides safeguards intended to ensure that OSHA will be fair and give all interested parties an opportunity to support or object to proposed new standards. In addition to promulgating new standards, OSHA is also permitted to revise old standards or even revoke them, following the same rulemaking procedure with prescribed safeguards.

"National Consensus"

The OSHA law recognized the existence of "national consensus standards" already in use prior to the enactment of the OSHA law. This is a very important part of the law because it gave OSHA the authority to *bypass the procedural safeguards* just mentioned and issue standards without consulting the public. The principle was that the standards by their prior existence had already been accepted by the public. OSHA's authority to issue national consensus standards expired in early 1973, two years after the effective date of the Act. Thus after two years OSHA was no longer permitted to "grandfather in" safety and health standards. The largest number of national consensus standards were developed by the two leading national standards-producing organizations, the American National Standards Institute (ANSI) and the National Fire Protection Association (NFPA).

Besides the national consensus standards, any "established federal standard" was also permitted to be adopted as a general standard by OSHA. The established federal standards had formerly applied only to limited groups such as construction industries or government contracts. With the OSHA law Congress permitted OSHA to extend these standards to virtually all industries.

The biggest issue with the national consensus standards was whether they ever really represented the "national consensus." Virtually all of these standards had never been enforced, and the language used made it obvious that the original drafters of many of the standards had never intended for them to be mandatory rules to be enforced with monetary penalties for offending employers. This issue will be examined further in the section "Public Uproar."

Structure

When discussing standards, this book will sometimes use the term *horizontal standard* or *vertical standard*. The Safety and Health Manager needs to understand these terms. Prior to OSHA, states enforced codes for safety and health by industry, issuing separate handbooks of rules for each industry to follow. These standards have come to be called *vertical standards*. OSHA's basic approach, by contrast, is to generalize and organize standards by hazard sources, regardless of industry. These are called *horizontal standards*. There are certain standards that OSHA limits in scope to a particular industry, but these are the exceptions; the basic structure of the standards is horizontal.

OSHA almost *had* to use the horizontal approach. The nature of the OSHA law itself was to authorize the new agency to generalize on specific standards. Imagine for a moment the morass of books that would have been generated if OSHA had attempted to write a separate standards book for each industry, sifting through the national consensus standards to determine which standards should be included in each. Unusual industry categories would inevitably "fall through the cracks" under a vertical scheme, characterizing themselves as being under the scope of neither this nor that standard.

By using the horizontal approach, OSHA left the job of sifting through the standards to the public. Thus a general book of standards entitled *General Industry*, Part 1910, was published to cover virtually all industries. A manufacturer of teacups must read out of the same book as does an airline, and both must determine whether they need to be concerned with a provision entitled "horse scaffolds."

The construction industry is one for which OSHA published a vertical standard entitled *Construction Standards,* Part 1926, but even these special categories of industries are also covered by the more general Part 1910 with respect to any hazard for which no standard exists in the narrower vertical standard. OSHA did make the job of compliance a little easier by subsequently publishing a larger Construction Standard which included those parts of the General Industry standard that OSHA might cite at construction sites. However, they did not abdicate the right to turn to General Industry standards, which were not included in even the larger version of the Construction Standard.

The concept of vertical versus horizontal standards extends down within the subdivisions and individual paragraphs of the OSHA standard. For instance, OSHA standard 1910.106 pertains to flammable and combustible liquids and is basically a horizontal standard to be applied to general industry. However, within the standard there is paragraph 1910.106(i), which is entitled "Refineries, chemical plants, and distilleries." Therefore, that portion of 1910.106 should be considered a vertical standard applying only to those industries named.

One problem in determining the scope of all standards and of vertical standards in particular is the infeasibility of an exclusive classification of industries and equipment. Consider the diagram in Figure 4.1 (called a Venn diagram), which relates two different classes of equipment. The circles represent the two classes, and the shaded area is the overlap between the two groups. The circles could also be considered safety standards, and thus equipment represented by the shaded area would be subject to both standards.

Another classification for standards divides them into "specification" standards versus "performance" standards. The specification type is easier to enforce because it spells out in detail exactly what the employer must do and how to do it. The performance type has its advantages, however, in that it permits the employer latitude in devising innovative ways to eliminate or reduce hazards. In short, the specifica-

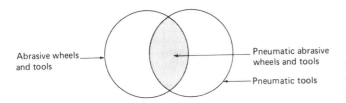

Abrasive wheels and tools — Pneumatic abrasive wheels and tools — Pneumatic tools

Figure 4.1. Venn diagram to represent the relationship of two different overlapping types of equipment.

tion standard emphasizes *methods* while the performance standard emphasizes *results*. The comparison is best shown by an example.

Example 4.1

(a) Example of a *specification* standard:

OSHA standard 1910.110   Storage and handling of liquefied petroleum gases

   (c) Cylinder systems

   (5) Containers and equipment used inside buildings or structures

   (vi) Containers are permitted to be used in industrial occupancies for processing, research, or experimental purposes as follows:

   (a) . . .

   (b) Containers connected to a manifold shall have a total water capacity not greater than 735 pounds (nominal 300 pounds LP-gas capacity) and not more than one such manifold may be located in the same room unless separated at least 20 feet from a similar unit.

(b) Example of a *performance* standard:

OSHA standard 1910.36   General requirements (under Subpart E—Means of Egress)

   (b) Fundamental requirements

   (2) Every building or structure shall be so constructed, arranged, equipped, maintained, and operated as to avoid danger to the lives and safety of its occupants from fire, smoke, fumes, or resulting panic during the period of time reasonably necessary for escape from the building or structure in case of fire or other emergency.

The standard in part (a) leaves no doubt in the reader's mind as to what must be done and specifies exactly the limitations of containers and manifolds within the room. There is no leeway for the employer to devise a better way—perhaps safer and cheaper—to minimize hazards. However, once the exact standards have been met, the employer can feel safe from citation.

In the performance standard stated in part (b), it can be seen that employers have all kinds of latitude to set up their buildings in ways to avoid "undue danger." If a new type of effective smoke alarm is developed and placed on the market, the employer has the choice of installing the new alarm or adopting a different strategy to enhance safety in the event of "fire or other emergency." But with the performance standard, a disagreement may develop between an enforcement inspector and the employer over the effectiveness of the methods selected by the employer to protect against "undue hazards."

Most standards do not exactly fit either type but do rather roughly fit one category or the other. Employers tend to prefer performance standards, while enforcement officials generally prefer specification standards. However, this issue does not clearly divide the two groups. Sometimes employers prefer specification standards because it is easier to determine whether a given facility or equipment meets a specification standard or not. Also, a few enforcement officials prefer the performance standard because then if an accident does occur, it tends to strengthen the case for a citation regardless of the specific steps the employer may have taken to prevent the hazard.

NIOSH

The National Institute for Occupational Safety and Health (NIOSH) was established by the OSHA law to carry on research and training. NIOSH is mentioned here because of its important role in recommending new standards to OSHA. OSHA has the sole authority to promulgate new standards, but NIOSH develops the criteria for new standards and conducts the research necessary to justify the need for new standards. NIOSH is often thought of as being concerned primarily with health and toxic materials, but it is important to remember that NIOSH also has responsibility for safety standards research and development.

Although the OSHA law infused new significance into the agency, NIOSH actually predated OSHA in existence. As early as 1914, NIOSH was part of the Department of Industrial Hygiene and Sanitation in Pennsylvania. In 1937 it became the Division of Industrial Hygiene as a part of the National Institutes of Health. Note the exclusion of the word *safety* in its title during this period. This orientation persisted for decades as the agency was made a part of the Bureau of State Services in 1944 and the Bureau of Occupational Health in 1968. Two years later the OSHA law elevated the agency to include safety and made it a part of the Department of Health, Education, and Welfare (HEW), which was subsequently changed to the Department of Health and Human Services (HHS).

ENFORCEMENT

Occupational safety and health ought to be topics of vital interest and importance regardless of OSHA, but most of us would not have realized this if OSHA had not been given the authority in the 1970s to inspect industries and issue citations with monetary penalties.

Inspections

The original OSHA law gave the OSHA official the right to enter a factory or other workplace without delay (at reasonable times) upon presenting credentials. Such credentials consisted of basic identification of the officer but no court-issued search warrant. The right of a government official to do this was later challenged by an Idaho businessman, and the U.S. Supreme Cout ruled in his favor in the famous "Barlow decision" in 1978. Employers may now invoke the Fourth Amendment to the US Constitution and require OSHA to obtain a search warrant to conduct an inspection.

Some company managers take the position that no time is a "reasonable time" to be visited by the OSHA inspector. They argue that their plant has such proprietary processes that a visit by the inspector would jeopardize their trade secrets. Congress anticipated this excuse and provided that any information gathered that might reveal a trade secret be kept confidential. Actually, this provision was essential to prevent OSHA from conflicting with existing laws to protect trade secrets.

OSHA inspections are generated according to the following priorities:

1. Imminent danger
2. Fatalities and major accidents
3. Employee complaints
4. High-hazard industries

"Imminent danger" would be a situation in which death or serious physical harm could be expected to occur immediately. Time is of the essence in these situations, and ordinary enforcement procedures may be too late to protect workers. OSHA may go to a US district court and get a temporary injunction to remove employees from the work area in an imminent danger situation. Such an action would be rare and must be reasonable in allowing some employees to remain to correct the condition and to permit a safe and orderly shutdown. If the process is continuous, a sudden shutdown might cripple the equipment and might even be unsafe. OSHA realizes this, and in the case of continuous process operations, the process does not have to be completely closed down; a pilot crew is permitted to remain and to maintain the capacity of the process to resume normal operations.

OSHA requires a telephone call or other notification within 48 hours upon occurrence of fatal accidents or accidents in which five or more persons are hospitalized. This notification will invariably trigger an OSHA inspection because this category of inspections is second in priority only to imminent danger inspections. In fact, at first fatalities were considered top priority; later it was decided that it would be better for OSHA to assign top priority to serious, perhaps fatal, hazards before the accident occurs rather than after the fact. OSHA has a policy of investigating both categories, however, within 24 hours of notification.

An employee can request that OSHA investigate a hazard by filing with OSHA a complaint describing a hazard believed to exist in the workplace. To be valid, the complaint must be signed by the employee. Many workers fear recrimination or other reprisal if they are subsequently identified by their signatures. But OSHA is bound by law to keep the origins of an employee complaint confidential if the employee so requests. However, the employee has additional safeguards against discrimination, as will be seen later in this chapter.

Next in priority to complaints come the inspections of industries that are shown by statistical records to be particularly hazardous. Early in the implementation of OSHA's enforcement program, a collection of industries were named "target industries." Subsequently, a "target health standards" group was named. Focus on the Target Industries Program (TIP) shifted later to a "special emphasis program," which named trenching and excavation cave-ins as the principal concern. Next came the "National Emphasis Program" (NEP), emphasizing foundries and metal stamping. The presidential administration of Ronald Reagan assigned a national priority to those firms whose lost-workday incidence rate was above the national average. Thus the old target industries and national emphasis program that emphasized whole *industries* were discarded in favor of a system that concentrates on *individual firms* that have poor safety and health records.

The shift in inspection strategy that occurred in the early 1980s permitted OSHA to redirect its efforts in the face of imperative budget cuts in agencies throughout the federal government. Since not enough OSHA field personnel were available to inspect firms frequently and thoroughly, the emphasis was shifted to those firms that need improvement the most. The revised strategy is a powerful signal to the Safety and Health Manager to achieve results in reducing lost-workday cases to avoid OSHA inspection. Although the total number of inspections has been reduced, the focus results in a greater chance of inspection if the firm's records show that it is really dangerous.

Citations

After the inspection, OSHA may issue a citation for alleged violations of the standards or of the General Duty Clause. The statutory limit is six months, so if no citation has been received within six months, the employer can be assured that a citation will not be forthcoming. A growing percentage of firms do not receive a citation upon inspection. OSHA has on occasion pointed to this percentage and designated these firms as "in compliance." The designation may be a misnomer, however, because the depth of inspections varies considerably. A really thorough wall-to-wall inspection of a comprehensive manufacturing plant of any size except the very smallest is almost certain to reveal some OSHA violations. The OSHA Compliance Officer may be reluctant to issue a citation and may overlook some minor items if it is felt that the employer has shown a good-faith effort to comply with the standards.

OSHA recognizes that some conditions, even though they represent a violation of the letter of the law, have no direct or immediate relationship to safety or health. In these instances OSHA may issue a "de minimus notice" in lieu of a citation. De minimus notices do not carry a monetary penalty.

If a citation is issued, it is not a very private matter. To the chagrin of the Safety and Health Manager, the citation must be prominently posted near where the violation occurred. Employees and management alike have the opportunity to read the citation and note the alleged violation. This increases the overall awareness of OSHA's enforcement powers and may spawn complaints by employees for other potential violations. As far as management is concerned, the Safety and Health Manager can prepare top management by explaining that many companies do receive a citation when inspected by OSHA; there is no disgrace to being found out of compliance with a few detailed provisions of the standards.

OSHA penalties can be quite severe, as Table 4.1 indicates, but actual fines are usually quite inconsequential. Penalties for nonserious violations are usually less than $100. OSHA has a formula for computing a reduced penalty, taking into consideration size of the company (small businesses tend to be assessed smaller fines), the history of previous violations, and "good faith" shown by the employer. By the time a given citation is issued, the formula for computing a possible reduction in penalty will have already been applied.

TABLE 4.1 OSHA Penalties (Statutory Maximums)

Offense	Maximum Penalty	Offender
Nonserious violations	$1,000	Employer
Serious violations	$1,000	Employer
Willful violations	$10,000	Employer
Repeat violations	$10,000	Employer
Failure to abate a violation	$1,000 per day	Employer
Willful violation resulting in death to employee	$10,000 and/or 6 months in prison	Employer
Second offense: willful violation resulting in death	$20,000 and/or 1 year in prison	Employer
Failure to post notices, citations, etc.	$1,000	Employer
Giving advance notice of an inspection	$1,000 and/or 6 months in prison	Anyone
Falsifying records or reports	$10,000 and/or 6 months in prison	Anyone
Killing an OSHA inspector	Life imprisonment	Anyone

Source: Ref. 73.

Although most OSHA penalties are small, it is possible for fines to become quite severe. Note that the "failure-to-abate" penalty is assessed for *every day* a violation remains uncorrected. Some "willful and repeat" violations have resulted in fines of hundreds of thousands of dollars assessed on a single firm, although these cases are rare.

In addition to the statutory penalty categories listed in Table 4.1, an additional category, "egregious violation," was established administratively in the 1980s by the agency. Worse than a "willful" violation, an "egregious" violation is a glaring or flagrant violation, which may invoke even higher penalties. Citation of egregious violations requires clearance from OSHA's national headquarters in Washington, DC.

The following questions are frequently asked in seminars about OSHA:

1. Who in the organization goes to jail when the "employer" is found to be guilty of a willful violation resulting in death?
2. Where does the money collected in OSHA fines go—to OSHA's budget to pay inspectors?

In answer to the first question, "employer" in a complex organization can be construed to represent the entire chain of supervision, from the supervisor of the employee victim to the chief executive officer of the firm. However, what few cases have been prosecuted indicate that OSHA will usually attempt to zero in on only one guilty person to go to prison. The rationale is that OSHA attempts to prosecute the manager with the highest authority who knew about the violation and had the authority to correct it but failed to correct the situation that resulted in the employee's death.

The other question, "Where does the money go?", apparently has its origins in some state and local enforcement schemes. It would hardly seem reasonable to deposit fine money into the coffers of those agency officials who assess the fines. But the misconception that OSHA is permitted to keep the money from fines collected seems to persist in the minds of the public. The origins of this idea appear to be the tradition of state boiler inspection agency regulations, which provide for money collected to be deposited into the boiler inspection division accounts. OSHA does not use this strategy, however, and all fines are deposited directly into the US Treasury and not earmarked for OSHA's use. OSHA fines would be insufficient to operate OSHA anyway, since the total amount collected annually is generally very low compared to OSHA's annual budget.

A big mistake made by some managers is to pay OSHA fines and to then consider the matter closed. Such a strategy ignores the far more important aspect of the citation—the prescribed abatement period. The OSHA fine itself may be inconsequential; the big impact of the OSHA citation, if it is accepted, is that each item listed in the citation must be *corrected*, regardless of cost. The cost of correcting violations is usually much greater than the amount of the OSHA penalty. Before accepting a citation, the Safety and Health Manager should stop and think about whether it is really feasible to correct the violation. There is only a 15-day period (working days) after receipt of the citation during which a decision must be made whether or not to contest the citation. Appeals can even be taken through the judicial processes all the way to the Supreme Court.

Appeals of citations are not to be confused with variances. Appeals are for employers who have already been cited. Variances are for employers who need time to comply or have an alternative to compliance which is more practical and still protects employees.

Prescribed abatement periods can be seen to be very important. It is easy to overlook abatement periods and concentrate on the nature of the violations and the proposed penalties. The abatement period can seem to pass quickly, and then the firm is subject to a possible reinspection and potentially severe penalties. OSHA is known to be very reasonable in extending abatement periods if the employer contacts OSHA and gives reasons for a request to extend. But if the employer takes no action and the OSHA Compliance Officer returns, a severe penalty is likely. It is easy for OSHA to be too optimistic about timetables for changes in facilities, allowing insufficient time for administrative approvals, delayed equipment deliveries, installation schedules, and, of course, Murphy's law. It is up to the Safety and Health Manager to bring up the subject of abatement and to ensure that OSHA sets reasonable periods.

Employee Discrimination

One penalty that is sometimes costly does not appear in Table 4.1. This is the penalty paid by the employer who is found to have discriminated against an employee because that employee filed an OSHA complaint or answered the OSHA Compliance

Officer's questions during an inspection or exercised any other right afforded employees under the OSHA law. Company management should be very careful to document reasons for discharging any employee, especially if that employee has a history of complaining about safety and health hazards. This type of complaining is no basis for dismissal, and the company that dismisses such an employee may find itself later reinstating the employee with back pay. No fine must be paid to the government for such an infraction, but if the case has lain idle for many months, the cost of reinstatement with back pay can be substantial.

Besides dismissal of the employee, OSHA recognizes other, more subtle means of discrimination. Any of the following employer actions is considered by OSHA to be illegal discrimination if it is used as punishment for the employee's exercise of rights under OSHA:

- Termination
- Demotion
- Assignment to an undesirable job or shift
- Denial of promotion
- Threats or harassment
- Blacklisting the employee with other employers

OSHA has even told employees that the employer may be in violation of the law for suddenly punishing an employee for doing something else wrong after the employee has protested a hazardous condition. It would be especially incriminating to single out a complaining employee for punishment for some unrelated action that goes unpunished for other employees. As can be seen, the whole matter can be extremely delicate, and the Safety and Health Manager should ensure that all persons in management, from first-line supervisors on up, be aware and careful not to discriminate either directly or indirectly against any employee for complaining about safety and health violations, either during the worker's employment with the company or in the future if and when the worker seeks employment elsewhere.

One questionable worker "right" is whether an employee can walk off the job because of unsafe or unhealthy conditions and still expect to be paid. The courts have ruled against employees who sue to be paid by their employer for not working when they have walked off the job due to unsafe or unhealthful conditions.

PUBLIC UPROAR

OSHA survived a very stormy first decade. Despite the validity of its purposes, OSHA quickly grew to be one of the most hated agencies the federal government had ever created. At times its demise seemed imminent, but it continued to survive.

At the root of public criticism of OSHA lie the OSHA standards. Much has been said about "nitpicking inspectors," "unjustified fines," and "gestapo techniques," but these criticisms would probably never have arisen if the standards themselves had been set up differently.

The original standards had a few ancient and obsolete provisions that have subsequently been deleted. Another problem with the standards was that advisory provisions containing language such as "should" were incorporated as mandatory rules with language such as "shall." The courts frowned upon this, and so did the public. The government has since worked hard to eliminate advisory provisions from the standards. Federal standards have also been criticized for their level of detail, vagueness, redundancy, and irrelevance.

Some federal standards have seemed to do more for industries manufacturing safety equipment than they have done to protect the worker. A good example was the original standards for fire extinguishers. OSHA now permits alternatives to fire extinguishers for many applications.

A favorite weapon of OSHA's critics is the old question "Has it done any good?" They usually feel comfortable with the question without any examination of the record because they believe that OSHA could not have any measurable beneficial effect. It is difficult to assess the impact of federal regulation on worker safety and health because even the statistical recordkeeping has changed somewhat since OSHA was created. It is not even agreed whether the changes in recordkeeping have made injury and illness rates look better or worse. Some believe that the presence of OSHA enforcement with OSHA Compliance Officers examining injury and illness records has tempted management to hide injuries and illnesses, making the overall record of injuries and illnesses under OSHA look better than it really is. Others believe that because injuries and illnesses requiring medical treatment are now required by law to be recorded, statistical summaries will show more injuries and illnesses and thus make OSHA look worse. There is no question, however, that in some areas OSHA has had an impact on worker safety and health. The most dramatic results are probably with respect to reduction of fatalities from trench and excavation cave-ins.

The concept of cost effectiveness continues to gain in importance. Thus the old question "Does a federal regulation do any good?" is changing to "Does the regulation do *enough* good to warrant the *cost* of compliance?" The cost of compliance is a much greater consideration than is the cost of monetary penalties assessed by regulatory agencies.

ROLE OF THE STATES

Prior to OSHA, occupational safety and health were generally considered to be concerns of the states, not of the federal government. The general feeling among supporters of OSHA was that the states historically had not been doing a sufficient job in establishing and enforcing adequate standards for occupational safety and health. Recognizing, however, that some states might develop effective occupational safety and health standards and enforcement programs, the OSHA law provides for state plans to be submitted to OSHA for approval.

Enforcement

It must first be said about state enforcement that OSHA itself has been given no authority to regulate state agencies, counties, or municipalities. Even federal agencies are exempt from regular OSHA enforcement—a sore point with some private employers—but the OSHA law does provide for federal agency coverage as a part of the responsibilities of the heads of the various federal agencies. Obviously, it would be impractical for the federal government to cite and penalize itself, so OSHA does make inspections but does not issue citations to its sister agencies. As far as states are concerned, if the federal government were to issue citations and assess penalties from state and local governments, there would be sovereignty problems, so that is prohibited.

But if a state submits a plan for occupational safety and health standards and enforcement, to be approved by OSHA the plan must contain a program applicable to employees of state agencies and political subdivisions of the state. In addition, the state standards and enforcement must be at least as effective as the corresponding federal standards and enforcement. About half of the states have achieved "as effective as" status and have had their state plans approved (see Appendix G). Safety and Health Managers operating in any of the state-plan states should consult with appropriate state officials for copies of standards and enforcement procedures. In most state-plan states, the standards are virtually identical to the federal OSHA standards.

Consultation

Besides standards and enforcement programs, OSHA also delegates to states authority and responsibility for consultative assistance in occupational safety and health to employers upon request of the employers seeking such assistance. OSHA has authority to make federal grants to the states to support enforcement, consultation, and other purposes of the OSHA law. Recent political tides have been in the direction of transferring federal agency authority and activity to the jurisdiction of the states. OSHA is pursuing this path with the state consultation programs. Cooperation with a state agency can even result in temporary immunity from OSHA citation in some cases. The possibility is worth checking into, and Safety and Health Managers should consider state consultation as a part of their overall strategies. There is no charge for state consultation, and as of this writing it was available in some form in every state.

Future Trends

Despite its problems, OSHA appears to be quite durable and has survived repeated attempts to repeal or modify it. Many thought that the change in administration in 1980 would spell the end of OSHA, but it did not turn out that way. And toward the end of the decade, OSHA enforcement appeared to intensify, with assessment of some exceptionally large penalties. This period also saw the inception of the concept of egregious violations.

The Barlow decision was an important one, but it really did not affect OSHA's enforcement operations as much as the public first thought it would. Those firms insisting on inspection warrants are usually inspected later; the procedure merely delays the process somewhat.

An attempt to exempt small businesses and family farms merely shifted enforcement emphasis. The same can be said of a strategy to inspect only those firms which have a lost-workday rate greater than the national average. An exemption of one category means emphasis on another, provided that the staffing and organization of the agency remain intact.

One indicator of OSHA's future is the congressional acts that have been patterned after the OSHA law and have been passed despite criticism directed at OSHA. The laws governing product safety and liability and mine safety were approved in the 1970s and were constructed and worded in a manner similar to the OSHA law. This illustrates that Congress has continued to endorse the concept of OSHA.

The evolution of OSHA standards is very slow. Once a standard has been adopted as the "national consensus," it becomes very difficult to revoke the standard at a later date by saying that the standard did not represent the "national consensus" after all. Such a step repudiates or at least reflects adversely on the thousands of citations enforcing the given standard in inspections prior to the revocation. Any revision of the standards that can be considered subjective or controversial can be challenged in the courts. All of this tends to preserve the status quo despite widespread controversy. OSHA has been able to overcome the inertia of existing standards by making some rather broad changes to some of the standards, the fire protection standard being the most notable example.

Positive Developments

Gilbert J. Saulter (ref. 81) summarized his assessment of public sentiment in the late 1980s by saying that OSHA was reaching "maturity." He cited successes in achieving public awareness by observing that the number of university academic programs in safety and health had tripled since 1970. Saulter emphasized the success of the program of exemption through consultation, whereby employers seek free consultation, usually provided by the states, and thus achieve limited immunity from citation. The Voluntary Protection Program (VPP) is a procedure for giving recognition to noteworthy safety and health programs in companies that are voluntarily doing a good job. There are two levels of recognition; "Merit" and "Star," the latter for the more outstanding programs.

One of the most recent OSHA emphases is upon the *management* of safety and health, an approach that is embraced by this book. OSHA has become increasingly interested in training and in the effectiveness of employee committees for safety and health. The stress is upon hazards analysis and feedback to correct safety and health problems so that the same injuries and illnesses do not recur.

Like the emphasis upon management is the attempt by OSHA to use a systems approach to dealing with hazards. This view of safety and health as a complex systems problem that impacts many other facets of the production system is a further sign that the agency is maturing. Sophisticated OSHA technical personnel now look beyond the citation and abatement of individual violations and seek solutions more basic to the problem.

Ergonomics

In no arena is the systems approach more evident than in the emphasis upon ergonomics. Ergonomics is the study of human capability in relation to the work environment, and solutions to ergonomics problems usually require a sophisticated analysis involving perhaps a redesign of the work station to fit the process. An example can be seen in the problem of "carpal tunnel syndrome," an injury to the wrist that occurs due to repetitive hand motion. The hand motion taken alone might be harmless, but when the same motion is repeated thousands of times per shift, the syndrome can develop. When a systems approach is used, work station analysis and redesign can result in dramatic improvement in safety and health as well as have a significant positive impact upon production.

OSHA's recent emphasis upon ergonomics has been highly visible to the public in the form of severe penalties, especially in the meat-packing industry. The trend is expected to continue and will have an impact upon production that goes beyond the issue of safety and health.

In general, the prospects for OSHA's future seem positive. Despite its problems, OSHA has had a tremendous impact on employers, workers, and the public alike by raising everyone's consciousness of safety and health hazards. For this it will never be forgotten and probably will never be repealed, at least not in its entirety.

EXERCISES AND STUDY QUESTIONS

4.1. What is NIOSH? What is its role?

4.2. By what procedure can an employer request time to comply with an OSHA standard before an enforcement inspection?

4.3. Describe the procedure established for employers who believe that OSHA has issued an unfair citation.

4.4. What legal rights are extended to employees by OSHA? (Identify at least three.)

4.5. What is the General Duty Clause?

4.6. What is a "national consensus standard" as defined by the OSHA law? Have any such standards been adopted in the 1990s? Why or why not?

4.7. Compare "performance" versus "specification" standards.

4.8. Compare "horizontal" versus "vertical" standards.

4.9. Explain the significance of the Barlow decision.

4.10. Explain the difference between "appeal" and "variance" as far as OSHA is concerned.

4.11. Suppose that you are a writer of new standards. Pick a familiar hazard and write a two- or three-sentence paragraph for a possible standard to protect against this hazard. Write the standard first in the style of a specification standard and then rewrite it in the language of a performance standard.

4.12. Compare advantages of "specification" standards versus "performance" standards from the viewpoint of the employer and then from the viewpoint of an enforcement agency.

4.13. List in order of priority four OSHA inspection categories.

4.14. What is the difference between a "repeat violation" and failure to correct a violation? How do the penalties differ?

4.15. Name some public criticisms of the OSHA standards.

5

Information Systems

There are two schools of thought regarding responsibility for risk in the workplace. The more ambitious of these two factions places full responsibility upon the employer not only to identify hazards but also to eliminate them so that the employee is assured a safe and healthful workplace regardless of the nature of the hazard. For the most part this is the approach used by the drafters of the OSHA law. It is true that the law contains a general-duty clause for employees as well as employers, but there is no question that the enforcement provisions of the law are to confirm compliance on the part of the employer, not the employee.

The second of these schools of thought is more conservative in that it recognizes the inability of the employer to completely eliminate some hazards and accordingly shifts some of this responsibility to the employee by requiring information systems to pass data to the employee specifying the nature and degree of hazard associated with the job. The theory of this school of thought is that the employee is thus given the necessary data with which to evaluate the risks and take action accordingly.

As OSHA entered its second decade of existence, it increasingly shifted toward this more conservative approach, reflecting the more conservative political climate that was ushered in by the change in government administrations in the United States in 1980. Even OSHA's critics acknowledged the fairness of a system of disclosing knowledge to employees of hazards to which he or she would be exposed and of which the employer had knowledge. Thus began the movement that became known as "right-to-know," along with regulations requiring "Material Safety Data Sheets" (MSDSs) and labeling for hazardous materials to which employees or the public might become exposed.

Though more conservative in concept, the right-to-know movement should not be interpreted as a weakening in the protection of the rights of the worker to a safe and healthful work environment. To the contrary, knowledge of the hazard can be a potent weapon in the employee's fight for improved safety and health, and the employer is aware of this power in the hands of the employee. The specter of future litigation for today's hazards is a powerful motivator for careful concern on the part of the employer, especially if the company is large and can be shown by attorneys to have a "deep pocket." Despite the immunities provided by Workers' Compensation laws, employers are increasingly exposed to litigation risk as a result of exposing employees and others to hazards, and information systems that proliferate knowledge of these hazards heighten that risk.

HAZARD COMMUNICATION

Action in the right-to-know movement was precipitated by OSHA in late 1983 with the promulgation of the Hazard Communication standard (29 CFR 1910.1200). A significant provision of this standard is the requirement that manufacturers and importers must label containers they ship and provide a MSDS for each hazardous chemical they produce or import. Employers in industries that make use of the hazardous substances also have responsibilities to maintain hazard communication programs to protect their employees.

Container Labeling

The Hazard Communication standard placed the labeling responsibility upon the manufacturer or importer of the substance. Just about any container one can think of is included, except pipes; the labeling of pipes is the concern of other OSHA standards. Recognizing that the labeling of some substances is governed by rules administered by another regulatory agency and for other reasons, some substances are excluded from the OSHA labeling requirement, as follows:

- Pesticides
- Food, drugs, or cosmetics
- Alcoholic beverages
- Substances covered by the labeling requirements of the Consumer Product Safety Commission (CPSC)
- Hazardous wastes
- Tobacco or tobacco products
- Wood or wood products
- "Articles"

The term *articles* is to be interpreted as manufactured items formed to a specific shape or design during manufacture, having end use dependent upon that shape or design, and which do not result in hazardous chemical exposures during normal use. The distinction really is whether the item is a material or a manufactured object. In some cases it may be difficult to distinguish between articles and materials. One example given is that of a desk, which is an article, versus a piece of lumber, which is considered a material, even though both the desk and the piece of lumber might be objects manufactured from wood.

Material Safety Data Sheets

In addition to labeling, chemical manufacturers or importers must provide MSDSs for hazardous substances. The standard lists specific categories of information that must be included in the MSDS. Figure 5.1 is an exhibit of a blank form that has been used as a generic format to comply with the standard.

A delicate issue with MSDSs is the problem of dealing with trade secrets. There is likely no industry more sensitive about trade secrets than the chemical industry, and the OSHA standard addresses this issue in detail. Basically, the manufacturer or importer, and in turn the employer, can withhold specific chemical identities from the MSDS, but only if they can justify this position according to specific criteria contained in the standard. Even then they must disclose the chemical identities to health professionals upon request in accordance with criteria specified in the standard. In nonemergency situations, the manufacturer, importer, or employer may require a confidentiality agreement. Disputes are anticipated as an inevitable consequence of conflicting interests between the manufacturers and persons claiming a need to know chemical identities. Accordingly, the standards provide for referrals to OSHA to weigh the evidence on both sides, and, if appropriate, subject the manufacturer, importer, or employer to OSHA citation.

Another problem to consider is how to deal with mixtures in which a hazardous substance is only an ingredient. What must be done depends upon circumstances best illustrated by a decision diagram (see Figure 5.2.).

Employee Hazard Communication Program

After the manufactured or imported chemical is distributed to others, responsibility for protection of employees from potential exposure becomes the responsibility of the employer in those firms that "use" (that is, "package, handle, react, or transfer") the hazardous substances. A principal requirement for such employers is that they have a written hazard communication program. The Safety and Health Manager should ensure that employees know about this program, because the workers themselves may be asked by federal enforcement inspectors. One required component in the written program is a list of the hazardous chemicals known to be present in the workplace. For each of the hazardous chemicals listed, an MSDS must be on hand and available to employees. If the substance was purchased prior to the era of right-to-know and no MSDS is on hand for a given substance, the employer is required to obtain or generate one. Safety and Health Managers can turn to current manufacturers or distributors, or perhaps write their own MSDSs. Sometimes a state

Material Safety Data Sheet

May be used to comply with
OSHA's Hazard Communication Standard,
29 CFR 1910.1200. Standard must be
consulted for specific requirements.

U.S. Department of Labor

Occupational Safety and Health Administration
(Non-Mandatory Form)
Form Approved
OMB No. 1218-0072

IDENTITY *(As Used on Label and List)*	Note: *Blank spaces are not permitted. If any item is not applicable, or no information is available, the space must be marked to indicate that.*

Section I

Manufacturer's Name	Emergency Telephone Number
Address *(Number, Street, City, State, and ZIP Code)*	Telephone Number for Information
	Date Prepared
	Signature of Preparer *(optional)*

Section II — Hazardous Ingredients/Identity Information

Hazardous Components (Specific Chemical Identity; Common Name(s))	OSHA PEL	ACGIH TLV	Other Limits Recommended	% *(optional)*

Section III — Physical/Chemical Characteristics

Boiling Point		Specific Gravity (H$_2$O = 1)	
Vapor Pressure (mm Hg.)		Melting Point	
Vapor Density (AIR = 1)		Evaporation Rate (Butyl Acetate = 1)	
Solubility in Water			
Appearance and Odor			

Section IV — Fire and Explosion Hazard Data

Flash Point (Method Used)	Flammable Limits	LEL	UEL
Extinguishing Media			
Special Fire Fighting Procedures			
Unusual Fire and Explosion Hazards			

(Reproduce locally)

OSHA 174, Sept. 1985

Figure 5.1. Material Safety Data Sheet (MSDS).

Section V — Reactivity Data

Stability	Unstable		Conditions to Avoid
	Stable		

Incompatibility (*Materials to Avoid*)

Hazardous Decomposition or Byproducts

Hazardous Polymerization	May Occur		Conditions to Avoid
	Will Not Occur		

Section VI — Health Hazard Data

Route(s) of Entry: Inhalation?   Skin?   Ingestion?

Health Hazards (*Acute and Chronic*)

Carcinogenicity: NTP?   IARC Monographs?   OSHA Regulated?

Signs and Symptoms of Exposure

Medical Conditions
Generally Aggravated by Exposure

Emergency and First Aid Procedures

Section VII — Precautions for Safe Handling and Use

Steps to Be Taken in Case Material Is Released or Spilled

Waste Disposal Method

Precautions to Be Taken in Handling and Storing

Other Precautions

Section VIII — Control Measures

Respiratory Protection (*Specify Type*)

Ventilation	Local Exhaust		Special
	Mechanical (*General*)		Other

Protective Gloves	Eye Protection

Other Protective Clothing or Equipment

Work/Hygienic Practices

☆ U.S.G.P.O. 1986-491-529/45775

89

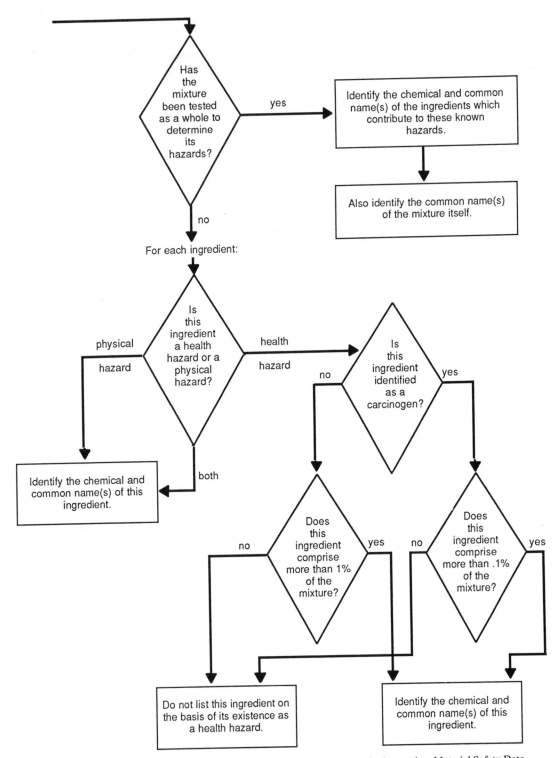

Figure 5.2. Decision diagram for reporting the content of mixtures in a Material Safety Data Sheet (MSDS). (*Source:* OSHA std 29 CFR 1910.1200(g)(2)(i)).

consultation agency, as described in Chapter 4, can be of assistance in identifying hazardous substances and in preparing the MSDS.

Federal standards permit the MSDS to be kept in any form, even within operating procedures. Sometimes it is more practical to address multiple hazards as a process, not as separate hazardous chemicals. In designing the right-to-know system, however, the Safety and Health Manager should ensure that the required information is available for each hazardous chemical present among a group of materials in a single process and that the information is readily accessible to employees of each work shift.

In addition to retaining and maintaining the MSDSs, the employer must maintain the labels provided by the manufacturer or importer of the substance, except that labels are not required for portable, in-house containers intended for immediate use.

Record Retention

Records are of ever-increasing importance, and the Safety and Health Manager should set up information systems that trace the identities, location of use, and time of use for hazardous substances, along with each employee exposure, for a retention period of *at least 30 years*. Employee medical records, except health insurance claims records maintained separately, must be preserved and maintained for the duration of employment *plus* 30 years. The reason for the very long retention period is to permit tracing of the cause of illnesses that may have extremely long latency periods after exposure to the hazardous substance. Such a responsibility, of necessity, presupposes a sophisticated, often computerized, information system.

Sale of the business or closing the business does not relieve the employer of the record-retention responsibility. Upon sale of the business, the successor employer is required to receive and maintain the records. If the business is closed permanently, the employer may be required to transfer the records to NIOSH, depending upon the requirements of specific standards pertaining to the hazardous substances in question.

ENVIRONMENTAL PROTECTION AGENCY

It was stated in Chapter 1 that the Safety and Health Manager often holds responsibility for compliance with Environmental Protection Agency (EPA) regulations. The dual roles of OSHA and the EPA were recognized in the enactment of the "Superfund Amendments and Reauthorization Act of 1986" (SARA) on October 17, 1986. OSHA responded with its standard 29 CFR 1910.120—hazardous waste operations and emergency response. The OSHA standard covers hazardous substance response operations under the Comprehensive Environmental Response, Compensation, and Liability Act (CERCLA) of 1980 and major corrective actions taken in clean-up operations under the Resource Conservation and Recovery Act (RCRA) of 1976.

One part of the SARA provisions, Title III, is the Emergency Planning and Community Right-to-Know Act of 1986 (Ref. 87). Title III establishes requirements for federal, state, and local governments and industry regarding emergency planning and reporting on hazardous and toxic materials.

At the outset of enforcement of the SARA law, the EPA published the initial list of 402 extremely hazardous substances. The list is subject to revision by the EPA at any time, but the latest available list at the time of this book's printing appears in Appendix E. It is essential that the Safety and Health Manager keep this important list on hand and up to date for reference so that the firm can be in compliance if it uses any of the materials on the list. There are reporting requirements for these substances, and there are federal standards for protecting workers who may become exposed to the substances.

Medical Surveillance

Medical surveillance requirements make up one of the elements of the OSHA standards for dealing with the EPA-listed hazardous substances. And wherever there is a medical surveillance program, there are medical records. Fundamental to the right-to-know concept is the employees' right of access to their personal medical records held by the company that employs them. A medical surveillance program is required for

1. All employees who may be exposed to health hazards at or above the established permissible exposure limits for 30 days or more a year, whether or not the employee uses a respirator for protection against the hazards.
2. All employees who wear a respirator for 30 days or more a year.
3. Employees designated by the employer to plug, patch, or otherwise temporarily control or stop leaks from containers that hold hazardous substances or health hazards (i.e., members of hazardous material [HAZMAT] teams).

The medical surveillance program serves both to determine whether the person is fit to work with the hazardous materials present and to recognize any adverse effects upon the worker arising from the exposure. It will be seen in Chapter 9 that various conditions, such as heart problems, the wearing of a beard, or even a perforated eardrum, could disqualify the employee from working in jobs requiring the use of a respirator.

Monitoring the adverse effects of exposure is the other, and perhaps more important, purpose of the medical surveillance program and the purpose that relates most to right-to-know. The worker wants to know everything the employer knows about his or her health and its possible deterioration due to hazardous exposures on the job.

Federal standards prescribe intervals at which medical examinations and consultations shall be made. The regular times are as follows:

1. Prior to assignment to duties that may require hazardous material exposure.

2. At least every twelve months during assignment to such duty.

3. Upon termination of such duty, unless the employee has had an examination within the last six months.

In addition to these regular times, an examination is required as soon as possible if an employee develops signs or symptoms indicating possible overexposure or if an unprotected employee becomes exposed in an emergency situation. Also, the examining physician might advise that increased frequency of examination is medically necessary, and in such instances, the employer is obliged to comply. All medical examinations and procedures are required to be performed by or under the supervision of a licensed physician, at a reasonable time and place, without cost to the employee, and without loss of pay for employee time lost.

It is the business of the Safety and Health Manager to be aware of the circumstances that require a medical surveillance program and to ensure that company management implements such a program if necessary. In presenting such a recommendation to management, the Safety and Health Manager can point out certain legal benefits to the company over and above the benefits of increased safety and health of employees and avoidance of OSHA fines. A record of the initial medical examination can be proven to be a valuable piece of evidence of preexisting conditions or symptoms in the event such conditions or symptoms surface during worker exposure. Management may already be aware of that, but what many management personnel have overlooked is the importance of the medical examination at the end of employment also, which serves the purpose of documenting conditions, symptoms, or lack of symptoms at the end of the period of risk. In this age of right-to-know, employees are usually aware that they can take legal action at a later date if symptoms related to the hazardous material exposure reveal themselves after the fact. At such time, the content of the medical examination at employee termination will have obvious value to the employer as well as to the employee.

Full disclosure in the spirit of right-to-know extends to the physician also. The employer is required to provide the examining physician with a copy of the federal standard covering medical examinations and must provide data pertaining to the employee's job, anticipated exposure levels, personal protective equipment to be used, and information from previous examinations.

It is the responsibility of the employer to obtain and furnish to the employee a copy of a written opinion from the examining physician containing the results of the examination, including any opinions or recommendations by the physician regarding increased risk or limitations upon the employee's assigned work. In the interest of privacy, however, the written opinion obtained by the employer is prohibited from revealing specific findings or diagnoses unrelated to occupational exposure.

Title III of the SARA act requires EPA to establish an inventory of toxic chemical emissions from certain facilities. This requirement is in effect a reporting requirement for manufacturing facilities (SIC[1] Codes 20xx through 39xx) that have ten or

---

[1]SIC refers to Standard Industrial Classification, and codes 20 through 39 are identified with manufacturing industries. Appendix F identifies the principal manufacturing categories of the SIC Code.

more employees and have manufactured, processed, or otherwise used a listed toxic chemical in excess of specified threshold quantities. By "listed" is meant one of the extremely hazardous substances listed by EPA (see Appendix E). Threshold quantities vary and were phased in over a three-year period in the late 1980s. Also, the threshold quantities are different depending upon whether the facility uses or manufactures the hazardous substance. Facilities in the manufacturing industries (SIC Codes 20xx through 39xx) that *use* listed toxic substances in quantities over 10,000 pounds in a calendar year are required to submit toxic chemical release forms (EPA Form R) by July 1 of the following year. For firms that *manufacture* or *process* these materials, the threshold quantity is 25,000 pounds per year, for amounts above which the firm is required to submit the toxic chemical release form.

For firms that both manufacture and use the same substance, if the threshold quantity is exceeded in either instance, then the firm must report, a point illustrated by Case Study 5.1.

Case Study 5.1   A fertilizer plant (SIC Code 2873) produces 22,000 pounds of ammonia, 16,000 pounds of which it uses within the plant. Is the firm required to report to EPA and, if so, in what fashion?
Solution: Yes, the firm has a manufacturing SIC (28xx through 39xx) and exceeds the "use" threshold and therefore must report to EPA using the Toxic Chemical Release Inventory Reporting Form R in accordance with Title III of the SARA Act of 1986. Since one of the thresholds was exceeded (the "use" threshold), the firm must complete a full report based on all activities and releases of ammonia from its facility, not just the releases from the use activity. (This study assumes that the firm had ten or more employees.)

Case Study 5.2   During a year of operations, a manufacturer of chemical coatings (SIC Code 2821) processes 20,000 pounds of cresol and uses 6,000 pounds of it within the plant. Is the firm required to report to EPA and, if so, in what fashion?
Solution: No, the firm is not required to report for either its processing or its use of cresol, because neither the "use" nor the "processing" thresholds were exceeded.

One other point should be made about threshold amounts. Stockpiling a material that is neither processed nor used within the plant does not count toward the computation of whether the threshold quantity has been exceeded. EPA may ask about total stockpiled amounts in the course of collecting data for a chemical for which a threshold has been exceeded, but stockpiling does not constitute either processing or use.

Firms that manufacture or process materials in excess of the threshold quantity in a year are required to submit the toxic chemical release form. The form is in four parts and is too lengthy to be included here, but it can be obtained by requesting "EPA Form R" from the regional or national office for the EPA or by contacting the designated state office ("Designated Section 313 Contact"). The form is designed to gather data for a national computer information system and includes entries disclosing basic data on the process within the plant, including maximum amounts on site, receiving stream or water body, if applicable, quantities of release into the atmosphere, off-site disposal locations, and waste treatment methods and efficiency.

In addition to the requirement for filing the toxic chemical release form, emergency notification is required by telephone, radio, or in person if there is a release of a listed hazardous substance that exceeds the reportable quantity for that substance. This includes substances listed by EPA as extremely hazardous (see Appendix E) and substances subject to the emergency notification requirements under CERCLA. Each such notification must be followed with a written confirmation including additional information on actual response actions taken, any known or anticipated data on chronic health risks associated with the release, and advice re-

Title III — Major Information Flow Requirements

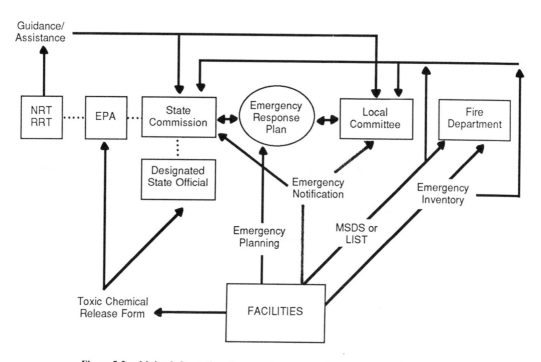

Figure 5.3. Major information flow requirements under Title III of the SARA act. (*Source:* Ref. 87).

garding medical attention necessary for exposed individuals. The requirements for information flow between private companies and the various governmental agencies that have a role in the control of hazards arising from toxic substances result in a complex information network that is best visualized by use of a diagram (see Figure 5.3).

COMPUTER INFORMATION SYSTEMS

In the previous section, reference was made to a national computer information system for toxic chemical data. Increasingly, Safety and Health Managers, both within and outside government, are turning to computerized data bases for quick access to detailed facts about the thousands of toxic substances and other workplace hazards. This development follows upon, and sometimes coordinates with, earlier computerized management information systems for administration and reporting under the recordkeeping requirements of OSHA as described in Chapter 2.

A significant advance in the computer technology was marked by the entry of high-speed, low-cost desktop models equipped with large memory capacity and tele-communication links to large mainframe computers holding vast databases. Prior to this development, the use of computerized databases was limited primarily to airlines reservations systems, government agencies, and large organizations. With the advent of the low-cost desktop computer, individuals and small firms became able to gain access to large-scale databases throughout the country. It is not unusual now for individuals to access stock exchange quotes or make airline reservations from their own desks. The possibilities for Safety and Health Managers who are able to take advantage of this technology are unlimited.

Artificial Intelligence and Expert Systems

One of the technologies that shows particular promise in the management of safety and health information systems is the general field of artificial intelligence, or more specifically, expert systems. Artificial intelligence is that general field of development that is attempting to make computers "think" or respond to problem-solving situations more in the manner in which humans do. Expert systems is the branch of artificial intelligence that encompasses computer systems that give advice based upon a knowledge base of logic rules provided by a human expert. Thus the computer is able to answer questions posed by a wide variety of problem situations and provide advice or assistance much as a human expert would. A key point to understand with these systems is that an exhaustive list of answers to specific questions has not been preprogrammed into the computer. Rather, the human expert has provided the basic logic or rules of thumb, and the computer is capable of gleaning what it needs to know from this knowledge base to answer specific questions in the future. Augmenting the logic of the human expert is the computer's power to rapidly pinpoint facts accessed from massive tables contained in on-line data systems. A prominent element in the advancement of expert systems is the evolution of "natural language" interfaces that empower the computer to understand requests expressed in ordinary

English, instead of in the rigid formats of computerese. Such interfaces are often called ''intelligent frontends,'' a reference to their appending to the front of existing computerized database management systems to make these systems more user-friendly and more capable of understanding human requests for answers to questions.

Industrial Chemical Hazards Database

Recent safety and health research at the University of Arkansas (ref. 68) has been conducted to develop a prototype computer system for answering questions about safety and health hazards. The system, identified as the Industrial Chemical Hazards Database (ICHD) is structured using the relational database tool R:BASE®. Queries to the database are made using the natural language access tool CLOUT®.[2]

The design of the relational database consists of a primary table with secondary tables for cross-reference. Each chemical in the database is keyed to its CAS number.[3] This identifier is used to key the chemical name to a variety of synonyms that might be used by various workers or users that might query the database via a computer terminal. The user need not know the CAS number itself; it is a cross-reference tool, internal to the database. Thus, using natural language, a user may inquire about ''coal tar naphtha,'' and the relational database will recognize that this is a synonym for benzene and will respond to the user's questions as if the questions were about benzene. The user can inquire about such particulars as protective eyewear required or recommended, if any, first aid instructions, or perhaps washing instructions. If the user does not know what chemical he or she is dealing with but has symptoms of exposure or can identify target organs affected, the relational database can use its cross-referencing capability to retrieve answers to inquiries in addition to identifying the chemical or groups of chemicals that match the symptoms. Figure 5.4 diagrams the relationships of the various files in the ICHD database available for retrieval in response to a user inquiry. The following case study exhibits an application of the ICHD relational database using an inquiry worded in natural language format acceptable to the system.

Case Study 5.3 Suppose a medical examination reveals that a worker has developed symptoms of kidney damage. The plant Safety and Health Manager becomes curious about whether any hazardous chemicals used in the plant could be associated with the kidney damage and poses the following question: ''Give me a list of all the chemicals that harm the kidney.'' The ICHD system would automatically scan all of the relevant tables in the relational database to find the desired information, merging information using the CAS number cross-reference key to provide the requested response.

---

[2]R:BASE and CLOUT are registered trademarks of Microrim, Inc.

[3]Chemical Abstracts Number, a well-known reference list for substances.

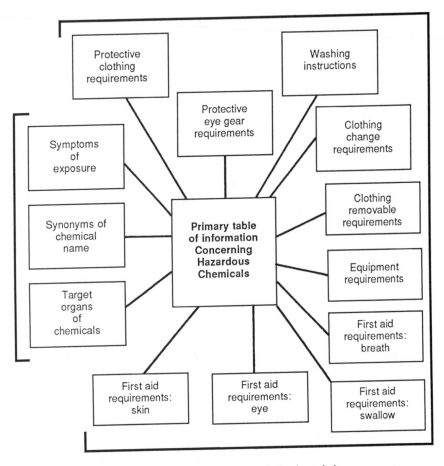

Figure 5.4. Industrial chemical hazards database design.

The CLOUT® natural language interface has built-in features to break down the structure of the sentence and interpret the meaning using a basic vocabulary of 300 words and capacity for the user to customize the system with another 500 buzz-words or common phrases used in the industry in which the ICHD system is being implemented. The system is dynamic in that the language can be updated at any time to accommodate new terminology adopted by users. A single user can teach the system new words to be used on a temporary basis and then discarded, or the system can be instructed to remember the new vocabulary indefinitely.

SUMMARY

This chapter has afforded a glimpse of current developments in managing and making accessible the large volumes of detailed data necessary to protect workers from hazards in the workplace. The evolution is the result of two recent developments:

the availability of low-cost computers with significantly larger memory and processing capacities, and increased interest on the part of workers and the public who are asserting their right to know about hazardous substances to which they are, or might become, exposed. Regulated by both EPA and OSHA, chemical hazards were the first to receive attention in the right-to-know movement, but it is projected that information system needs for safety and mechanical hazards will follow.

EXERCISES AND STUDY QUESTIONS

5.1. With regard to responsibility for risk in the workplace, describe the change in emphasis that occurred in the early 1980s.

5.2. Does "right-to-know" represent a strengthening or weakening of workers' powers in the fight for improved safety and health? Explain.

5.3. What type of container is expressly exempted from the Hazard Communication standards for labeling?

5.4. For purposes of hazard communication, what is the difference between an "article" and a "material"?

5.5. How may chemical companies protect their trade secrets despite requirements to furnish MSDSs?

5.6. Are MSDSs required for mixtures in which some ingredients are hazardous and others are not? Explain.

5.7. Is an MSDS required for a chemical that was purchased by a plant prior to the Hazard Communication standard?

5.8. How long must records be kept to trace the use of hazardous substances? Why?

5.9. How long must employee medical records be preserved?

5.10. What should the Safety and Health Manager do about records if a firm decides to go out of business?

5.11. What do the acronyms SARA, CERCLA, and RCRA represent, and how do they relate to OSHA?

5.12. What is the function of a HAZMAT team?

5.13. Under what circumstances is a medical surveillance program required for a given employee? How frequently are medical examinations required for such an employee?

5.14. If an emergency notification of an accidental release of a toxic chemical is required, who should be notified?

5.15. What are "expert systems"?

5.16. What commercially available computer tools for relational databases are used by the Industrial Chemical Hazards Database (ICHD) system?

6

Buildings and Facilities

 13%

Percent of OSHA
general industry citations
addressing this subject

This book now turns to the business of examining hazards in various categories, highlighting applicable standards and suggesting methods of bringing about change to eliminate or reduce hazards. Most enterprises begin with a building in which to conduct operations, and this is an appropriate beginning for the examination of hazards. The enterprise is fortunate if its management considers hazards and applicable safety and health standards during its building design stages.

Safety standards for buildings, whether they be municipal, state, or federal, are usually called "codes." For the most part, building codes apply to the construction of new buildings or to their modification. Thus, although building codes change constantly, those buildings built or remodeled before a particular change in the code are not required to be torn down and rebuilt or remodeled according to the new code. Needless to say, most buildings in existence do not meet the latest provisions of current building codes.

Some federal standards for buildings have been applied to all buildings, regardless of age. The standards have included matters of relative permanence such as floors, aisles, doors, numbers and locations of exits, and stairway lengths, widths, riser design, angle, and vertical clearance. Industry's objection has been that such standards are unfair, not only because they apply to existing buildings but also because they are vague and generally worded. But despite these problems, industries have undertaken a large number of retrofit programs to update their buildings and facilities to satisfy federal standards.

In the defense of federal standards for buildings and facilities stands the fact that some of the most frequent categories of worker injuries and even fatalities arise from improper building design, lack of guardrails, and problems with exits. Building equipment is designed and built without sufficient thought about the worker who must have access to this equipment to clean, maintain, repair, replace lightbulbs, or otherwise service the buildings or facilities. Some workers are even in locations where they would be unable to escape in event of fire. Aisle widths are often set up arbitrarily without giving thought to clearance between moving machinery and personnel.

Federal standards pertaining to buildings and facilities include the following categories:

- Walking-working surfaces
- Means of egress
- Powered platforms, manlifts, and vehicle-mounted work platforms
- General environmental controls

A few provisions of these standards generate almost all of the problems. This chapter will now single out those individual provisions and analyze each to determine what the Safety and Health Manager should do to alleviate hazards and conform to standards.

WALKING AND WORKING SURFACES

One does not normally call a floor a "walking or working surface," so why do the standards writers select such a complicated bit of jargon? This question is answered by reflecting on the hazardous locations in which people work. To be sure, many accidents, especially slips and falls, occur on floors, but consider the other "walking and working surfaces"—for instance, mezzanines and balconies. Then there are the even more hazardous platforms, catwalks, and scaffolds. Not to be forgotten are ramps, docks, stairways, and ladders.

Guarding Open Floors and Platforms

The most frequently cited standard in the walking-and-working-surfaces subpart is indeed one of the most frequently cited standards in all of the OSHA standards. It is repeated here in its entirety, due to its importance.

> OSHA standard 1910.23—Guarding floor and wall openings and holes
> (c) Protection of open-sided floors, platforms, and runways
> (1) Every open-sided floor or platform 4 feet or more above adjacent floor or ground level shall be guarded by a standard railing (or the equivalent as specified in paragraph (e)(3) of this section) on all open sides, except where there is entrance to a ramp, stairway, or fixed ladder. The railing shall be provided with a toeboard wherever, beneath the open sides,
> (i) Persons can pass,
> (ii) There is moving machinery, or
> (iii) There is equipment with which falling materials could create a hazard.

To many people, 4 feet seems an innocuous height, but the reason for this illusion is that they are thinking of jumping, not falling. Almost everyone has *jumped* without injury from elevations of more than 4 feet, but few adults experience a *fall* from that height without injury. A surprising number of *fatalities* result from falls at heights of only 8 feet.

The illusion of safety of platforms in the range 4 to 14 feet has led to complacency on the part of building and facility designers as well as of Safety and Health Managers. To stand on the extreme point of a precipice overlooking the Grand Canyon without a guardrail would seem foolhardy to the average person. But many "average persons" would think nothing of standing on the unguarded top of a tank 10 feet high. An unexpected event in such a situation can easily result in the worker taking a reflex action that results in a fatal fall.

The top of a tank, in the example just cited, was considered a "working surface" to be reckoned with by the facility designer. This is an important thing to remember. Even if the surface is the top of a tank, or even the top of a piece of equipment that is in the process of being manufactured, it may act as a temporary walking and working surface. But in unusual temporary situations there are other ways to provide temporary fall protection for workers, as will be seen in Chapter 15.

Another problem area is the protection of personnel from falls from loading docks. A few loading docks are slightly less than 4 feet high, and the issue is thereby avoided. Some Safety and Health Managers have had the area immediately adjacent to the dock built higher to comply with standards, but this is of dubious value in preventing injuries. Others have installed temporary, removable railings, and some have used chain-type gates. The chain-type gates do not qualify as "standard railings" but have sometimes been accepted as a practical substitute in an attempt to deal with the hazard in the face of a difficult situation.

A rooftop is, of course, a walking and working surface for roofing workers. This raises the question of whether guardrails, fall protection, or some other type of protection is needed for perimeters of rooftops where persons are working. A provision of the OSHA Construction Standard specifies "catch platforms," unless

1. The roof has a parapet.
2. The slope of the roof is flatter than 4 inches in 12 inches.
3. The workers are protected by a safety belt attached to a lifeline.
4. The roof is lower than 16 feet from the ground to the eaves.

The term "standard railings" mentioned in the platform guarding standard is further clarified by federal standards, and salient features are shown in Figure 6.1. Certain reasonable deviations from the specifications in Figure 6.1 are permitted. For instance, some other material in place of the midrail is permitted, provided that the alternative material provides protection equivalent to the midrail. Also, the height of the railing does not have to be exactly 42 inches high. Railings of 36 inches are acceptable if they do not present a hazard and are otherwise in compliance. This avoids

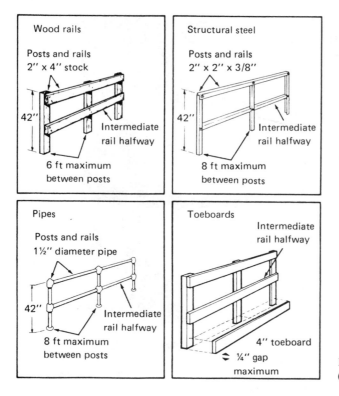

Figure 6.1. Standard railings. (From NIOSH.)

the necessity of tearing down good railings and rebuilding to meet current codes, as has actually been done by some employers, as shown in Figure 6.2.

One railing that is definitely not standard but still can afford reasonable protection in some cases is lines of parallel ductwork or pipes along the otherwise exposed sides. In many cases parallel ductwork or pipes completely eliminate the hazard of the otherwise open-sided floor, platform, or runway. In fact, one might argue that such a floor, platform, or runway is not even "open-sided."

In OSHA standard 1910.23(c)(1) quoted earlier, three conditions are enumerated, any one of which calls for a "toeboard." Toeboards are vertical barriers along the exposed edges of the walking or working surface to prevent falls of *materials*. A standard toeboard is 4 inches high and leaves no more than a ¼-inch gap between the floor and the toeboard. If the toeboard is not of solid material, its opening must not be greater than 1 inch.

Although 4 feet is the maximum height for open platforms in general, special situations require protection *regardless* of the height of the walking or working surface. Examples are open pits, tanks, vats, and ditches. When the hazardous opening size is small, it may be more practical to place a removable cover over the opening than to place a guardrail around it. Some Safety and Health Managers have saved their firms a great deal of money by calling attention to this alternative. Another

Figure 6.2. New guardrails. Some employers tore down their guardrails and rebuilt them 3 or 4 inches higher just to meet OSHA's standard in the early years of OSHA enforcement. Note the paint marks on the wall where the old guardrail had been attached. OSHA has since relaxed its rule somewhat.

special situation is walking and working surfaces adjacent to dangerous equipment, where standard railings are appropriate.

Floors and Aisles

The most important consideration for floors and aisles is not how they are built, but how they are maintained. Federal standards for housekeeping require areas to "be kept clean and orderly and in a sanitary condition." The obvious question is, how is the Safety and Health Manager to know what constitutes "clean and orderly and in a sanitary condition"? There is no clear-cut answer to this question, but some information can be gleaned from past cases of OSHA citations:

Example 6.1

In the railroad yard area of a steel manufacturing industry, piles of debris such as railroad ties and cables were lying about close to the track and presented tripping hazards to employees who must work in the area of the tracks. A complicating factor was that some of the debris was hidden by weeds.

Example 6.2

Dangerous accumulations of grain dust were found in many locations in a grain elevator. The dust was sufficiently concentrated so as to not only be a health hazard to cleanup personnel but also to present a serious explosion hazard. Since the condition had been cited before, the citation was established as "serious and repeat," and the penalty was set at $10,000.

Example 6.3

A cluttered workshop had obstructions that some of the employees had to step over or go around to do their jobs.

Example 6.4

Leaking hydraulic cylinders dripped oil on the floor in a work area. No one was responsible for cleaning up the leaked oil.

Example 6.5

Cutting oil was spilled on the floor of a machine shop and was not cleaned up.

Some conditions might *look* bad but not represent a hazard. In one example an OSHA Compliance Officer wrote a citation for piles (approximately 2 feet high) of sawdust around a planer and jointer in a woodworking shop. But the company resisted the citation because few employees would be in the area, and those few would be more inconvenienced than endangered. The company won its case.

It is also recognized that it is necessary in the course of many jobs for objects and materials to lie about in the work area. This is especially true of repair and disassembly work, and also of construction. What is generally considered excessive is material accumulation in the immediate work area in quantities in excess of what is actually required to do the given job, or scrap material lying hazardously about the work area in excess of a day's accumulation.

A prime measure of whether a housekeeping program is inadequate is the number of accidents such as trips and falls occurring in the area. Note that the preceding sentence was worded negatively in that too many accidents suggest an inadequate program, but "no accidents on record" does not prove that the work area is *free* of hazards. For every accident that is on file in the plant, the wise Safety and Health Manager will have documented what steps have been taken to remove or reasonably reduce the offending hazard, whatever that hazard has been determined by analysis to be.

Finally, Safety and Health Managers can look to their own industries for guidance. Any reasonable person knows that the definition of "clean and orderly and in a sanitary condition" is different in a foundry than in a pharmaceuticals plant. If the industry has any work practice guidelines, as put forth by trade associations, for instance, these guidelines would be very helpful in illustrating what is reasonable for a given industry.

Water on the floor is a problem in many industries, and constant surveillance becomes necessary to assure that the floor is kept clean and dry. In the design phase of buildings and facilities, attention to the problem of wet processes will suggest slopes and floor drainage systems to alleviate the problem. Another helpful device is to use a false floor or mat to provide a dry standing place for the worker who must work in a wet process.

Some buildings are not built to facilitate cleaning. The floors in others are not only poor from a cleaning standpoint but present trip hazards such as protruding nails, splinters, holes, or loose boards.

Trip hazards due to uneven floors can result in serious injury. In one case, a steel mill, a difference of 1 to 3 inches existed at a point where a grating and plate came together in the floor. The employee who worked in the area had the job of guiding red-hot steel butts from the end of a roller line. A fall could have brought the employee into contact with the red-hot butts of steel.

Aisles are important, and standards for aisles specify that permanent aisles be kept clear of hazardous obstructions and that they be appropriately marked. Ironically, the more clearly the aisles are marked, the more noticeable will be clutter or materials allowed to accumulate in the aisles. Conversely, the more materials or obstructions are kept clear for forklifts or other travel, the more conspicuous will be such a corridor if it is not appropriately marked as an aisle.

There is a lesson to be learned from the irony just described. There is a tendency among Safety and Health Managers to go out all over the plant, indiscriminately laying out aisles, and then to take pride in marking them according to appropriate standards. It is human nature for Safety and Health Managers to want to show everyone that they are taking positive action to enhance safety and comply with established standards. The unfortunate result is that aisles can become too regimented, and so much time is spent policing aisles that production efficiency is lost. In fact, the effect on safety and health can thus become negative, with employer and employee alike wanting to say, "Do we really need that Safety and Health Manager in the plant?" Aisle-marking procedures are a prime illustration of the principle that overzealous safety and health action can do more harm than good. Every time a decision is made to mark an aisle, the Safety and Health Manager should stop and ask the question, "Can we keep this aisle free of material or other obstructions?"

Conspicuously absent from federal standards for aisles is a specific minimum width for aisles. State codes which reflect the National Fire Protection Association's Life Safety Code often specify a minimum exit access width of 28 inches. But the federal code is silent on this point except to use the language ". . .sufficient safe clearances shall be allowed . . ." Aisles in forging machine areas receive special attention: ". . .sufficient width to permit the free movement of employees . , ." But this standard also does not specify an actual minimum width dimension. These two standards are therefore good examples of "performance standards."

The OSHA aisle-marking standard has left a mark of its own upon industry. The term "appropriately marked" originally meant black or white or combinations of black and white for aisles. Untold thousands of dollars were spent by industries switching from the yellow stripes widely used for aisles to OSHA-acceptable white stripes. However, the white striping used was not as durable, and frustrated facilities superintendents would return in desperation to the more durable yellow stripe in order to maintain any stripe at all. The black-and-white rule became more and more unpopular until OSHA finally revoked the aisle color code as superfluous. In summary, the Safety and Health Manager should be sure that aisles are well marked, but the choice of color for the marking stripe is of little importance. Thus the aisle-marking standard was formerly a specification standard but is now a performance standard.

The discussion about floors thus far has concerned guarding of openings, surfaces, maintenance, and appropriate markings. Nothing has been said about the structural design of the floor itself or whether the floor can withstand the loads applied to it. One is reminded of the eight-story vehicle parking building in which the eighth floor collapsed upon the seventh. The seventh floor was well constructed but was unable to withstand the shock load of the falling eighth floor, and from then on, all the floors came down like dominoes. An even more grim reminder was the Kansas City hotel tragedy of 1981, mentioned in Chapter 3. Almost no one really pays attention to whether a floor is overloaded. Everyone has heard of some structural collapse somewhere, but the few incidents that do occur seem so remote and rare that few of us ever worry about the problem.

Federal standards require marking plates for floor loads approved by the "building official." One of the chief complaints from industry Safety and Health Managers is that the floor loading standard does not explain its use of the term "building official." Confusion over this term has led to phone calls to various agencies in attempts to identify some *public* official to come to the facility to make an engineering determination to use as a basis for the floor-load-marking plates. Since the term "building official" is not defined, a better course of action for the Safety and Health Manager is to secure the services of a competent Professional Engineer either inside or outside the company. This would show a good-faith effort to comply with the standard and would virtually eliminate the possibility of a hazard.

Floor-load-marking plates are of little value if they are ignored by employees. Adherence to floor load limits is an administrative or procedural matter and thus calls for vigilance in order to maintain compliance. A good system of records and inventories can make weights and locations part of the database of an overall management information system. If this system is computerized, the computer can monitor loads at all times and print out a warning message in the unlikely event that a distribution of inventories exceeds a floor load limit.

Such a computerized system may seem to be an overreaction to the problem, and in fact it would be an overreaction for most companies medium to small in size. But in a large-scale computerized warehouse system, the floor load monitoring system would represent such a small sideline to the overall computerized operating monitor that it would require only a few seconds of computer time per month. The system could be programmed to print out status reports monthly or on an exception basis whenever loads exceed or perhaps approach limits. The monthly status reports would be preferable to the exception reports because the monthly status reports would provide evidence of positive adherence to approved standards.

A word of caution is in order here. If there actually exists a violation of floor load limits in the plant, no fancy computerized information system is going to hide it. Manual calculations can be used to determine whether a floor load limit has been exceeded, and symptoms such as crushed, bent, or cracked floor members can be as embarrassing as they are hazardous.

The whole problem of floor and aisle design and maintenance is magnified by the use of mechanical handling equipment such as forklift trucks. Forklifts both aggravate the problems and at the same time become more of a hazard themselves whenever problems with floors exist. This problem will be revisited in Chapter 11.

Stairways

Building codes and standards for stairs are well established, but many factories and businesses have stairways that do not meet code. One or two steps are usually not a problem, but if the stairways have four or more risers, they need standard railings or handrails and must be kept clear of obstructions. Note the use of the words *handrail* and *railing. They are not the same.* A handrail, as used in this standard, is a single bar or pipe supported on brackets from a wall to furnish a handhold in case of tripping. A railing, however, is a vertical *barrier* erected along the exposed sides of stairways and platforms to prevent falls. A handrail and railing may of course be combined into the same unit, but the two terms are not interchangeable.

The OSHA standard is very flexible on the subject of the placement of stairway landings, using such language as "avoid" (long flights of stairs), and "consideration should be given to. . ."

The placement of stairway landings is a safety consideration. Most people think that the purpose of stairway landings or platforms is to give the climber a chance to rest, and it is true that this is a supplemental purpose. However, the main purpose of the stairway landing is to shorten the distance of falls, and thus landings play an important role in building and facilities safety. Extremely long flights of stairs are obviously more dangerous than stairs interrupted by landings. To be effective they must be no less than the width of the stairway and a minimum of 30 inches in length measured in the direction of travel.

Ladders

Ladders are not the simple devices most people think they are. Design is critical because the construction should be neither too strong nor too weak. A weak ladder is obviously dangerous, but a ladder that is overdesigned is difficult or impossible to handle safely. A long, heavy ladder can be as much hazard when it is carried as when it is climbed. It is easy to see why ladders must be manufactured to exacting standards.

Most firms buy ladders from regular ladder manufacturers, and the Safety and Health Manager can be quite sure that the ladders were constructed properly in the first place. What is more important to industrial safety is how the ladders are used and maintained. Defective ladders must be either repaired or destroyed, and while awaiting either fate they must be tagged or marked "Dangerous; Do Not Use." Personal interviews with Safety and Health Managers have indicated that they often saw defective ladders in half rather than risk the possibility of a defective ladder being returned to service. When a job needs to be done and there is no regular ladder in sight, the temptation is too great for maintenance or other personnel to remove the danger tag and make immediate use of a defective ladder. At best, a repair job on a portable ladder usually does not look very good and will likely arouse suspicion even if the ladder is safe. It takes a good engineer to convince anyone that the repaired ladder is just as good as new, even if the engineer has been able to convince himself or herself.

Portable metal ladders share common hazards with portable wood ladders; a major difference, however, is the fact that <u>metal ladders conduct electricity</u>. Most workers are aware of the increased hazards of electrocution present when using metal ladders. Rubber or otherwise nonconducting feet are a good precaution on metal ladders, but the hazard is still present, and the rubber feet should not be allowed to give rise to complacency.

Whether the portable ladder is made of wood or metal, it is <u>the way the ladder is used that will chiefly determine its safety</u>. Almost everyone knows the admonition that it is unsafe to ascend or descend a ladder with the climber facing away from the ladder. Less obvious is the fact that portable ladders are typically not designed to be used as platforms or scaffolds, and they become very weak if used at angles close to horizontal.

A common error in the use of ladders is to use ladders that are too short. For instance, when accessing a roof, the ladder needs to extend at least 3 feet above the upper point of support. For ordinary stepladders, the top should not be used as a step. Another foolish practice is to place ladders on boxes, barrels, or other unstable bases to obtain additional height. Some people even try to splice short ladders together to make longer ones.

The first consideration in the use of a portable ladder is its condition, especially the rungs. Next to the hazard of broken rungs, the greatest hazard probably is a ladder that slips or tips because it is insecurely positioned. The proper slant is 4 feet vertical to 1 foot horizontal. A safe practice is to tie off the ladder at the top so that it cannot tip or slide down. This is not always practical, however, and alternative solutions can solve the slipping problem. Sometimes a ladder can be made stable by positioning it where the structure of the wall or building limits its movement and makes the ladder safe. Another solution is to use nonslip bases, but this method may not work on some surfaces, such as oily, metal, concrete, or slippery surfaces. Metal ladders in particular may be subject to slipping and need to be equipped with good safety shoes. Safety shoes are as much to prevent slipping as they are to prevent electrocution. Even spikes may be useful on some surfaces to make the footing secure. If the ladder is used without safety shoes on a hard slick surface, the ladder needs a footladder board to prevent slipping.

Fixed Ladders

From a safety and health standpoint, fixed ladders require a somewhat different approach from portable ladders. With portable ladders the emphasis is on care and correct use, as just discussed. With fixed ladders, the emphasis is on design and construction. It is beyond the scope of this book to specify all of the details of fixed ladder design. If fixed ladders need to be constructed, the designer should follow the detailed specifications laid out in established standards, not this book.

This book seeks only to alert Safety and Health Managers to problems they may encounter with existing fixed ladders. Some fixed ladders may have been constructed long ago or perhaps without the benefit of standards. It is the duty of the Safety and Health Manager to alert the company if fixed ladders are found to be incorrectly constructed. The Safety and Health Manager can then recommend that design details be followed when the ladders are rebuilt.

Some knowledge of fixed ladder design principles is necessary, however, if the Safety and Health Manager is to know how to recognize common problems found. For brevity, this book selects some easy-to-observe problems with fixed-ladder design, illustrated in Figure 6.3. The rung offset is to prevent the foot from sliding off the end of the rung. Other means, such as siderails, would also be acceptable.

When the fixed ladder is more than 20 feet long but less than 30 feet long, ladder cages are needed, but many Safety and Health Managers do not realize that there is an alternative to installing ladder cages. On tower, water tank, and chimney ladders over 20 feet in unbroken length, ladder safety devices may be used in lieu of cage protection. Ladder safety devices usually take the form of a combination of fixed equipment on the ladder and personal equipment worn by the climber. One type uses a trolley that moves along a fixed rail. The trolley is attached to the climber's belt. The trolley moves easily up or down the rail when the climber ascends or descends, but a brake engages which stops the trolley in event of a fall. One thing to watch carefully with a rail-and-trolley device is to ensure that the device is equipped to deal with ice on the rail if the ladder is located in a climate where ice is a pos-

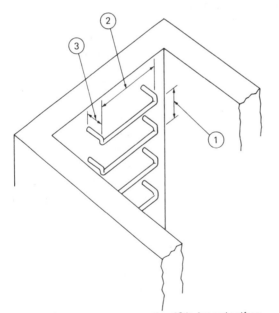

1. Distance between rungs not more than 12 inches and uniform.
2. Minimum rung length is 16 inches.
3. Back clearance minimum is 7 inches. This dimension is the most frequently violated, according to frequency of OSHA citation.

Figure 6.3. Easy-to-recognize problems with fixed ladders; all three are required by OSHA. Note the rung offset to prevent the foot from slipping off. (Based on OSHA standards.)

sibility (most outdoor locations in the United States). The safety problem with ice on the rail is somewhat indirect; the trolley simply cannot move along the ice-covered rail, making the system useless. But this is when the safety device is most needed. The problem is not without solution because there are de-icing devices for trolley systems. Safety and Health Managers should choose their protection systems with care, however, being sure to consider all factors, including the problem of icing, in order to avoid later embarrassment and wasted investment on the part of the company. An example rail-and-trolley ladder safety device system is illustrated in Figure 6.4.

Dockboards

A dockboard, or bridge plate, provides a temporary surface over which loads can be transported, particularly during loading or unloading of a cargo vehicle. One of the main safety hazards with these boards is that they may shift while in use. It is also possible that the surfaces connected by the dockboard can shift, such as when the cargo vehicle itself moves. Finally, the dockboard itself may not be strong enough to carry the load.

Figure 6.4. Example of ladder safety device system.

EXITS

Exits are usually considered doors to the outside and from a safety standpoint are considered a means of escape, especially from fire. Such thinking is accurate but incomplete. The Safety and Health Manager should enlarge the concept by using the more general term "means of egress" to include

1. The way of exit access
2. The exit itself
3. The way of exit discharge

By considering means of egress instead of simply exits, the Safety and Health Manager can analyze the entire building to determine whether every point in the building has a continuous and unobstructed way of travel to a public way. In this way, one must think of such building facilities as stairways, intervening rooms, locked interior doorways, and limited-access corridors. Outside the building one must think of yards, exterior storage of materials, fences, courtyards, and shrubbery. One thinks of shrubbery and landscaping as affecting neither safety nor health. However, employees might escape a burning building (or perhaps a building with a ruptured chlorine gas line) through an exit door, only to find outside that the exit empties into a courtyard tightly confined by a fence, dense shrubbery, or other obstruction.

Almost every Safety and Health Manager at some time during his or her career encounters the embarrassment of an exit that is *locked*. This can be a two-edged problem because many Safety and Health Managers are also responsible for plant security. In many cases the only practical solution is to provide panic bars or other mechanisms for locking doors from outside entry while maintaining free and unobstructed egress from the inside. Where unauthorized exit can be as much a security problem as unauthorized entry, the automatic-alarm-sound type of door may be the only alternative. Facilities designers are turning more and more to the use of unlocked, automatic-alarm-sounding emergency exit doors. Even more frequently encountered than locked exits are exits that are cluttered or blocked by obstructions or impediments. Stacks of material obstructing the door or way of travel defeats the purpose of the exit.

Exits must be marked by a readily visible sign. Sometimes exits are marked, but the way to reach the exit may not be apparent. Intermediate signs along the access to the exit may be necessary to show occupants how to reach the exit.

A line of reasoning similar to the reasoning described earlier for aisles and aisle marking applies to exits. The Safety and Health Manager should not be too eager to go out and buy signs from the exit-sign sales representative. It is important to weigh carefully a decision to name a door an "exit." Once a sufficient number of exits have been identified, the Safety and Health Manager should be careful about putting exit signs on any additional doorways. An exit sign is an invitation to start checking for obstructions, locked doors, inadequate signs along the exit access, incorrect sign dimensions, obstructions outside the exit, and even lighting. Provided that a sufficient number of marked exits have been provided, there is no law that says that an employee shall not escape the building by an unmarked passage or doorway to the outside. Such doorways should not even be called "exits."

ILLUMINATION

The subject of lighting was mentioned in the last paragraph. Lighting, or the lack of it, can be a safety hazard, but there is no code for minimum safe lighting except for specialized areas. For instance, if forklift trucks are operated in the plant area, the minimum general lighting level is 2 lumens per square foot unless the forklift trucks themselves have lights. Every exit sign should be suitably illuminated by a reliable light source giving a value of not less than 5 footcandles on the illuminated surface. This is not to say that the exit sign must be the kind that is *internally* lighted. An alternative to be considered is artificial lighting *external* to the sign. Also, there is nothing wrong with relying on *natural* illumination (sunlight) on the exit sign in an amount not less than 5 footcandles. Natural illumination can be a problem, however, if the area is accessed on second or third shifts. Incidentally, 5 footcandles is not very much illumination. Most plant areas are normally illuminated by much greater levels of illumination.

MISCELLANEOUS FACILITIES

Maintenance Platforms

The importance of planning for maintenance activities when constructing a new building was pointed out earlier in this chapter. Many modern buildings have built-in, safe suspension systems for exterior window cleaning and other exterior maintenance. Maintenance workers for buildings not so equipped are less fortunate and typically work from suspended scaffolds of the same type as construction scaffolds, which are discussed in Chapter 15. Not only are the maintenance workers less fortunate, but so are the employers of these workers and the Safety and Health Managers who must worry about the safety of the scaffolds, the proper securing of scaffolds on the roof of the building, and other items governed by applicable standards.

Safety and Health Managers who do find that their buildings are equipped with powered platforms for exterior maintenance should direct most of their attention to how these platforms are being used and maintained, not how they are made. The manufacturer of such equipment would normally be very careful to adhere strictly to standards when fabricating the powered maintenance platform. Typical problems with these platforms are missing guardrails, missing toeboards, missing side mesh, disabled safety devices, and inadequate inspections or records of inspections.

Regarding the equipment itself, some companies have been tripped up for not having load-rating plates on the platform. The load rating must be stated in letters at least ¼ inch in height. The wire rope suspending the platform must also be marked with a metal tag stating its maximum breaking strength and other data, including the month and year the ropes were installed.

Workers on some types of powered platforms need to wear safety belts; on other types they are safe without the belts. A platform supported by four or more wire ropes can be so designed that the working platform will maintain its normal position even if one rope fails. However, many powered platforms are suspended by only two wire ropes and will tip dangerously if one of the ropes fails. One of the more dangerous types of platform is known as "type T," and workers on these platforms must wear safety belts attached by lifelines. If the platform qualifies as type T, it will upset with a single wire rope failure, but it will not fall to the ground. Therefore, the lifeline may be attached either to the building structure or to the working platform. Compare this to construction industry standards (Chapter 15) which require that lifelines be secured to an anchorage or structural member instead of to the scaffold.

Public-utilities workers and tree trimmers often use platforms that are vehicle mounted, such as aerial baskets, aerial ladders, boom platforms, and platform-elevating towers. Again the majority of accidents arise from improper use of the platform rather than from equipment failure or design. This is even more true of the vehicle-mounted platforms than of the building-mounted models discussed earlier.

The most serious hazard with vehicle-mounted platforms is contact with high-voltage power lines, and this kills workers every year. The hazard is so severe that a safety distance must be maintained at all times, except, of course, in the case of electric utility companies who by the nature of their work must approach closer. For safety, the utility companies must insulate aerial devices that work closer than the standard safety distance. For nonutility companies, the accepted standard is a 10-foot distance in the case of a 40-kilovolt line, for example. Different line voltages may need higher or lower safety distances, and these distances are covered in more detail in Chapter 15 in the discussion of mobile cranes.

Sometimes special sensors called "proximity warning devices" are installed on the boom to warn the operator when the basket is too close for safety. However, these warning devices do not provide positive protection and thus should not be considered an excuse for moving the boom closer to the line than authorized minimums.

Workers in aerial baskets often fail to wear a body belt and lanyard attached to the boom. Adding to the fall hazard is the possibility of unexpected contact with an object that might strike and perhaps sweep the worker out of the basket or off the platform. Echoing the hazards avoidance principles of Chapter 3, such unusual hazards point to the importance of training for personnel who work in aerial baskets. Other unsafe procedures are failure to secure the aerial ladder before traveling, climbing or sitting on the edge of the basket, or improvising a work position other than the floor of the basket.

Elevators

Elevators are everywhere, but when can you remember one falling? The catastrophic fall of an elevator is such a horrifying thought that the public long ago set up regulations for safe elevators. Jurisdiction was placed within the states, and most states administer elevator inspections through "labor" or "labor and industry" commissions. Next time you ride an elevator, look at the certificate of inspection posted inside the elevator car.

Elevators must be inspected both when new (or altered) and periodically thereafter. Many states even require construction permits from the authorized elevator inspection agency before elevator construction is begun. Elevator operating permits and fees are also required by some states. Not every inspector must be an agency official, but state licensing procedures for elevator inspectors may be applicable.

Manlifts are used as elevators, but unlike elevators, there are federal standards for manlifts. Manlifts are much cheaper and more efficient than elevators for many plant operations and are thus sometimes used instead of elevators. As can be seen in Figure 6.5, however, a manlift is inherently more hazardous than an elevator. It is ironic that elevators—which are safer than manlifts—are governed by strict state inspection, licensing, permits, and approvals. With manlifts, though, it is up to the Safety and Health Manager to interpret general standards and identify hazards. It is apparent from Figure 6.5 that the biggest hazard with manlifts is getting on and getting off. Exit is essential because the belt is continuous, and to stay on the belt past the top or bottom floor would be either impossible or extremely hazardous.

Boilers

Steam boilers and pressure vessels are so safe today that most people do not even think about them. It has not always been that way. Although steam boilers are not as popular today for building heating systems, many are still in use; in addition, industrial processes use hundreds of thousands of boilers and pressure vessels. So the lack of familiarity with boiler accidents is not because the boilers themselves are rare. It is the accidents that are rare, not the boilers, and when an accident does occur, the energy released by the explosion is so devastating that it usually is a catastrophe.

The extreme hazard of an unsafe boiler led to the early regulation and safeguarding of these vessels. As with elevators, the historical development of boiler codes has placed their jurisdiction within the states. State control has been very effective in keeping boiler accidents to an absolute minimum.

The Safety and Health Manager needs to ensure that boilers and pressure vessels in the plant are being inspected and that state procedures are being followed. One question that immediately comes to mind is, What pressure vessels are covered by the regulations? Most states exempt containers for liquefied petroleum gases (LPG), as these are covered by other regulations. The same can generally be said of vessels approved by the Department of Transportation for public highway transportation of liquids and gases under pressure. Some states also exempt vessels used in connection with the production, distribution, storage, or transmission of oil or natural gas. Note, however, that the foregoing does not exempt refineries or chemical plants that produce petroleum *products*. In the industry, the term *oil production* refers to the drilling and extraction, or "mining," of petroleum, not its refining. The only way to be sure about exemptions in a given state is to check with that state's agency of authority.

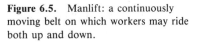

Figure 6.5. Manlift: a continuously moving belt on which workers may ride both up and down.

The time to stop and think about regulations for boiler and pressure vessel safety is whenever such a vessel is to be purchased, installed, modified, moved, or sold. Welding on such vessels may weaken them and is scrutinized carefully by the inspector, although welding is not absolutely prohibited. Even hot water storage containers should be installed or reinstalled by persons properly licensed to do the work.

SANITATION

The sanitation of lunchrooms seems straightforward and obvious, but sanitation decisions can be trickier than they appear. If a decision is made to allow employees to eat in the plant, principles of hygiene must be observed. The Safety and Health

Manager should be sure that a sufficient number of waste receptacles is provided to avoid overfilling. But before going overboard, the Safety and Health Manager should realize that too *many* waste receptacles can be provided also. If too many receptacles are provided, maintenance personnel will become lax about emptying containers that receive little use, resulting in additional sanitation problems.

The presence of toxic materials complicates the whole problem of food service, consumption, and storage. Certainly, food and beverages must not be stored in areas where they will be exposed to toxic materials. This rule may seem obvious, but the Safety and Health Manager should consider not only the plant cafeteria or lunchroom but also the employee who brings snacks from home and stores them in areas in which they will be exposed to toxic materials.

Some toxic materials, such as lead, are particularly susceptible to exposure by ingestion during food consumption. Some toxic materials, such as vinyl chloride and arsenic, are of such concern that there are strict, specific standards for their control. Toxic substances will be discussed in more detail in Chapter 7.

SUMMARY

Safety and Health Managers who are willing to plan ahead can save their companies a great deal of money by heeding building and facility codes *before* commencing construction or expansion of plant space. Planning ahead is the key to compliance with standards for floors, aisles, exits, and stairways. Guardrails, ladders, and platforms may be added, but they, too, deserve some advance consideration to ensure that installations meet requirements.

Safety and Health Managers need to be careful not to get too enthusiastic and provide for too many aisles or exits. An extra aisle or exit that is improperly maintained or marked can easily lead to problems and indeed is not even in the interest of safety.

The subject of buildings and facilities may not be the most exciting topic for the Safety and Health Manager's attention. However, even ordinary matters such as housekeeping and sanitation deserve careful consideration and judgment to promote safety and health at reasonable cost.

EXERCISES AND STUDY QUESTIONS

6.1. What is the height of a standard railing for walking and working surfaces?

6.2. A portable ladder is needed to climb onto a rooftop 14 feet high. How long should the ladder be?

6.3. What are the two principal federal requirements for aisles in industrial plants?

6.4. Explain why the title for a major OSHA subpart is "Walking and Working Surfaces" instead of simply "Floors." Name 10 different walking and working surfaces.

6.5. What is deceptive about the danger of an open-sided floor or platform only 4 feet high?

6.6. Suppose that in your plant, welders must stand on top of a 10-foot-high tank to complete manufacturing operations. What should be done, if anything, to protect them from falling?

6.7. What is the purpose of a toeboard?

6.8. Explain under what conditions, and how, roofing workers should be protected from fall hazards.

6.9. When must an open-sided floor or platform be equipped with a railing?

6.10. When would OSHA require guarding for a service pit for use in vehicle maintenance? How would you guard such a pit?

6.11. Explain how poor housekeeping could result in an OSHA citation that classifies the violation as "serious."

6.12. As a Safety and Health Manager, how would you deal with the problem of housekeeping?

6.13. Explain the danger of marking too many permanent aisles.

6.14. Explain the difference between a "handrail" and "railing."

6.15. Describe the purposes and requirements for stairway landings.

6.16. Describe important points to consider in safety with portable ladders.

6.17. Under what conditions are ladder cages specified? When may alternate ladder safety devices be used? What are the advantages and disadvantages of these devices?

6.18. Define the term *means of egress*.

6.19. What are the biggest problems with exits?

6.20. What are vehicle-mounted work platforms? What are their chief hazards?

6.21. Explain how either too many or too few waste receptacles can lead to sanitation problems.

7

Health and Environmental Control

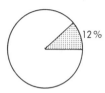 12%

Percent of OSHA
general industry citations
addressing this subject

Health hazards carry a great deal of impact because the potential harm to exposed employees is great, and the cost of correction of a single health hazard can run into millions of dollars.

Industrial hygienists have been saying for many years that health hazards deserve more attention, and in response to this pressure, a shift from safety toward health activities has been in evidence almost since OSHA commenced operations. At first OSHA did not have a sufficient cadre of qualified health professionals to assess health hazards, and the natural focus was on safety. But the proportion of health specialists in OSHA has increased a great deal since the early 1970s.

Chapter 1 discussed the bases of competition and even friction between safety professionals and health professionals. The two groups have common objectives, but their backgrounds are usually vastly different, which results in conflict. The more successful professionals on both sides are busy emulating the characteristics typical of the opposite side, and thus the differences between the two professions are beginning to dissolve.

Health hazards will always tend to be more subtle to detect than safety hazards, by the very definitions of *health* and *safety*. As was noted in Chapter 1, health deals with the long-term chronic exposure effects, whereas safety deals with the more obvious acute effects that do their damage immediately.

BASELINE EXAMINATIONS

Almost everyone has taken a preemployment physical examination, but few understand the importance of the physical exam to the overall safety and health program. Each employee's baseline health status is established by this examination. This status is important in placing the employee in the right job and in detecting any health deterioration due to job exposures. Occupational health exposures are perhaps the most important reason for preemployment physical examinations because of the chronic nature of health hazards. If an employee already has emphysema or other lung disorders, it is essential that this fact be established at the time of employment. The same can be said of hearing impairment, as will be discussed later in this chapter.

TOXIC SUBSTANCES

Exposure to toxic substances is the classic "health problem," and this topic will be used to model the entire subject of health and environmental control. Choice of words can be important here. For instance, the term *hazardous materials* is sometimes used to refer to toxic substances, but the term *hazardous* is a much more general term that would include such safety hazards as presented by flammable and combustible liquids and explosives. This book follows popular convention, which tends to associate the term *materials* with safety hazards and the term *substances* with health hazards. In fact, the title of Chapter 8 is "Hazardous Materials," and the chapter deals almost exclusively with safety hazards.

The Safety and Health Manager needs to have a general knowledge of what various types of toxic substances can do to the body. Such general knowledge will be useful in convincing workers and management alike that toxic substances must be controlled for the health of the workers as well as to avoid OSHA citation. This discussion will describe some of the various types of toxic substances by considering their effects on the body.

Irritants   *what type is ammonia?*

Irritants inflame the surfaces of the parts of the body by their corrosive action. Some irritants affect the skin, but more of them affect the moister surfaces, especially the lungs. Even a weak irritant to the upper respiratory tract will be easily detectable by the victim, but irritants to the lower respiratory tract may go unnoticed.

When the irritant is some type of dust, the lung disease that results is called *pneumoconiosis*. This is a general term that includes reactions to simple nuisance dusts as well as fibrosis, a more serious reaction that includes the development of fibrous scar tissue which impairs the efficiency of the lungs. Examples of pneumoconioses include siderosis (from iron oxide dust), stannosis (from tin dust), byssinosis (from cotton dust), and aluminosis (from aluminum dust). The more dangerous fibroses are asbestosis (from asbestos fibers) and silicosis (from silica).

Everyone is familiar with the strong odor of ammonia. Ammonia gas and moisture on the mucous membranes of the body combine to form ammonium hydroxide, a strongly caustic agent. It is easy to understand why this would irritate and injure the delicate tissues of the nose, trachea, lungs, and other parts of the body that it might contact. By similar logic, any of the gases that combine with water to produce acids would be irritants, as would airborne particles of the acids themselves.

Plating operations easily impart acid mists to the air since the plating tanks are often splashing, hot, and acidic. Particularly offensive is chromic acid mist, which causes an ominous-sounding malady, "chrome holes." Chromic acid also destroys the nasal septum, which separates the two nostrils.

Another well-known irritant is chlorine gas, a widely used industrial chemical. Chlorine's halogen relatives, fluorine and bromine, are also irritants, especially fluorine, the strongest of all halogens. Even the soluble *salts* of fluorine are poisonous. Less well known are those substances that are irritants deep within the lungs, such as the oxides of nitrogen and phosgene. Phosgene is best known as a chemical warfare gas, which is evidence of its toxicity. But phosgene can also be generated inadvertently in the workplace when chlorinated hydrocarbon solvents are exposed to welding radiation.

Chronic exposure to irritants over a long period can cause scar tissue to develop in the lungs. Some of these substances do not produce an appreciable immediate irritant effect but are dangerous in the long run. The most notorious of these scarring agents is asbestos fibers. Coal dust is also a scarring agent. Scarring agents are in the form of tiny solid particles, and their action on the lungs is *mechanical*, as contrasted with the systemic poisons discussed in the next section.

Systemic Poisons

More insidious than the irritants are the poisons that attack vital organs or systems of organs, sometimes by toxic mechanisms that are not understood. The chlorinated hydrocarbons common in solvents and degreasers, for example, are blamed for liver damage.

Perhaps the best known systemic poison found in occupational exposures is lead. Lead is disappearing from paint pigments because of its reputation as a poison, but it still exists in tetraethyl lead used in leaded gasoline. The author of this book formerly worked in a tetraethyl lead plant, and decades ago the workers in this plant had a high awareness of what lead can do to the body. Lead attacks the blood, the digestive system, and the central nervous system, which includes the brain. Autopsies have also shown damage to kidneys, liver, and reproductive systems, but these results are inconclusive. Other toxic metals are mercury, cadmium, and manganese. Magnesium, sometimes confused with manganese, is less toxic.

Another important systemic poison is carbon disulfide. Carbon disulfide is unusual in that its hazards are extreme from the standpoint of both safety (fire and explosions) and health. It is widely used in industry as a solvent, a disinfectant, and an insecticide. As a systemic poison, carbon disulfide attacks the central nervous system. Methyl alcohol (methanol), a popular solvent, also is a systemic poison to the central nervous system but is a much milder poison than carbon disulfide. Indeed, methanol is even acceptable in small quantities as a food additive! Methyl alcohol is also a fire and explosion hazard.

Depressants

Certain substances act as depressants or narcotics on the central nervous system and as such can actually be useful as medical anesthetics. Unlike the systemic poisons discussed earlier, the effect of the action of depressants on the central nervous system is usually temporary. However, some substances, such as methyl alcohol, are both systemic poisons and depressants. Besides affecting health, depressants can have an adverse effect on safety because they interfere with the concentration of workers who operate machinery.

The most familiar depressant is ethyl alcohol (the "drinking" variety of alcohol), sometimes called ethanol in industry. Its harmful effects as an industrial hazard are minimal compared to its effects from drinking. In fact, the greatest hazard from ethanol on the job is without doubt from "voluntary ingestion" from bottles brought on the premises by employees. Ethanol is not as toxic as methanol.

Acetylene, the most widely used fuel gas for welding, is a narcotic, but its health dangers are minimal compared to its safety hazard from fires and explosions. Acetylene has been used in medical anesthesia.

Benzene is a very popular industrial chemical used principally as a solvent. Benzene is a depressant on the central nervous system, an irritant, and a systemic poison, and has recently been recognized as causing leukemia. In addition, benzene is a dangerous fire and explosion hazard. OSHA has a movie that dramatically films the hospital-bed testimony of a young benzene worker who is dying from leukemia.

Asphyxiants

Asphyxiants prevent oxygen from reaching the body cells, and in the general sense, any gas can be an asphyxiant if there is enough of it to crowd out the essential proportion of oxygen in the air. Many people have committed suicide by breathing natural gas, which is essentially methane. But methane is merely a simple asphyxiant, in that it displaces the proportion of oxygen in inhaled air. Methane can be encountered in industrial environments, as it is a product of fermentation. Other simple asphyxiants commonly encountered are the inert gases such as argon, helium, and nitrogen used in welding gases.

It may seem incorrect to classify nitrogen as an air contaminant and an asphyxiant when it is the principal constituent (78%) of normal air. But too much nitrogen will reduce the normal proportion of oxygen (21%) in the air. Any proportion of oxygen less than 19.5% is considered oxygen deficient. Oxygen deficiency is very dangerous, a more serious condition than most people think. The following case study is an accident description quoted from federal records of workplace fatalities (ref. 24).

Case Study 7.1   A contract employee was assigned to sandblast the inside of a reactor vessel during turnaround activities at a petrochemical refinery. Instead of relying on the contract company's own air compressors in accordance with the contractor's policy, the contract foreman connected the employee's supplied air respirator to a hose containing what he thought was plant air but was actually nitrogen. Both hoses were identical except for markings at the shutoff valve. The sandblaster entered the vessel, descended to the bottom, placed the respirator hood on his head, and was overcome.

The cause of death in this accident case history was oxygen deficiency, and the irony is that the worker was breathing nearly pure nitrogen, the primary constituent of normal breathing air.

Carbon dioxide is one of the most important simple asphyxiants, although it is a harmless constituent in air in normal quantities. Fire is the primary source of dangerous industrial concentrations of carbon dioxide. Carbon dioxide is heavier than air, which causes it to accumulate in low, confined spaces, increasing its hazard potential. Confined spaces have a great deal of hazard potential, not only for carbon dioxide but for all air contaminants.

The asphyxiants discussed so far are *simple asphyxiants,* essentially nontoxic substances which replace the essential oxygen content in the air. The other type of asphyxiant is the *chemical asphyxiant,* which interferes with the oxygenation of the blood in the lungs or later interferes with the oxygenation of the tissues. The most notorious chemical asphyxiant is carbon monoxide, a substance for which blood hemoglobin has a stronger affinity than it does for oxygen (over 200 times the affinity). The resulting compound, carboxyhemoglobin, is a very stable substance which prohibits the life-critical exchange of oxygen and carbon dioxide by their vehicle, blood hemoglobin.

Another well-known chemical asphyxiant is hydrogen cyanide, an industrial insecticide but better known as the gas used in prison execution chambers. The gas is formed by dropping sodium cyanide pellets into a small container of acid. Some workplaces have the potential for becoming execution chambers. In a workplace inspection in California, a laboratory was found in which strong acids in glass bottles were stored on shelves located directly above sodium cyanide salts!

Carcinogens

Carcinogens are substances that are known to cause or are suspected to cause cancer. A great deal of attention has been given to carcinogenesis since the advent of OSHA, but the source of the emphasis is not OSHA alone. NIOSH, the Consumer Product Safety Commission (CPSC), and other agencies have concentrated on carcinogens. Public awareness is now high, and many workers and consumers alike are becoming very cautious about their exposures to carcinogens. One of the frightening things about carcinogens is that the cancer has such a long latency period. Sometimes a lapse of 20 or even 30 years occurs between exposure and appearance of the cancerous tumors.

New carcinogens are being labeled every year, and many of the substances indicted are very commonly used industrial materials, such as benzene and vinyl chloride. The familiar saying "Everything I do is either illegal, immoral, or fattening" might be modified to include "or causes cancer." So many substances have been indicted by laboratory tests on animals that it is possible that a sort of complacency could be developing in the minds of the public.

Vinyl chloride, mentioned earlier as an example carcinogen, is extremely dangerous from many aspects. It is a severe explosion hazard, and when it burns it is very difficult to extinguish. Added to this is the hazard of highly toxic phosgene being liberated during vinyl chloride fires. Acute skin exposures can result in injury from skin freezing due to rapid evaporation of vinyl chloride. Chronic inhalation is now recognized to cause a form of cancer of the liver, angiosarcoma. Vinyl chloride is also believed to be a teratogen, a type of substance to be explained in the next section.

Before going to teratogens, a distinction should be made between the extremely hazardous vinyl chloride and polyvinyl chloride (PVC). PVC is a type of plastic synthesized by polymerizing the unstable vinyl chloride monomer into the stable polyvinyl chloride polymer. Figure 7.1 illustrates the arrangement of the atoms in the dangerous vinyl chloride monomer and in the harmless and very stable polyvinyl chloride polymer.

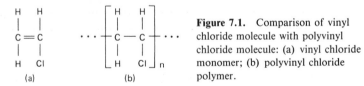

Figure 7.1. Comparison of vinyl chloride molecule with polyvinyl chloride molecule: (a) vinyl chloride monomer; (b) polyvinyl chloride polymer.

Teratogens

Teratogens affect the fetus, so their toxic effect is indirect. Women should be careful about exposures to certain substances during pregnancy, especially in the first trimester (first three months). Teratogens should not be confused with mutagens, substances that attack the chromosomes and thus the species instead of the individual. Teratogens do their damage after conception but before birth, whereas mutagens do their damage before conception. Mutagens can affect the chromosomes of either potential fathers or potential mothers.

A tricky legal question is whether an industry can prohibit women of childbearing age from working in jobs in which they may be exposed to teratogens. The question is whether this constitutes sex discrimination. The affirmative contingent says that the teratogens must be controlled by the industry so that pregnant women can have a safe and equal opportunity to work in the same jobs as those offered to men. The negative contingent says that it makes more economic, safety, and social sense to employ women in other jobs during pregnancy.

Routes of Entry

The term *toxic substance* can be considered synonymous with the term *poison*, a word familiar to all of us who were taught as children not to eat or drink poison. Poison is even in our fairy tales, such as "Snow White and the Seven Dwarfs." It may be true that the greatest danger from poisons in the home is from ingestion (swallowing), but on the job the greatest danger is from breathing. In fact, it has been said that the order of importance of routes of entry to the body for toxic substances is the exact opposite on the job from the order at home, as shown in Figure 7.2.

Routes of entry	Importance At home			At work	
Ingestion	Most important			Least important	
Skin contact	Of moderate importance			Of moderate importance	
Inhalation	Least important			Most important	

Figure 7.2. Toxic substances' routes of entry to the body. Note reverse sequence of importance at work versus at home.

The various routes of entry to the body for toxic substances are more related than most workers realize. Inhalation of toxic substances results in accumulations on the mucous membranes; mucus is later coughed up, and some of it is inevitably swallowed. Skin contact with toxic materials can also result in ingestion as substances become embedded beneath fingernails and on hands which later come into contact with food. Toxic dusts in the air are also picked up in the hair and are later deposited on the pillow during sleep and thus gain entry to the body indirectly.

From knowledge of routes of entry for toxic substances it is easy to see the importance of sanitation. Some of the principles of sanitation discussed in Chapter 6 take on more importance when toxic substances are considered. Proper storage of food, and shower and washing facilities, may be essential in the control of the quantities of toxic substances entering the worker's body.

Air Contaminants

The greatest concern with toxic substances is with air contamination, and this is as it should be (as shown in Figure 7.2). Air contaminants take on many physical forms, and most people confuse these forms in everyday language. The Safety and Health Manager should know the difference, for instance, between vapors and fumes. Although air is essentially gases, the *contamination* of that air can consist of any of the three states of matter: solids, liquids, or gases.

Gases easily contaminate the air because air consists of gases, and gases readily mix. The most familiar toxic gas is carbon monoxide. Also dangerous in the industrial environment are hydrogen sulfide and chlorine. Even "harmless" gases, such as carbon dioxide and inert nitrogen, can become dangerous if allowed to accumulate in large quantities so that they become asphyxiants by crowding life-giving oxygen out of the breathing zone.

Vapors are also gases, but these substances are normally liquids or even perhaps solids that release small quantities of gases into the surrounding air. Some of our most useful industrial liquids, such as gasoline and solvents, have a strong tendency to release these vapors.

Mists are really tiny droplets of liquids, so small that they remain suspended in the air for long periods, as in clouds. Since liquids are heavier than air, they eventually settle out or coalesce into larger droplets which fall in the manner of rain. Long before that happens, however, mists can be inhaled by the worker. Fine mists are generated as vapors condense into clouds. Coarse mists are produced by splashing or atomizing operations, such as from machine tool cutting oils or from electroplating. Pesticide spray is also usually a mist.

Dusts are recognized as solid particles. Technically speaking, dust particles range in size from 0.1 to 25 micrometers (0.000004 to 0.001 inch) in diameter. Everyone is exposed to dust, and some dust is relatively harmless. Dangerous dusts include asbestos, lead, coal, cotton, and radioactive dusts. Silica dust from grinding operations is also becoming recognized as a hazard, although ordinary earth dust is principally silica. Asbestos dust particles are fiber shaped instead of round, and this contributes to their hazard.

Fumes are also solid particles but are generally too fine to be called dusts. Actually, the sizes for fume particles and dusts overlap, as can be seen in Figure 7.3. Whereas dust particles are typically finely divided by mechanical means, fumes are formed by the resolidification of vapors from very hot processes such as welding. Chemical reactions can also produce fumes, but gases and vapors also produced by chemical processes are not to be confused with fumes. Metal fumes are the most dangerous, especially those of the heavy metals.

Particulates are really a general classification which includes all forms of both solid and liquid air contaminants (i.e., dusts, fumes, and mists). Particle sizes thus can vary greatly; some are visible to the naked eye, but most are not. Figure 7.4 displays some example particle sizes over the range from tiny visible particles down to submicroscopic large molecules.

It is probably already obvious that industry and technology do not eliminate exposures to toxic substances; they merely control these exposures to keep them within acceptable levels. It is both naive and unnecessary for the Safety and Health Manager to adopt a strategy of complete elimination of worker exposure to toxic substances.

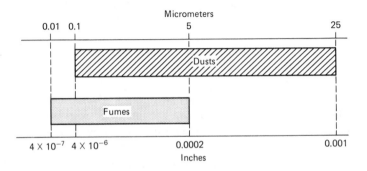

Figure 7.3. Dust and fume particle sizes compared (not to scale).

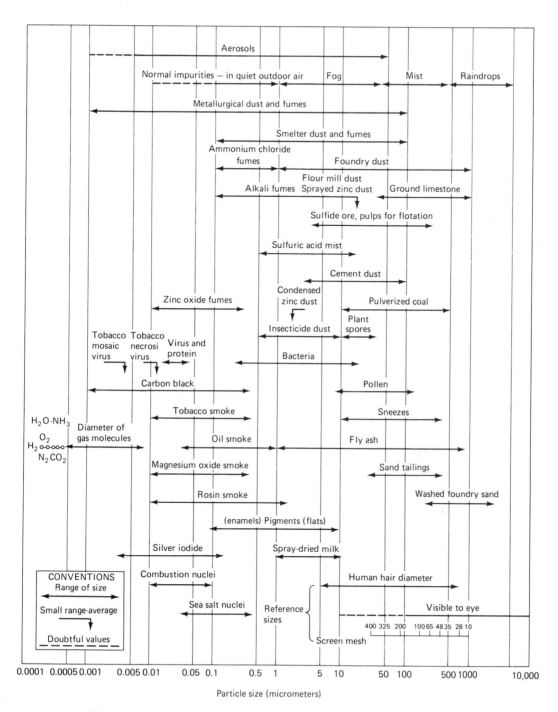

Figure 7.4. Size of airborne contaminants. (Note the logarithmic scale.) (*Source:* Courtesy of the Mine Safety Appliances Company.)

There is no known poison to which a human being cannot be exposed without significant harm, provided that the exposure is tiny enough and distributed over a long enough interval for the body to assimilate or counteract it. On the other hand, even the mildest of poisons can become lethal if the worker is constantly exposed to it in massive doses. And just such massive doses are the kind that can be found in industrial exposures, more than in any other environment.

Threshold Limit Values

Since no poison is lethal in small enough doses, and all poisons are lethal in large enough doses, no clear-cut line separates the harmful from the benign worker environment. But some line has to be drawn to serve as a basis for action to control toxic substances. Particularly with regard to airborne contaminants, it becomes necessary to identify some level of concentrations below which one need not worry about worker exposures. Thus the term *threshold limit value* (TLV) evolved to mean that level of concentration to which the worker could be exposed during the entire workday without significant harm. The TLV varies drastically with the toxicity of the contaminant, of course, and every toxic substance has its own TLV.

So far this discussion of TLVs is in a generic sense, and thus every toxic substance has a TLV even if no one knows what it is and even if no one knows that the particular substance is toxic. But for *known* toxic substances there is a "listed TLV," which is a value agreed upon by a committee of the American Conference of Governmental Industrial Hygienists (ACGIH) and is listed in the "TLV booklet." When the acronym TLV is used instead of spelling out the words "threshold limit value," the ACGIH listed value is usually the meaning intended.

Just because a committee decides on the TLV for a given substance, it does not mean that employers must control worker environments to comply with that level. Some TLVs are based on hard scientific data and are very robust criteria for action. Others are based on much more sketchy data, and thus professional judgment is necessary to determine the actions that should be taken to control the worker environment. The TLVs themselves even change from year to year as more information becomes available. Thus to be strictly correct, one should refer to the date of the list when naming a given TLV—for example, "the 1983 TLV for carbon monoxide was. . . ." Incidentally, when TLVs change, their levels usually decrease as new information about hazards is uncovered.

Permissible Exposure Limits

The focus of this book is on what enforcement agencies require, not what ACGIH lists, but the two are definitely related. At OSHA's beginning, when it was permitted to adopt "national consensus standards" without formal promulgation, the agency adopted hundreds of TLVs, most of which represented the TLV levels published in 1968 by ACGIH. Since the OSHA-published list assumed a regulatory character, the term *permissible exposure limit* (PEL) was used to distinguish between the OSHA-prescribed level and the ACGIH term "TLV." For the most part, PELs have remained static, as OSHA has had difficulty getting the public to accept stricter and stricter levels of control. However, the TLVs remain dynamic, as ACGIH continues to revise the list whenever the collective professional judgment of the committee determines that a new TLV should be added to the list or an old one adjusted, usually downward.

For about thirty substances considered notoriously hazardous, OSHA promulgated full standards instead of merely publishing PELs in a table. This effort became known as the "standards completion project," a subject that will be covered in more detail in a later section in this chapter. Except for these few substances, though, the PEL table remained virtually unchanged for nearly two decades. Then, on January 19, 1989, eighteen years and one month after the passage of the OSHA law, OSHA published a completely revised table that changed nearly all of the PELs to reflect the latest threshold limit values published by ACGIH and to correspond to the findings of research studies conducted by NIOSH. This was the boldest action by OSHA in its two decades of existence. Many new substances were added to the tables, and whole industries were immediately and profoundly affected. The wood products industry, for example, primarily concerned with safety hazards and machine guarding in the early days of OSHA, was suddenly faced with a stiff new standard for airborne concentrations of wood dust, a contaminant formerly considered harmless but now believed to be a carcinogen.

The entire table of PELs, as revised in 1989, is included in this book as Appendices A.1, A.2, and A.3, corresponding to OSHA's designations Z.1, Z.2, and Z.3. Appendix Table A.1 is the main table and contains most of the PELs, listed alphabetically by substance name, with "CAS No." (Chemical Abstracts) for reference. A common mistake is to ignore the hazard when the substance is not listed in the main table, but a few of the most hazardous and frequently encountered substances in industrial exposures are those found in the second table, Appendix Table A.2. This table has its origins in a version formerly (prior to OSHA) published by ANSI for certain substances. Appendix Table A.3 is for mineral dusts, which are considered separately because solid particles are sampled and measured by different means than are toxic gases, mists, and vapors.

MEASURES OF EXPOSURE

The tables of Appendix A.1 are long and complicated, and the many columns specified for each toxic substance warrant explanation. The reason for this complication is that it is difficult to measure levels of contamination of the atmosphere in the workplace. The problem is compounded by the various physical states—solid particles, liquid droplets, mists, gaseous molecules—in which the contaminant can exist in the atmosphere. Further, the medical data incriminating a particular poison may point to danger in a single short-term exposure or may suggest deleterious effects from long-term exposures. Finally, there is the legal complication of introducing strict new standards fairly and affording industries reasonable time periods to bring their facilities into compliance. We will address this legal complication first.

Transitional Limits

The table of limits in Appendix A.1 shows a column of *Transitional Limits* that represent basically the old values for the PELs before the massive revision occurred in 1989. The strategy was to keep a floor of protection during the period of moving into compliance with the newer *Final Rule Limits*. In Chapter 3 of this book we studied the three lines of defense against health hazards: engineering controls, administrative or work practice controls, and personal protective equipment. The federal standard recognizes these three lines of defense by requiring the employer to seek first to eliminate the hazard by engineering or work practices controls "whenever feasible." When such controls are infeasible to achieve full compliance, the federal standard permits the use of personal protective equipment to bring the workplace into full compliance with the PELs listed in Appendix A.1. The federal standard requires that the three lines of defense priorities be applied in controlling air contaminant exposures to the levels specified in the Transitional Limits column. However, for the stricter Final Limits values, during a preliminary transitional period (1989 through 1992[1]), the federal standard permitted the employer to use *any combination* of the three lines of defense to control employee exposures to contaminants within limits. This was a significant economic issue because it permitted the employer to dispense, and required employees to use, personal protective equipment in lieu of designing, purchasing, and installing expensive engineering control equipment to eliminate the contamination of the workplace beyond atmosphere beyond limits. Thus, only during the transition period, and only for the revised stricter exposure limits, was the employer permitted to seek first the lower-cost, or at least more expedient, solution to the problem—that is, the provision and use of respirators or other personal protective equipment to control exposures within the limits specified in Appendix A.1.

It should be acknowledged that a program of acquiring, maintaining, and requiring employees to use personal protective equipment is not always a cheaper alternative to the use of engineering controls to eliminate the hazard. Employers often underestimate the continuing costs of sustaining an effective respirator program, with appropriate provisions for employee training and maintenance of the equipment. The engineering solution may have a very large initial cost, but after the initial outlay the lower cost of maintaining the engineering solution may make it more economical than the personal protective equipment approach. However, the point here is that for the Transitional Limits, the employer has no choice: engineering and work practices controls must be applied first whenever feasible. But for the stricter Final Rules, the employer was extended the freedom to choose whatever method is cheapest or otherwise most desirable during the transition period.

There is another facet to the strategy of promulgating a list of both Transitional and Final Rule Limits. The federal OSHA agency anticipated the possibility that industry groups would file court challenges to the stricter levels set forth in the re-

---

[1]The transition period is extended through 1993 if the provisions in the standard for the use of engineering controls, administrative or work practices controls, and personal protective equipment do not undergo final rulemaking proceedings.

vised limits. So, the standard stated that, if any of the revised limits were "administratively or judicially stayed or vacated," then the existing limits specified for those substances in the Transitional Limits columns would apply. This was a clever strategy to prevent a loss of protection in the face of court challenges to the stricter revised limits.

Time-Weighted Averages

The most popular measure of air contaminant exposures, both for Transitional Limits and for Final Rule Limits, is the *time-weighted average (TWA)*. Appendix A.1 specifies TWA as one of the measures of the Final Rule Limits in the column headings. But even for the Transitional Limits, the PELs are understood to be TWAs unless otherwise specified. The TWA is a computed weighted average concentration over an 8-hour shift. Such a calculation recognizes that concentrations of air contaminants change over time and that it is sometimes permissible for a workplace concentration to exceed the threshold limit value (TLV) if at other times during the workday the exposure is sufficiently lower than the TLV, such that the average exposure for the workshift is lower than the specified level.

The following formula is used to compute the TWA:

$$E = \frac{\sum_{i=1}^{n} C_i T_i}{8} = \frac{C_1 T_1 + C_2 T_2 + \cdots + C_n T_n}{8} \qquad (7.1)$$

where
E = equivalent 8-hour time-weighted average concentration
C_i = observed concentration of the contaminant in time period i
T_i = length of time period i
n = number of time periods studied

The calculation will now be illustrated by a case study.

Case Study 7.2 Calculate the 8-hour, full-shift TWA for the concentrations shown:

Time period, i	Observed concentration, C_i	Length of period, T_i (hr)	$C_i \times T_i$
1	2	1 1/2	3
2	4	2 1/2	10
3	7	1	7
4	5	2	10
5	3	1	3
Total		8	$33 = \sum_{i=1}^{5} C_i T_i$

Solution:

$$E = \frac{33}{8} = 4.125$$

The Case Study 7.2 calculation shown is quite adequate if there is only one toxic substance present in the industrial atmosphere. However, mixtures present a different problem. Suppose, for instance, that an industrial atmospheric TWA concentration of nitric acid was barely below the specified PEL of 5 milligrams per cubic meter. But suppose that the same atmosphere showed TWA concentrations of just under the prescribed limits of 1 milligram per cubic meter for sulfuric acid and 25 milligrams per cubic meter for acetic acid *at the same time*. Taken separately, none of these three acid concentrations violates the standard, but common sense says that the three concentrations present at the same time are dangerous.

The synergistic effect of combinations of toxic substances is a complicated subject. Most research has concentrated upon direct effects of substances acting alone. Some mixtures of contaminants would tend to neutralize each other and be beneficial. For instance, caustics mixed with acids may produce benign salts. In the example just described, however, the three acids working together are bound to have a combined effect. For some mixtures the combined effect may be far worse than the sum of each of the individual effects.

OSHA takes a moderate approach by requiring simple combinations of toxic substances to be considered but generally ignoring the complex synergistic effects. The method is to sum the ratios of concentrations of each substance to its own PEL. The resulting sum should not exceed unity. The following formula summarizes the computation:

$$E_m = \sum_{i=1}^{n} \frac{C_i}{L_i} = \frac{C_1}{L_1} + \frac{C_2}{L_2} + \cdots + \frac{C_n}{L_n} \tag{7.2}$$

where E_m = calculated equivalent ratio for the whole mixture
        C_i = concentration of contaminant i
        L_i = permissible exposure level (PEL) for contaminant i
        n = number of contaminants present in the atmosphere

E_m may not exceed 1.

Case Study 7.3 An industrial process produces exposure in accordance with the following table.

	Nitric Acid	Sulfuric Acid	Acetic Acid
Contaminant, i	1	2	3
Concentration, C_i	4	0.9	22
Limit, L_i	5	1	25

Solution:

$$E_m = \sum_{i=1}^{3} \frac{C_i}{L_i} = \frac{4}{5} + \frac{0.9}{1} + \frac{22}{25} = 2.58$$

Since 2.58 > 1, the concentration of the mixture exceeds the PEL, even though the individual PELs are not exceeded.

Ceiling Levels

Most of the PELs listed in the main table (Appendix A.1) are to be considered TWAs, but for some substances the concern is for short-term exposures. A "ceiling" value, sometimes abbreviated *C* or *MAC* for *maximum acceptable ceiling,* is an exposure limit that should never be exceeded. Another convention is to specify an *STEL, short-term exposure limit,* recognizing the danger of acute exposures, but allowing short excursions above a level that, on an eight-hour shift basis, would clearly be hazardous. The STEL states a maximum concentration permitted for a specified duration, usually fifteen minutes. For instance, Table A.2 lists the following PELs for toluene:

Toluene

TWA	200 ppm
MAC	300 ppm
STEL	500 ppm for 10 minutes

Note that the STEL for toluene is much higher than the MAC. One would think that the STEL would lie somewhere between the TWA and the MAC, but the standards consistently list STELs higher than the MACs. This suggests that if the exposure duration is shorter than the duration specified by the STEL, there is no limit to the allowable concentration! This seems to be a contradiction of the definition of the MAC, but in reality it is not feasible to measure the concentration for so short a time period except for "grab samples," a very unreliable method of measurement. Indeed, even the STELs are impractical to check with current state-of-the-art instruments, so STELs are sometimes ignored by industry and enforcement officials alike.

Units

Regardless of the type of limit against which the exposure is being measured, the analyst must be concerned with the units of the measure. For most of the substances in Appendix A.1 the table lists for each limit a pair of values, which are really two different measures of the same limit, expressed in different units. Gases are usually more conveniently measured by volume, and thus the first column, labeled p/m (parts per million), is usually used for these substances. Liquids and some solids are more conveniently measured by weight, and thus the second column, labeled mg/m³ (milligrams of particulate per cubic meter), is preferred for these substances. If the molecular weight of the substance is known, conversion can be made by using the following formula:

$$p/m = \frac{mg/m^3 \times 24.45}{MW}$$

where MW is the molecular weight of the substances. Parts per million is sometimes abbreviated *ppm* rather than *p/m*.

Action Levels

One other level, the action level (AL), deserves mention. If control measures are taken only after TLVs are exceeded, it may be too late to prevent serious harm and also perhaps too late to prevent citation by authorities. ALs are somewhat of a lock-the-barn-before-the-horse-runs-away measure, which anticipates the problem before TLVs or other measures are exceeded. ALs are set arbitrarily at ½ PEL. A great deal of statistical and instrument variation prevents precisely accurate surveys. The difference between the AL and the PEL provides a margin for error to ensure that worker exposures do not exceed the PEL.

STANDARDS COMPLETION PROJECT

As was mentioned earlier, some toxic substances have whole standards devoted to their control. The standards for these substances have been formulated as a series, and all the standards in the series have been issued several years after the passage of the OSHA act. This means that the standards were scrutinized by the public promulgation procedures and survived the controversy of opposing factions. Several of the substances experienced stormy promulgation, including the invoking of "emergency temporary" promulgation procedures and subsequent court challenges. Much of the background research to justify the issuance of these separate standards for individual substances was conducted by NIOSH. The effort has been called the "standards completion project" because it has been envisioned by some that every toxic substance should eventually have its own unique standard instead of simply being included in a list of TLVs. A list of standards completed, along with the associated PEL for each substance covered, appears in Table 7.1.

TABLE 7.1   OSHA Standards for Specific Toxic Substances ("Standards Completion Project")

	Permissible Exposure Limit	
	TWA	
1910.1001 Asbestos[1]	2 fibers/cm³ (longer than 5μm)	CEILING: 10 fibers/cm³ (longer than 5μm)
1910.1101 Asbestos[2]	2 fibers/cm³ (longer than 5μm)	
1910.1002 Coal tar pitch volatiles	.2 mg/m³	
1910.1003 4-Nitrobiphenyl	Extremely hazardous. Refer to standard.	
1910.1004 alpha-Naphthylamine	Extremely hazardous. Refer to standard.	
1910.1005 [Reserved]		
1910.1006 Methyl chloromethyl ether	Extremely hazardous. Refer to standard.	
1910.1007 3,3'-Dichlorobenzidine (and its salts)	Extremely hazardous. Refer to standard.	
1910.1008 bis-Chloromethyl ether	Extremely hazardous. Refer to standard.	
1910.1009 beta-Naphthylamine	Extremely hazardous. Refer to standard.	
1910.1010 Benzidine	Extremely hazardous. Refer to standard.	
1910.1011 4-Aminodiphenyl	Extremely hazardous. Refer to standard.	
1910.1012 Ethyleneimine	Extremely hazardous. Refer to standard.	
1910.1013 beta-Propiolactone	Extremely hazardous. Refer to standard.	
1910.1014 2-Acetylaminofluorene	Extremely hazardous. Refer to standard.	
1910.1015 4-Dimethylaminoazobenzene	Extremely hazardous. Refer to standard.	
1910.1016 N-Nitrosodimethylamine	Extremely hazardous. Refer to standard.	
1910.1017 Vinyl chloride	Extremely hazardous. Refer to standard.	
1910.1018 Inorganic arsenic	10 μg/m³	
1910.1025 Lead	50 μg/m³	
1910.1028 Benzene	1 ppm	STEL: 5 ppm/15 min
1910.1029 Coke oven emissions	150 μg/m³	
1910.1043 Cotton dust	200–750 μg/m³ [3]	
1910.1044 1.2-Dibromo-3-chloropropane	1 ppb (parts per billion)	
1910.1045 Acrylonitrile	2 ppm	CEILING: 10 ppm/15 min
1910.1046 Exposure to cotton dust in cotton gins	(Refer to standard.)	
1910.1047 Ethylene oxide	1 ppm	EXCURSION: 5 ppm/15 min
1910.1048 Formaldehyde	1 ppm	STEL: 2 ppm/15 min

[1]Includes tremolite, anthophyllite, and actinolite.
[2]Includes chrysotite, amosite, crocidolite, tremolite, anthophyllite, and actinolite. 1910.1101 is an alternate standard to apply while 1910.1001 is partially stayed.
[3]Depending upon process (see standard for details).
Source: OSHA Safety and Health Standards (29 CFR 1910).

As the table shows, some of these substances are extremely hazardous and no PEL is specified. For these substances, the standard is very specific regarding procedures to be followed, respirators to be used, and other protective measures.

DETECTING CONTAMINANTS

It is nice to have a list of toxic substances with permissible exposure levels for each, but more is needed to determine whether a problem exists. There are simply too many substances on the list to keep tabs on all possibilities. Safety and Health Managers need to have knowledge of the processes within their plants so that they know where to look or at least whom to ask. Air sampling and testing are the way to determine concentrations as accurately as possible, but before the test is made, some suspicion of possible contamination needs to be provided by other evidence.

One of the most common ways of first detection of a potential problem is by sense of smell. People seem to believe that they can smell an air contaminant, and they *usually* can smell either the toxic substance or some odorous agent that usually accompanies the toxic substances. But sense of smell is not enough to detect some of the most dangerous contaminants. The most notorious example is carbon monoxide, but carbon dioxide, nitrogen, and methane are also essentially odorless and can be dangerous simply by displacing the oxygen in the air. Some readers will question the statement that methane is odorless because they know that methane is the principal ingredient in natural gas. But the odor in "natural" gas is from a stenching agent deliberately introduced as a safety precaution so that people will detect leaks by sense of smell. Even hydrogen sulfide, a gas that is both dangerous and has a very strong rotten odor, cannot be detected reliably by sense of smell. The odor is so foul that it quickly overcomes the olfactory system, and the victims begin to block out the smell sensation so that they are not aware of the degree of their exposure.

Another approach is to examine technical literature to determine what industries might release what substances. Table 7.2 provides some information from NIOSH literature regarding possible contaminants in various industries.

TABLE 7.2 Potentially Hazardous Operations and Air Contaminants

Process types	Contaminant type	Contaminant examples

Hot operations
  Welding
  Chemical reactions
  Soldering
  Melting
  Molding
  Burning
	Gases (g)
	Particulates (p)
	  (dust, fumes, mists)
		Chromates (p)
		Zinc and compounds (p)
		Manganese and compounds (p)
		Metal oxides (p)
		Carbon monoxide (g)
		Ozone (g)
		Cadmium oxide (p)
		Fluorides (p)
		Lead (p)
		Vinyl chloride (g)

TABLE 7.2 Continued

Process types	Contaminant type	Contaminant examples
Liquid operations		
Painting	Vapors (v)	Benzene (v)
Degreasing	Gases (g)	Trichloroethylene (v)
Dipping	Mists (m)	Methylene chloride (v)
Spraying		1,1,1-Trichloroethylene (v)
Brushing		Hydrochloric acid (m)
Coating		Sulfuric acid (m)
Etching		Hydrogen chloride (g)
Cleaning		Cyanide salts (m)
Dry cleaning		Chromic acid (m)
Pickling		Hydrogen cyanide (g)
Plating		TDI, MDI (v)
Mixing		Hydrogen sulfide (g)
Galvanizing		Sulfur dioxide (g)
Chemical reactions		Carbon tetrachloride (v)
Solid operations		
Pouring	Dusts	Cement
Mixing		Quartz (free silica)
Separations		Fibrous glass
Extraction		
Crushing		
Conveying		
Loading		
Bagging		
Pressurized spraying		
Cleaning parts	Vapors (v)	Organic solvents (v)
Applying pesticides	Dusts (d)	Chlordane (m)
Degreasing	Mists (m)	Parathion (m)
Sandblasting		Trichloroethylene (v)
Painting		1,1,1-Trichloroethane (v)
		Methylene chloride (v)
		Quartz (free silica, d)
Shaping operations		
Cutting	Dusts	Asbestos
Grinding		Beryllium
Filing		Uranium
Milling		Zinc
Molding		Lead
Sawing		
Drilling		

Source: NIOSH (Ref. 60).

Yet another approach is to analyze the processes within the plant to determine potential leaks to the industrial atmosphere. Such an analysis can become quite technical and involves not only an understanding of the machines, pumps, valves, sumps, and tanks, but also a knowledge of the materials used, quantities, intermediate phases, volatilities, and other characteristics that affect the amount of the contaminants that become airborne. A chemical engineering qualitative flow process chart and perhaps a quantitative flow process chart may be necessary to determine whether air contaminant potentials are present. The Safety and Health Manager should not hesitate to involve a chemical engineer in the analysis of potential air contaminations. A technical analysis at this phase may preclude a lot of expensive and laborious sampling experiments later.

Measurement Strategy

Once an air contamination risk has been determined to exist, a procedure is needed to go about taking samples, measuring employee exposures, and instituting controls. NIOSH recommends a strategy for this purpose, and this strategy is displayed in the form of a decision chart in Figure 7.5.

Measurement Instruments

Federal regulation by both OSHA and EPA of permissible exposure levels for toxic substances in the atmosphere has stimulated the electronics and instrumentation industries to develop new and more precise instruments for determining concentrations. Interest in parts per million is shifting toward closer scrutiny to detect parts per *billion*. Such demands are pressing the physics of the instruments, and large-scale inaccuracies are the result.

With such demands by high technology on atmospheric measurement devices, one would think that air contamination monitoring would be a new field. But historically there have been other, more crude, means for monitoring breathing air. Animals formerly were used to test for toxic gases or oxygen deficiency. A canary or a mouse in a cage was often carried into mines. If the canary or mouse died, the workers were alerted to the hazard. A flame safety lamp was used to test for oxygen deficiency; the flame would die if the oxygen proportion in the atmosphere was too low. A brighter-burning lamp was supposed to be an indication of the presence of methane.

These methods were crude, but they did provide some essential indications—for *acute* exposures. With the recognition of threshold limit values and the rise in importance of chronic exposures, the canary test or flame test became totally inadequate. By the time the canary developed cancer or scarred lung tissue, the workers would also be victimized. Furthermore, animal life spans are too short to deal with the chronic effect to which humans might be susceptible.

There are basically three approaches to measuring air contaminant exposures:

1. Direct-reading instruments
2. Sampling and subsequent laboratory analysis
3. Dosimeters

Direct-reading instruments are used principally for oxygen deficiency, flammable gases or vapors detection, and certain commonly encountered contaminants. Direct-

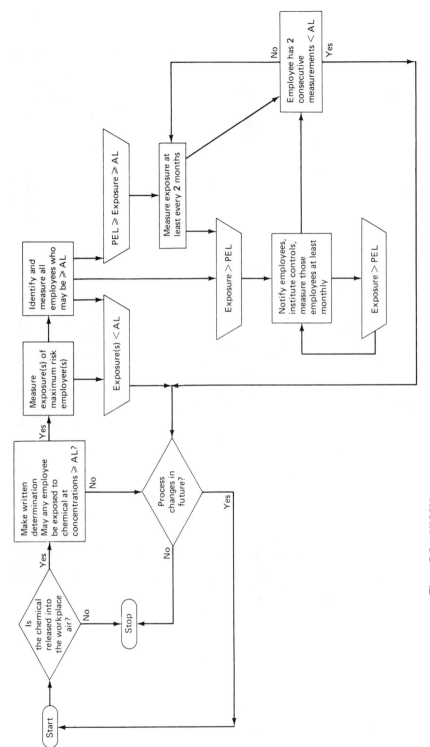

Figure 7.5. NIOSH exposure measurement strategy. Each individual substance health standard should be consulted for detailed requirements. AL, action level; PEL, permissible exposure limit. (*Source:* From Ref. 61.)

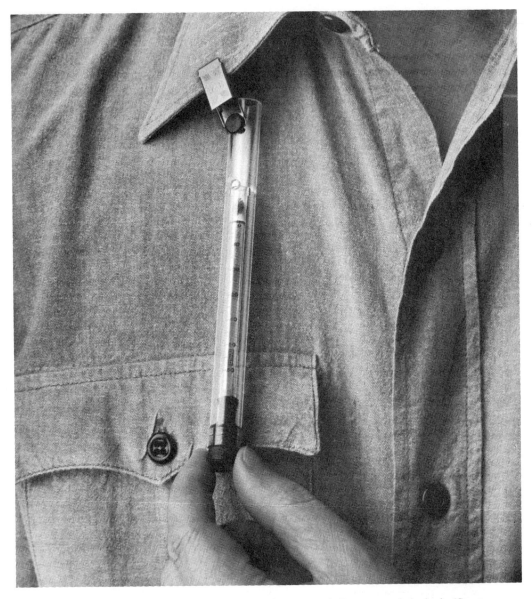

Figure 7.6. Dosimeter for gathering TWA data automatically on a cumulative basis. (Courtesy of the Mine Safety Appliances Company.)

reading instrument are, of course, more convenient than sampling and analysis, so instrument manufacturers are trying to meet demands for these devices for as many toxic substances as possible. The only way an atmosphere can really be tested for a dangerous acute exposure just prior to entry is by a direct-reading instrument.

For more obscure contaminants and for rarer concentrations, sampling devices must be used. Either these devices pump a prescribed quantity of air past a filter or sorbent which collects the contaminant, or they merely collect a precise volume of the air itself. The filter, the sorbent, or the air sample is then sent to a laboratory for analysis.

Dosimeters are the most convenient device of all, especially for gathering TWA data. A dosimeter is a small collector worn on the worker's body or clothing and which collects a time-weighted average exposure over a specified time period, such as a full shift. Unfortunately, accurate dosimeters are still beyond the state of the art for most toxic substances. Figure 7.6 shows one type of dosimeter for air monitoring.

VENTILATION

Ventilation is a good way to deal with toxic air contaminants, but it is not the best way. Before investing in an expensive ventilation system, the Safety and Health Manager should first ask some probing questions about the sources of the contamination.

The most desirable way to deal with an air contaminant is to change the process so that the contaminant is no longer produced. This is so obvious that it is sometimes overlooked. It may be that the process cannot be changed, but if it can, there may be tremendous gains in store, not only in health and safety, but in production cost and efficiency as well. For instance, it may be found that machined parts can be machined dry, avoiding cutting oil exposures to the machinist's skin as well as solvent contamination of the air as the parts are later cleaned. Forging, die casting, or powdered metal technology may eliminate several machining processes, changing the process in beneficial ways. Some of these ideas may generate more disadvantages than advantages in given situations, but each application should be checked for potential benefit. There is no better way for the Safety and Health Manager to win recognition from top management than by generating a clever idea that cuts costs or increases production while it enhances safety or health.

For some more idea generators, consider the hazards of toxic air contaminants from welding operations. Sometimes the principal source of the contaminant is the surface coating on the metal to be welded. Perhaps as a process change, this surface coating can be removed prior to the commencement of welding. Better yet, perhaps the material does not require welding at all. Possibly a crimping operation could produce an effective joint, eliminating the need for welding or soldering.

Chemical processes are classified as batch or continuous. The choice between the two usually involves many considerations, including investment cost, length of production run expected, and volumes to be produced. But an important factor to be considered among these is air contamination. Continuous processes generally reduce the exposure of materials to the air because open handling is reduced, and with continuous processing, batches of materials are not sitting idle awaiting processing. However, the mechanical handling equipment used for continuous processes may increase the contamination levels. Each situation must be studied to determine the best solution, but the safety and health aspects need to be considered.

One way a process can be changed is to isolate it or enclose it. If a particularly contaminating process is in the plant, perhaps it should be located in a separate building so that it does not contribute to the overall ventilation problem.

A slight variation to changing the process is to change the materials used. Carbon tetrachloride has been found to be a health hazard, so other solvents have been substituted. The chlorinated hydrocarbon solvents substituted, such as trichloroethylene and perchloroethylene, are also being found to be hazardous, but fortunately not as hazardous as carbon tetrachloride. New solvents may be found to reduce hazards even further.

In another example, silica sand is often used for blasting to improve surface characteristics. But airborne silica causes the lung disease called silicosis. Perhaps the silica sand could be replaced by steel shot, removing the silica contamination.

A classic example of changing materials to reduce hazards is the switch from hazardous lead-based paints to substitute materials such as iron oxide pigments. Another classic was the switch from Freon to propane as a propellant for aerosol cans. In this case the materials switch was intended to protect the environment (the ozone layer), but the solution may be more hazardous to the individual because propane is a flammable gas.

Design Principles

If the process cannot be changed or materials substituted, a well-designed ventilation system may be the best solution to the problem. OSHA has a standard that deals with this subject, but it must be emphasized that ventilation is a very technical subject, and the Safety and Health Manager may want to turn to a professional engineer to design an adequate ventilation solution to an air contamination problem. Exhaust ventilation is not the same as ordinary heating and air conditioning, and design errors can be made if this difference is not considered. An example is shown in Figure 7.7. Most heating and air-conditioning ducts have right-angle bends which may be fine for gases but greatly impair the ability of the ducts to transport particulates.

Another questionable ventilation system is an ordinary household fan used to blow away smoke from a contaminant source. It is true that a fan can dilute the con-

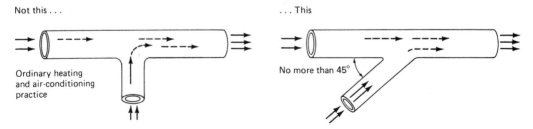

Figure 7.7. Avoid sharp angles at duct entry.

centration of a contaminant in a given place, and dilution ventilation is a recognized method of reducing concentrations to levels lower than the PEL. But the question is, Where is the fan blowing the contamination *to*? It is adding to the overall background level of air contamination in the plant and may later have to be dealt with if other processes are also producing contaminations.

A basic objective of exhaust ventilation is to isolate and remove harmful contaminants from the air. The more these contaminants are concentrated into limited plant areas, the easier they are to separate from the air. Dilution ventilation, although expedient in some cases, is counterproductive to the goal of removing the contaminant. Dilution ventilation can be likened to "sweeping the dirt under the rug."

Focus is an important consideration for ventilation systems. Sheer volume or flow velocity is not enough. Ventilation technology is producing some very fine local exhaust systems that focus the intake right on the contaminant, much as a vacuum cleaner is designed to do. Even if sufficient flow could be achieved with a general ventilation system, the flow could be a nuisance because such a wind might blow away papers and other materials, making the job awkward and inefficient.

The mention of a vacuum cleaner in the previous paragraph might give some Safety and Health Managers the wrong idea. Particle size for fume or other contaminants will usually be too small to be effectively trapped in the bag of an ordinary vacuum cleaner. If the process does not trap the contaminant, it is merely blowing it around, perhaps increasing exposure.

The best exhaust ventilation systems are the "pull" types, not the "push" types. Even within the exhaust duct, the fan should be placed at the end of the duct if possible, as shown in Figure 7.8. Leaks in the duct then merely draw in more air rather than pumping contaminated air back into the plant environment.

Makeup Air

With an exhaust ventilation system or systems, some source of "makeup air" is essential. Open windows and doors usually serve this purpose. But with the advent of the energy crisis in the 1970s, ensuring an adequate supply of makeup air has be-

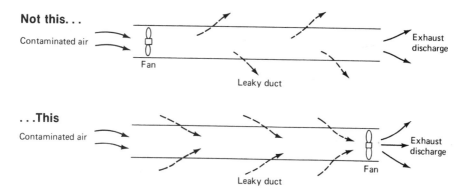

Figure 7.8. Fan location: keep negative pressure within duct.

come a real problem. The increased cost of energy has greatly enhanced the attractiveness of recirculating the exhaust air after filtration and decontamination. Not only does such a solution save energy, but it also reduces external atmospheric pollution, a very important point considering regulations of the Environmental Protection Agency (EPA) and the general environmental concerns of the public.

Figure 7.9 diagrams a recirculating system in which the objective is to remove dust by means of a high-efficiency filter. It is essential to recognize the importance of the condition of the filter to the overall effectiveness of the system. The filter will clog up over time if it is doing its job, and thus it must be serviced, cleaned, or changed. Note the bypass damper, which permits selective proportioning of the system from full recirculating to full exhaust. This bypass can save energy when weather conditions are mild and recirculation is unnecessary. Also note the inclusion of a manometer

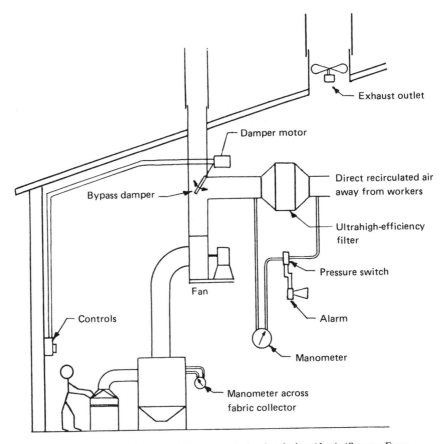

Figure 7.9. Example of recirculation from air cleaning devices (dust). (*Source:* From the American Conference of Governmental Industrial Hygienists Committee on Industrial Ventilation.)

to detect pressure differential across the filter as well as an alarm to sound if this differential becomes too great. Both are intended to provide indication or warning to the operator that the filter has become fouled and needs service.

Safety and Health Managers are warned to monitor recirculation systems closely. Frequently in industry a sophisticated ventilation system is installed for a process, and then it is ignored. Some filter alarms are merely red lights, which operators often ignore. Even the audible alarms such as buzzers and horns are sometimes disabled by disconnecting electrical leads. Both operations and maintenance personnel have been found guilty of disabling these devices, which are designed to protect the workers' health.

Besides using recirculating systems, there are some other ways around the problem of energy losses due to the introduction of makeup air into the building. One method is to introduce the makeup air right at the point at which the contamination is taking place. With this strategy the makeup air may need no air conditioning—no cooling in summer and no heating in winter. The exhaust system will merely suck up the unconditioned makeup air together with the contaminants, and workers will have little exposure to either.

Another solution to the problem is to use a heat exchanger to recapture the energy of the exhaust air and transfer it to the incoming makeup air. It is difficult to make this solution practical, however, because the heat differential between makeup air and exhaust air is usually too low to make the heat exchanger effective. Also, the heat exchanger approach necessitates passing the makeup air duct close to the exhaust air duct, which introduces the possibility of cross-contamination. Finally, heat exchanger systems can be expensive, both to install and to maintain effectively—too expensive in many cases to amortize by means of energy cost savings over the life of the system.

Besides the energy problem, another problem with supplying makeup air is the presence of contaminated air from the outside. This is an unusual problem, but it has on occasion presented itself. At one plant, the makeup air inlet was adjacent to a major freeway, and carbon monoxide and other automobile exhaust emissions were being drawn into the building. In another poor design, the makeup air inlet was so close to the exhaust system discharge that contaminants were being drawn back in and recirculated around the plant. That way if workers managed to escape breathing the contaminated air the first time through, they got another chance to become exposed!

A quick check to determine whether makeup air supply is sufficient is to check atmospheric pressure both inside and outside the plant. The pressure inside should be only slightly lower than the pressure outside. If the pressure inside is substantially lower, the makeup air supply is insufficient. The basic relationship between makeup and exhaust is illustrated in Figure 7.10. The cross-sectional area of makeup openings multiplied by the velocity of flow through those openings must equal the cross-sectional area of the exhaust openings multiplied by the velocity of flow through the exhaust.

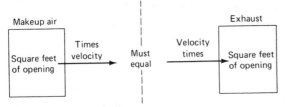

Figure 7.10. Balancing makeup and exhaust air.

The provision for adequate makeup air and a sufficient volume of general exhaust ventilation is sometimes the only practical solution to the problem of reducing air contaminant exposures to specified levels. The following case study will illustrate the principle of this solution to the problem.

Case Study 7.4   An industrial process liberates 2 cubic feet of chlorobenzene per hour into a room that measures 20 feet by 40 feet and has a ceiling height of 12 feet. What minimum general exhaust ventilation in cubic feet per minute is necessary to prevent a general health hazard in this room?

Solution: A subtle facet of this problem is that for a continuously operating process, the dimensions of the room are really irrelevant to the solution. It is true that for a short-duration exposure, the size of room will affect the dilution of the chlorobenzene within the confines of the room. But to deal with a continuous process, one must provide sufficient ventilation to yield an ample supply of makeup air to continuously dilute the chlorobenzene to levels within limits, *regardless of room size*.

The PEL for chlorobenzene is 75 ppm.

Let X = the total ventilation necessary to dilute the chlorobenzene. Then

$$\frac{2}{X} = \frac{75}{1,000,000}$$

$$X = \frac{2 \times 1,000,000}{75} = 26,667 \ \text{ft}^3/\text{hr}$$

$$= \frac{26,667}{60} \ \text{ft}^3/\text{hr}$$

$$= 444 \ \text{ft}^3/\text{min}$$

Purification Devices

If the exhaust air is clean enough to meet external standards, no filtration or purification may be necessary once the air gets outside the plant. But often some type of purification device is necessary on the outside as is required indoors for recirculating systems, particularly for removal of particulates. The following paragraphs describe some of the basic types of particulate removal devices.

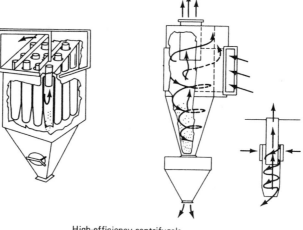

High-efficiency centrifugals

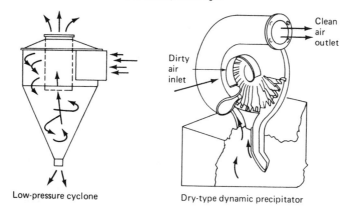

Low-pressure cyclone   Dry-type dynamic precipitator

Figure 7.11.   Cyclone and other dry-type centrifugal collectors for removal of particulates from exhaust air. (*Source:* ACGIH Committee on Industrial Ventilation.)

Centrifugal devices, often called ''cyclones'' (see Figure 7.11) take advantage of the mass of the contaminant particles, causing them to collect on the sides of the cyclone in the swirling air and then slide to the bottom and settle in the neck of the funnel, where they can be periodically emptied. Another type of centrifugal device causes the dirty air to strike louvers, whereupon particles separate from the air. A typical application for cyclones is grain dust removal for grain elevators and mills. Cyclones are also used for woodworking sawdust, plastic, dusts, and some chemical dry particulates.

Electrostatic precipitators place a very high (e.g., 50,000 volt) electrical charge on the particles, causing them to be attracted to an electrode of opposite electrical charge. The collecting electrode can consist of plates, rods, or wires, all of which can be agitated to shake off the collected dust and cause it to settle in the bottom of the chamber. Figure 7.12 illustrates one type of high-voltage electrostatic

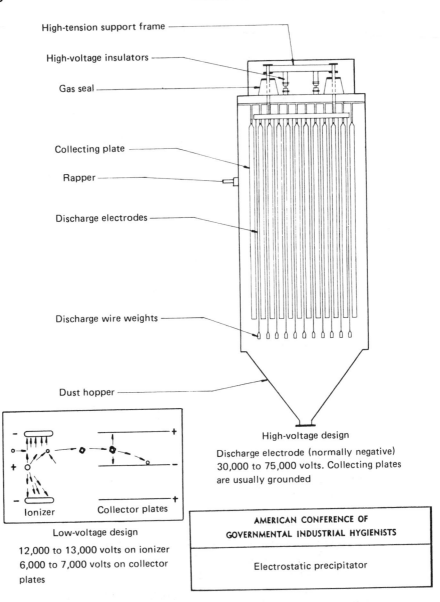

High-tension support frame

High-voltage insulators

Gas seal

Collecting plate

Rapper

Discharge electrodes

Discharge wire weights

Dust hopper

High-voltage design

Discharge electrode (normally negative)
30,000 to 75,000 volts. Collecting plates
are usually grounded

Ionizer   Collector plates

Low-voltage design

12,000 to 13,000 volts on ionizer
6,000 to 7,000 volts on collector
plates

AMERICAN CONFERENCE OF
GOVERNMENTAL INDUSTRIAL HYGIENISTS

Electrostatic precipitator

Figure 7.12. Electrostatic precipitators for removing particulates. (*Source:* ACGIH, Committee on Industrial Ventilation.)

precipitator. Electrostatic precipitators are used in the steel, cement, mining, and chemical industries. Electrostatic precipitators are also used in smokestacks to reduce fly ash.

Wet scrubbers include a wide variety of devices that employ water or chemical solution to wash the air of particulates or other contaminants. Some types pass the dirty air through standing water or solution. Another type forces dirty air to rise

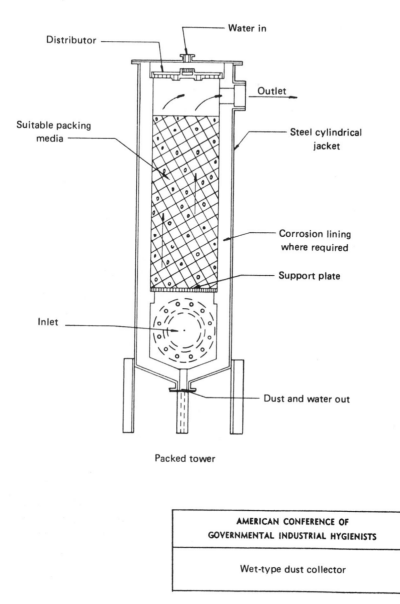

Water in

Distributor

Outlet

Suitable packing
media

Steel cylindrical
jacket

Corrosion lining
where required

Support plate

Inlet

Dust and water out

Packed tower

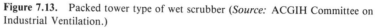

AMERICAN CONFERENCE OF GOVERNMENTAL INDUSTRIAL HYGIENISTS
Wet-type dust collector

Figure 7.13. Packed tower type of wet scrubber (*Source:* ACGIH Committee on Industrial Ventilation.)

in a tower packed with a filler media through which water falls, as shown in Figure 7.13. Note in the figure that the cleanest water is at the top of the tower, where the exiting air is at its cleanest. The bottom of the tower is the dirtiest, where the water vehicle is exiting the packing and is contacting the dirtiest, untreated exhaust air. Wet centrifugal types, like dry centrifugal types, take advantage of the mass of the particles by causing them to impinge on blades, plates, or baffles. Wet scrubbers are

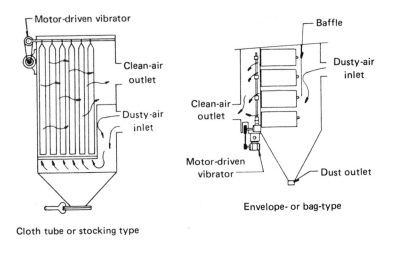

Cloth tube or stocking type

Envelope- or bag-type

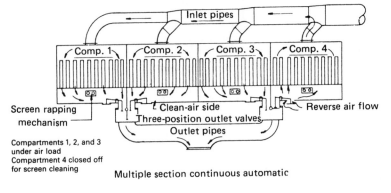

Multiple section continuous automatic

| AMERICAN CONFERENCE OF |
| GOVERNMENTAL INDUSTRIAL HYGIENISTS |
| Fabric collectors |

Figure 7.14. Fabric filters for exhaust air. (*Source:* ACGIH Committee on Industrial Ventilation.)

seen in the chemical industries, where they are able to remove gases and vapors in addition to particulates. Other industries that use wet scrubbers are rubber, ceramics, foundries, and metal cutting.

Fabric, or bag-type, filters are essentially like a vacuum cleaner bag. Some are huge and are located in separate buildings called "bag houses." Figure 7.14 shows three types of fabric filters. Fabric filters are used in the refining of toxic metals such as lead and in woodworking, metal cutting, rubber, plastics, ceramics, and chemical industries.

INDUSTRIAL NOISE

Noise exposure is another classic health problem because the chronic exposures are the ones that typically do the damage. A single acute exposure can do permanent damage, and in this sense noise is a safety problem, but noise exposures of this type are extremely rare. As with other health hazards, noise has a threshold limit value, *TLV* and exposures are measured in terms of time-weighted averages. To understand the units of these measures, some background is needed in the physical characteristics of sound.

Characteristics of Sound Waves

Noise can be defined as unwanted sound. In the industrial sense, noise usually means excessive sound or harmful sound. Sound is generally understood as a pressure wave in the atmosphere. In liquids, sound is also a pressure wave; in rigid solids the sound takes the form of a vibration.

Two basic characteristics of sound waves important to the subject of noise control are

1. The amplitude, or pressure peak intensity, of the wave
2. The frequency in which the pressure peaks occur

Our sense of hearing can detect both of these characteristics. Pressure intensity is sensed as loudness, whereas pressure frequency is sensed as pitch. Figure 7.15 illustrates the wave form of sound and also graphs the relationship between pressure and time. Note in Figure 7.15b that the "period" is the length of time required for the wave to complete its cycle. In the graph it is measured at the point at which the pressure differential becomes zero and is starting to become negative. However, it could have been measured from peak to peak, valley to valley, or at any other convenient reference point in the cycle. These "periods" are always short, too short to count their occurrences while listening to the sound. If we could count the occurrences of these wave cycles, the resulting count per unit time would be the *frequency*, usually measured in numbers of cycles per second (called *hertz*). A typical sound is at a frequency of 1,000 cycles per second—that is, 1,000 hertz (abbreviated 1,000 Hz). Obviously, we could never count 1,000 pulses of pressure in a single second, but our ears have a surprising sensitivity to variations in this frequency count. The sensation is known as *pitch*, and skilled musicians have trained their ears to hear very slight variations in sound-wave frequency. Frequency is important in analyzing the sources of occupational noise exposure.

Even more important than pitch in industrial settings is the pressure intensity of the sound wave. High peaks of pressure in the waves can do permanent damage

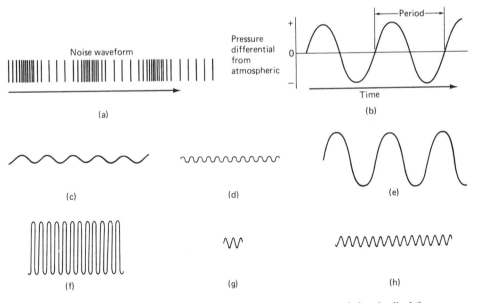

Figure 7.15. Characteristics of sound waves: (a) pressure wave is longitudinal (in the direction of travel of the sound); (b) relationship between pressure and time at a given point of sound exposure; (c) low pitch, soft sound; (d) high pitch, soft sound; (e) low pitch, loud sound; (f) high pitch, loud sound; (g) short-duration sound; (h) sustained sound.

to the delicate mechanisms in the human ear and cause permanent hearing loss. The ear must be delicate to pick up the tiny pressures of the faintest audible sounds, but it is able to withstand an incredibly large range of pressures. The human ear can withstand without damage a sound pressure 10,000,000 times as great as the faintest sound it can hear! A necessary result of this incredible range of pressures is that the ear is not very sensitive to shades of differences in these pressures, especially as the pressures get into the upper part of the range. In other words, as sounds get fairy loud, the human ear cannot readily detect a large increase in the intensity, even a doubling or tripling of the intensity.

Decibels

It is difficult to talk sensibly about such a large range of audible pressures, and it is especially difficult to set standards. Imagine a noise meter reading in the millions. The situation is further complicated by the lessening of the ability of the human ear to detect pressure differences as sounds get louder. To deal with these problems, a unit of measure called the *decibel* (dB) has been devised to measure sound pressure intensity. The decibel has a logarithmic relation to the actual pressure intensity, and thus the scale becomes compressed as the sound becomes louder, until in the upper

Noise sources	Sound level (dB)	Speaking effort required
Pneumatic chipper (at 5 ft)	115	Nearly impossible to communicate by voice
Chain saw (at ear) Teenage rock n roll band Riveting machine Nail machine Casting shakeout area Wood planer Punch press Forging hammer	110	Very difficult to communicate by voice
	105	Shout with hands cupped between mouth and other person's ear
Pneumatic air hoist: 4000 lb	100	Shout at 1/2 foot
Tumbler 6 in. x 3 in., small castings		
Automatic screw machine, Nut blanking	95	Shout at 1 foot
Boiler room Arc welder	90	Normal voice at 1/2 foot, shout at 2 feet
Milling machine (at 4 ft)	85	Normal voice at 1 foot, shout at 4 feet
Pneumatic drill	80	Normal voice at 1-1/2 feet, shout at 6 feet
Inside a car (50 mph)	75	Normal voice at 2 feet, shout at 8 feet

(Telephone use impossible — spanning from 80 to 110 dB)

Figure 7.16. Decibel noise levels of familiar sounds. (*Source:* From NIOSH.)

ranges the decibel is only a gross measure of actual pressure intensity. But this is appropriate because as was mentioned before, the human ear only hears gross differences anyway when the sound gets very loud. Figure 7.16 relates the decibel to familiar sound levels.

The logarithmic decibel scale is convenient, but it does give rise to some problems. If a machine in the plant is very loud, putting a second machine just like it right beside it will make the sound twice as loud—right? Wrong! Remember that the range of sound pressures is tremendous and that the human ear hears only a slight increase in loudness, when the actual sound pressure may have doubled due to the addition of the extra machine. The decibel scale recognizes the addition of the new machine as an increase in noise level of only 3 decibels. Conversely, if the noise level in the plant exceeds allowable standards by very much, shutting off half the machines in the plant—an obviously drastic measure—may have very little effect in bringing down the total noise level on the decibel scale. Table 7.3 provides a scale for combining decibels to arrive at a total noise level from two sources. If there are three or more sources, two sources are combined and then treated as one source to be combined with a third, and so on, until all sources have been combined into a single total. An example is helpful in illustrating the method.

TABLE 7.3 Scale for Combining Decibels

Difference between Two Decibel Levels to be Added (dB)	Amount to be Added to Larger Level to Obtain Decibel Sum (dB)
0	3.0
1	2.6
2	2.1
3	1.8
4	1.4
5	1.2
6	1.0
7	0.8
8	0.6
9	0.5
10	0.4
11	0.3
12	0.2

Source: NIOSH (Ref. 44).

Case Study 7.5 Suppose that noise exposure at a work station is essentially due to four sources, as follows:

Machine A	86 dB
Machine B (identical to machine A)	86 dB
Machine C	82 dB
Machine D	78 dB

First the two identical noise sources, machines A and B, are combined to produce a noise level of 89 dB. Then machine C is added as follows:

$$\text{dB difference} = 89 \text{ dB} - 82 \text{ dB} = 7 \text{ dB}$$

From Table 7.3, a difference of 7 dB between two sources results in the addition of 0.8 dB to the larger source. Therefore, the combined sound of machines A, B, and C is

$$\text{combined sound (A, B, C)} = 89 \text{ dB} + 0.8 \text{ db} = 89.9 \text{ dB}$$

Adding Machine D, we have

$$\text{dB difference} = 89.8 \text{ db} - 78 \text{ dB} = 11.8 \text{ dB} \approx 12 \text{ dB}$$

> Returning to Table 7.3 a difference of 12 dB between two sources results in the addition of 0.2 dB to the larger source. Therefore, the combined sound of all machines is
>
> combined sound (A, B, C, D) = 89.8 dB + 0.2 dB = 90.0 dB

Figure 7.17 diagrams the computation of Case Study 7.5. It should be noted that the measurement of both the total combined sound and the individual machine contributions toward that total are made from the position of the operator. Otherwise, distance factors would affect the results. The computation of combined noise levels is useful in considering the potential benefits of removing machines, enclosing machines, or changing the process.

We have been emphasizing sound intensity measured in decibels, but frequency or pitch also plays a role in noise control. Industrial noise is typically a combination of sound frequencies from each of several sources. The total range of sound frequencies audible to the human ear is from about 20 Hz to about 20,000 Hz. The ear is more sensitive to some of these frequencies than others, particularly the upper middle range from about 1000 Hz to about 6000 Hz. Thus sound-level meters have been devised to bias the decibel reading slightly to emphasize the frequencies from 1000 to 6000 Hz. This biased reading is called the A-weighted scale,[2] and resulting readings are abbreviated dBA instead of simply dB. OSHA recognizes the A-scale, and OSHA PELs are expressed in dBAs.

OSHA Noise Standards

Noise is unusual in that it is a hazard for which OSHA has set both a PEL and an AL. The best known is the PEL, which is set at 90 dBA for an 8-hour TWA. The action level (AL) was established in the early 1980s at 85 dBA for an 8-hour TWA, about 10 years after the 90-dBA PEL was established. It is widely recognized that workers can tolerate short periods of noise higher than the 8-hour TWA without damage, so OSHA specifies a range of decibel exposure levels for various exposure periods. The range of OSHA PELs for noise exposure is given in Table 7.4.

The range of permissible exposures in Table 7.4 makes possible a computation of a time-weighted average exposure, relating each exposure time to the limit permitted for that sound level. The procedure is very similar to the calculation used earlier where multiple contaminants are present in the atmosphere. The formula used is as follows:

$$D = 100 \sum_{i=1}^{n} \frac{C_i}{T_i} = 100 \left(\frac{C_1}{T_1} + \frac{C_2}{T_2} + \cdots + \frac{C_n}{T_n} \right)$$

(7.3)

[2] There are also a B-scale and C-scale, but these scales are seldom used.

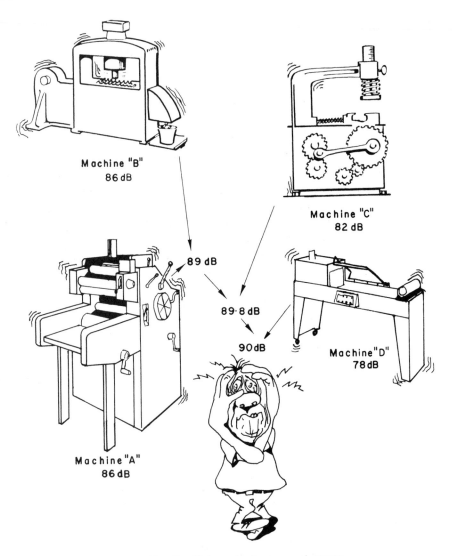

Figure 7.17. Combining noise from several sources.

where D = total shift noise exposure ("dose") as a percent of PEL
      C_i = time of exposure at noise level i
      T_i = maximum permissible exposure time at noise level i (from Table 7.4)
      n = number of different noise levels observed

An interesting computation is for total shift exposure exactly at the AL of 85 dBA. Using equation (7.3), the computation is as follows:

$$D = 100 \sum_{i=1}^{n} \frac{C_i}{T_i} = 100 \left(\frac{8}{16}\right) = 50\%$$

Thus the AL is computed to be 50% of the maximum permissible PEL. However, the reader should note from the earlier discussion of sound intensity that 85 dBA represents less than one-half of the absolute sound intensity of noise at 90 dB.

TABLE 7.4   OSHA's Table of PELs for Noise

A-Weighted Sound Level	Reference Duration Time (hr)
80	32
81	27.9
82	24.3
83	21.1
84	18.4
85	16
86	13.9
87	12.1
88	10.6
89	9.2
90	8
91	7.0
92	6.2
93	5.3
94	4.6
95	4
96	3.5
97	3.0
98	2.6
99	2.3
100	2
101	1.7
102	1.5
103	1.4
104	1.3
105	1
106	0.87
107	0.76
108	0.66
109	0.57
110	0.50
111	0.44
112	0.38
113	0.33
114	0.29
115	0.25
116	0.22
117	0.19

TABLE 7.4 (Continued)

A-Weighted Sound Level	Reference Duration Time (hr)
118	0.16
119	0.14
120	0.125
121	0.110
122	0.095
123	0.082
124	0.072
125	0.063
126	0.054
127	0.047
128	0.041
129	0.036
130	0.031

Source: Code of Federal Regulations 29 CFR 1910.95.

As a second illustration of the computation of noise dose from equation (7.3), a more comprehensive example is undertaken next.

Case Study 7.6 Noise-level readings show that a worker exposure to noise in a given plant is as follows:

8:00 A.M.–10:00 A.M.	90 dBA
10:00 A.M.–11:00 A.M.	95 dBA
11:00 A.M.–12:30 P.M.	75 dBA
12:30 P.M.–1:30 P.M.	85 dBA
1:30 P.M.–2:00 P.M.	95 dBA
2:00 P.M.–4.00 P.M.	90 dBA

Adding up the noise durations for each level, we obtain

At noise level 90 dBA:	2 + 2	= 4 hours
At noise level 95 dBA:	1 + ½	= 1½ hours
At noise level 75 dBA:		1½ hours (ignore)
At noise level 85 dBA:		1 hour
Total		8 hours

The reason that the 1½-hour exposure at 75 dBA was ignored is that 75 dBA is below the range of Table 7.4. In other words, workers may be exposed to noise levels of 75 dBA for as long as desired with no adverse effects, at least as far as safety standards are concerned.

Computing the ratios at each level and summing, in accordance with equation (7.3), yields

$$D = 100 \sum_{i=1}^{n} \frac{C_i}{T_i} = 100 \left[\frac{4}{8} + \frac{1\frac{1}{2}}{4} + \frac{1}{16} \right]$$

$$= 100 \, (0.5 + 0.375 + 0.0625)$$

$$= 93.75\%$$

Since 93.75% is less than 100%, the PEL is not exceeded. However, since 93.75% is greater than 50%, the AL of 85 dBA (8-hour TWA) is exceeded.

Sometimes noise is percussive or intermittent so that, technically speaking, there are tiny intervals of silence between sharp reports. Some employers have followed the scheme that these tiny intervals of silence can be counted toward the quiet time, reducing the observed duration of noise in excess of 90 dBA, but this interpretation is incorrect. Any variations in noise levels where the maxima are less than 1 second apart are to be considered continuous. The "slow response" scale of modern noise-level meters tends to ignore such tiny interval variations, and thus "slow response" is specified in noise-level metering.

The standards do have a specification for peak impulse or impact noise at 140 dBA, but this, of course, is much higher than the PELs for continuous noise. Thus the 140 dB OSHA specification can be considered a ceiling, or C, value. The 140-dB ceiling should be considered a limit for acute exposure and is accordingly a safety hazard. However, such exposures are so rare and difficult to measure after the fact that ceiling violations are virtually never cited. Ordinary sound-level meters are not very effective in measuring impact noise. Even for continuous exposures, measurement can be a problem which will now be addressed.

Noise Measurement

As a first check for potential noise problems, the Safety and Health Manager should take a walk through the plant and listen. As a rule of *thumb,* if you can reach out and touch someone with your *thumb* but still cannot hear and understand that person's conversation (without his or her shouting), either your hearing is already damaged or defective, or there is excessive noise in the area. If the noise is continuous throughout the working shift but is no louder than a continuously running vacuum cleaner, there is probably no violation of standards. But if the noise is as loud as a subway passing through a station continuously throughout the full shift, a violation probably exists. If subways are unfamiliar to you, imagine a fast-moving freight train passing within 20 feet; such a noise level would easily constitute a violation if the exposure were continuous for a full 8-hour shift. Some exposures louder than the train might be permissible if they are of short duration, as was obvious in the calculation of average exposures in the preceding section. Anything between the vacuum cleaner and the train in sound level would be a gray area which ought to be measured with accurate meters.

Accurate measurement of sound levels requires instruments such as the sound level meter (SLM) illustrated in Figure 7.18. The meter registers the sound intensity

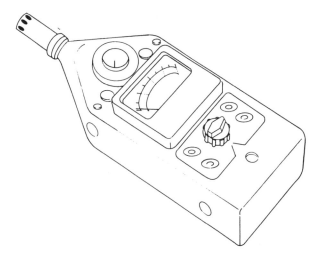

Figure 7.18. Example of sound-level meter.

in decibels. Sound-level meters are delicate instruments and must be handled carefully. Accuracy is a problem, and the Safety and Health Manager should not expect performance better than $\pm$ 1 dB. Calibration is extremely important, and no sound-level meter is complete without a calibration device (known sound source) nearby. Variations in battery level must be compensated, and humidity and temperature conditions can cause distortions.

Some skill is required in using the sound-level meter to obtain reliable readings. Naturally, the microphone receiver of the instrument must be held in the vicinity of the subject's ear in order to be representative of the exposure. However, the instrument should not be held too close, as the subject's body may affect the reading. The instrument's microphone receiver should be shielded from wind currents, and the instrument itself should not be subjected to direct vibrations. The Safety and Health Manager may find the sound-level meter most useful in comparing various locations in the plant and in comparing different machines and operating modes.

Determining the full-shift, time-weighted average exposure with a sound-level meter is tedious and requires a great deal of effort. A convenient substitute is a cumulative device called a "dosimeter," which is worn on the person of the subject under study. Dosimeters may seem to be a panacea, but they have their drawbacks. The wearer can easily bias the dosimeter by holding it close to a loud mechanism,

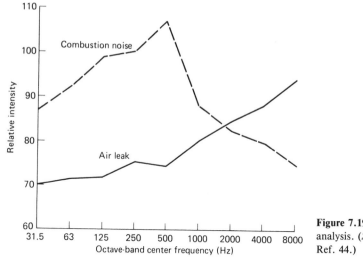

Figure 7.19. Example of octave-band analysis. (*Source:* From NIOSH, Ref. 44.)

or by simply rubbing, tapping, or blowing on the microphone. They are useful survey devices, however, and are used to monitor exposures if the firm's noise levels have exceeded the AL of 85 dBA for an 8-hour TWA.

Once it is determined that a problem exists, a more sophisticated measurement of noise level may be necessary to isolate the sources of the undesirable noise. An octave-band analyzer permits decibel readings to be taken at various frequencies over the audible range as shown in the example in Figure 7.19. Various noise reduction media have characteristic frequencies at which they are most effective. The octave-band analysis will help to delineate the problem frequencies as well as provide evidence to identify the problem sources. Finally, octave-band analysis is useful in determining the frequency characteristics of a particular industry's noise sources so that damage from these sources can be distinguished from damage from noise exposures that have occurred off the job.

Engineering Controls

Once the instruments have proven that a problem exists, the Safety and Health Manager needs physical solutions to the problems. If noise levels exceed the PEL, federal standards require that feasible engineering or administrative controls be used. If these measures fail to reduce noise exposures to within the PEL, personal protective equipment must be provided and used to reduce sound levels to within the PEL. Engineering controls should be considered a more thorough and permanent solution to the problem.

As with the control of toxic substances, the simplest solutions may be so obvious that they are overlooked. Process modification or elimination should always receive consideration. Another very simple solution, if feasible, is merely to move the operator away from the primary source of the noise. This idea has more merit than it would

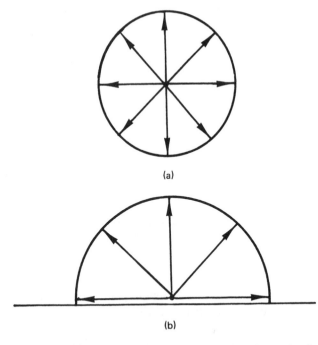

(a)

(b)

Figure 7.20. Distribution of sound intensity over the surface of a sphere as it radiates from a single source. (a) Sound emanates from a point source in all directions distributing over the surface of a sphere, the area of which is calculated by the formula $4\pi r^2$. Thus the sound intensity is reduced as the square of the radial distance from the source. (b) Sound emanates from a point source located on the floor or other surface. The floor either absorbs or reflects the sound, but the resultant sound is still distributed over a hemisphere, the area of which is $2\pi r^2$. The squared-distance relationship still holds approximately.

intuitively appear because the noise intensity from a given source goes down as the square of the distance, in the absence of reflective walls and other distorting factors. The reason for this relationship can be seen in Figure 7.20.

It must be remembered that it is absolute sound intensity, not decibels, that varies inversely with the square of the distance from the source. The logarithmic decibel scale results in a 3-dB change whenever the sound intensity is changed by a factor of 2 (doubled or halved). This leads to a rule of thumb for distance. Since sound intensity varies as the square of the distance from the source, a doubling of distance results in a *fourfold* reduction in sound intensity, which in turn reduces the decibel level by 6 dB.

Case Study 7.7 A worker's machine is located at a distance of 2 feet from the operator and produces a noise exposure of 95 dB to the operator. How much is to be gained by moving the operator to a position 4 feet from the machine? How much reduction could be achieved by a move to 8 feet?

Solution: A move from 2 feet to 4 feet is a doubling of distance and results in a 6-dB reduction in sound level. The resultant level would be

$$95 \text{ dB} - 6 \text{ dB} = 89 \text{ dB}$$

which would probably be within the 8-hour PEL of 90 dB even after considering reflections and other sources, if these sources are not very significant.

A move to 8 feet would be a second doubling, resulting in a reduction of another 6 dB to a resultant 83 dB, ignoring reflections and other sound sources. This would reduce noise exposures to less than the AL of 85 dBA for an 8-hour TWA.

It is not likely to be feasible to move operators away from their own machines, and even if it is, to move away from one machine may place the operator close to a neighboring machine. Distance factors work best in separating operators from noise arising from adjacent machines or other processes in the area. A general spreading out of the plant layout can be beneficial in this regard.

If spreading out the plant layout is infeasible or too costly, the installation of sound-absorbing barriers between stations can increase their virtual separation as far as noise is concerned. The gain to be achieved by such barriers is variable and complicated to estimate in advance. An acoustics expert is recommended for advice in this area, and even the expert is likely to experiment with various temporary barriers, measuring the "before" and "after" sound levels with each. Heavy materials absorb sound vibrations, a fact that makes curtains or shields containing lead a popular choice.

Sheet metal surfaces on machines are susceptible to mechanical vibrations and may act as sound-amplifying surfaces on machines. Metal gear contact in the drive mechanism can sometimes be eliminated by substitution of nylon gear wheels for metal gears or by the use of belt drives instead of gears. Simpler still would be a stepped-up preventive maintenance schedule to lubricate the gears more often, perhaps reducing noise levels. Useful principles for engineering controls to reduce noise are illustrated in Figures 7.21 to 7.26.

More expensive than any of the engineering control approaches discussed so far would be the isolation of the offending machine by means of an enclosure. The effectiveness depends on the type of material used to construct the enclosure and also depends to a surprising degree on the number and extent of openings or leaks in the enclosure. Figure 7.27 on p. 167 shows the relationship between opening size and loss of effectiveness for an example noise enclosure which has a capability of a 50-dB reduction if there are no leaks. Note that most of the effectiveness of the enclosure is lost if there is a hole in the enclosure wall of less than one-half of 1% of the total area of the enclosure.

Administrative Controls

It was stated earlier that either engineering or administrative controls are specified where feasible for excessive noise levels and that engineering controls are preferable. However, the administrative controls alternative was left unexplained. Administrative-

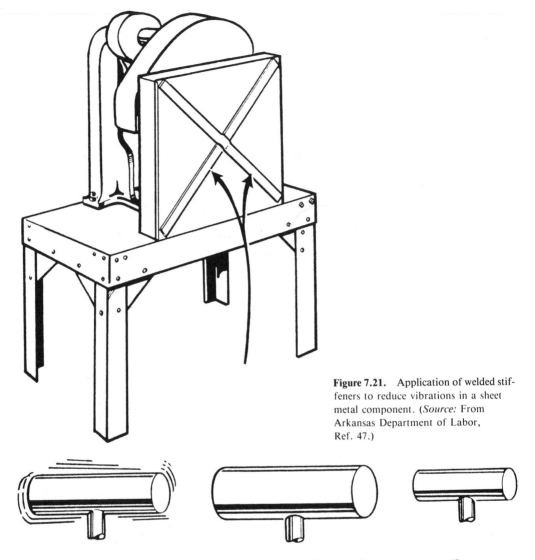

Figure 7.21. Application of welded stiffeners to reduce vibrations in a sheet metal component. (*Source:* From Arkansas Department of Labor, Ref. 47.)

Figure 7.22. Enlarge or reduce size of a part to eliminate vibration resonance. (*Source:* From Arkansas Department of Labor, Ref. 47.)

ly, management can schedule production runs so that noise levels are split between shifts and individual workers are not subjected to full-shift exposures. Other tricks are to interrupt production runs with preventive maintenance to give workers quiet time. During normal shift breaks, workers can be removed to a quiet rest area. Sometimes workers can share a loud job and a more quiet one by trading jobs at midshift. All of these practices can be used to bring down noise exposure levels to within the PEL for the given time of exposure as determined from Table 7.4. Since

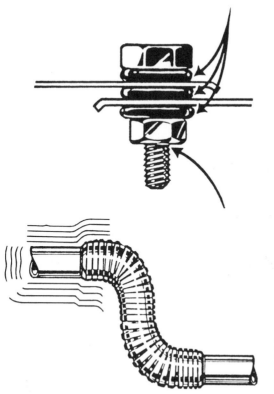

Figure 7.23. Rubber cushions on both sides of vibrating sheet metal surfaces where they are joined. (*Source:* From Arkansas Department of Labor, Ref. 47.)

Figure 7.24. Flexible section in rigid pipe isolates vibrations. (*Source:* From Arkansas Department of Labor, Ref. 47.)

the term *administrative control* is somewhat vague, the term *work practices control* has become preferable to refer to the various methods of shifting employee exposures to comply with Table 7.4.

Hearing Protection and Conservation

Personal protective equipment is required when engineering and administrative controls both fail to reduce noise to legal levels. Specifically, hearing protection is to be provided to all employees exposed to the 85-dBA TWA AL. In addition to providing protection, employers must permit employees to select protectors from a variety of suitable types and must train employees in the proper use and care of the protectors. One aspect of proper use is fit, and employers must ensure proper fit.

The actions just described are mandatory when ALs are exceeded, but the reader may have noted that the actual wearing of the protectors by the workers is not among the mandatory steps to be taken if ALs are exceeded. However, there are conditions under which workers must actually wear the protectors. These are as follows:

1. Whenever worker exposures are greater than the PELs (see Table 7.4)
2. Whenever worker exposures are greater than the AL of 85 dBA (TWA) *and* the worker has experienced a "permanent significant threshold shift"

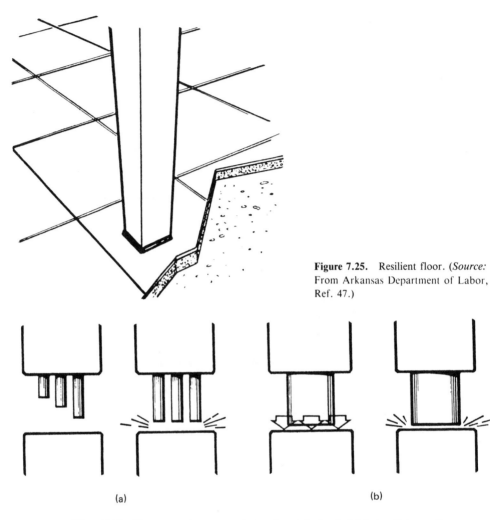

Figure 7.25. Resilient floor. (*Source:* From Arkansas Department of Labor, Ref. 47.)

(a)   (b)

Figure 7.26. Power press improvements to reduce noise levels: (a) design die to result in a dull, crunching sound instead of a sharp bang; (b) replace the sharp impact action of a mechanical press with the relatively quiet pressure action of a hydraulic press.

When testing shows a "threshold shift" in a worker's hearing, the implication is that the worker's hearing has been damaged and needs special protection. In these cases the noise levels are required to be reduced by the protectors to 85 dBA (TWA), not 90 dBA (TWA).

Personal protective equipment should not be considered a final solution because elimination of the source of the noise provides a more satisfactory work environment. Sometimes workers are lax in the wearing of personal protective equipment, and injurious exposures result. The careful selection and fitting of various types of personal hearing protective equipment is discussed in Chapter 9.

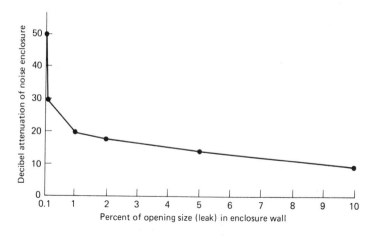

Figure 7.27. Loss in effectiveness of a noise enclosure due to leaks. For a sample enclosure that has an ideal (airtight) noise reduction (attenuation) capability of 50 dB, leaks in the walls reduce the decibel attenuation by remarkable amounts, as shown.

Whenever the AL of 85 dBA (8-hour TWA) is exceeded, a "continuing, effective hearing conservation program" should be administered, including audiometric testing, noise monitoring, calibration of equipment, training, warning signs for noisy areas, and recordkeeping of audiometric tests and equipment calibration.

Audiometric testing involves a small, perhaps portable, listening station in which a subject listens to recorded sounds and an audiologist measures the subject's hearing acuity at various frequencies. Audiometric testing can be very useful in determining sources of hearing loss, or more particularly in providing data to determine whether hearing loss is due to occupational or off-the-job exposure. If plant noise levels are high, it is foolhardy to hire new employees without first testing their "baseline" hearing acuity. Without any evidence of hearing deficiency at the time of employment, any hearing deficiencies that show up after employment appear strongly to be work related.

When observing the frequency profile of a worker's hearing acuity, the audiologist often looks for a "4,000-Hz shift" for evidence of occupational exposure. Experience has shown that much industrial noise occurs in the 4,000-Hz frequency range, which causes audiologists to suspect occupational exposure when hearing acuity becomes deficient in the 4,000-Hz range.

RADIATION

A natural progression from the subject of noise is to the subject of radiation. Noise is, in fact, a form of radiant (wave) energy, but the term *radiation* is generally considered to mean electromagnetic radiation such as X rays and gamma rays, or high-speed particles such as alpha particles, protons, and electrons. Federal standards separate radiation into the categories "ionizing" and "nonionizing."

Ionizing radiation is the more dangerous of the two types and is the type most associated with atomic energy. By far the most important category of ionizing radiation, from the occupational exposure standpoint, is the X ray. X rays are no longer the exclusive domain of the medical and dental professions. X rays are being widely used in manufacturing operations, especially in inspection systems.

Nonionizing radiation is somewhat of a misnomer but applies to a more benign type of radiation in the electromagnetic spectrum. Included are radio and microwave frequencies. These phenomena are also increasingly important in industrial applications.

COMPUTER TERMINALS

One of the most dramatic developments in the workplace in the 1980s was the increased use of computer terminals. It is estimated that more than half of all office workers in the United States make use of computer terminals in some way in their jobs. Occasional use is not a concern, but workers who sit at a computer terminal all day are reporting eye strain, headaches, backaches, and neck and shoulder pain. Chapnik and Gross (ref. 8) report a Wisconsin study by Sauter that shows significantly greater incidence of these conditions among computer terminal operators than among other office workers. Consistently higher percentages of the 250 exposed subjects studied reported discomfort compared with the 84 workers who did not use computer terminals in their jobs.

The problem of eye strain reported in studies of computer terminal operators is no surprise. What has been unexpected is the level of musculoskeletal aches and pains associated with computer terminal operation. Chapnik and Gross (ref. 8) have reported increased incidence of repetitive strain injuries such as tenosynovitis, tendonitis, and carpal tunnel syndrome. OSHA's increased attention to ergonomic hazards, coupled with the dramatic growth in the use of computer terminals, ensures that this subject will be of vital concern to the Safety and Health Manager in the remaining years of this century.

SUMMARY

Health hazards, although subtle, can be very serious. Worker health has historically taken a backseat to safety, but recent attention to long-term effects of worker exposures to toxic substances, noise, and radiation has led to rapid development of the field of industrial hygiene.

Threshold limit values for worker exposure to chronic health hazards are continually being studied and updated by the American Conference of Governmental Industrial Hygienists. TLV values do not necessarily become law, but federal standards have been strongly influenced by ACGIH's list of published TLVs, especially in the initial publication of permissible exposure levels. Many toxic substances and even noise have acute effects and should be considered safety hazards as well as health hazards. There are ceiling limits for many of these exposures, but some of the most common and dangerous substances have no ceiling values. Carbon monoxide is a notorious example, having a TLV for health considerations but no ceiling value for safety.

Fundamental to the assessment and control of health hazards and the industrial environment are instruments capable of measuring conditions. These instruments can be very sophisticated, and the underlying physics limit the accuracy that can be obtained. Calibration and care of the instruments are extremely important.

The Safety and Health Manager will probably find it beneficial to employ experts to take time-weighted averages with appropriate meters and sampling instruments. Subsequent control measures, such as ventilation systems for toxic substances and acoustic panels for noise control, may require the design capabilities of experts in their respective fields. Dealing with these experts demands a degree of understanding of their methods and terminology but does not require the Safety and Health Manager to duplicate the experts' capability in each field. Such an understanding of methods, terminology, and basic principles of occupational health and environmental control are what this chapter has attempted to provide.

EXERCISES AND STUDY QUESTIONS

7.1. What is the definition of the word *fumes*?

7.2. How much carbon *mono*xide in the air is permissible, given a normal (0.033%) concentration of carbon *di*oxide (and no other contaminants)?

7.3. Air samples show an industrial atmosphere to contain 0.001% methyl styrene during the morning half of the shift and 0.015% during the afternoon half. What is the TWA? Assuming that no other contaminants are present, does the exposure exceed the PEL? Does it exceed the AL?

7.4. Air samples show the following contaminant concentrations in an 8-hour shift (from 8:00 A.M. to 4:00 P.M.):

1. trifluoromonobromomethane: one-tenth of 1% from 11:00 A.M. to 2:00 P.M.
2. propane: 0.05% all day
3. phosgene: one part per million at 2:00 P.M., with a duration of 15 minutes

   (a) Assuming that no other contaminants are present, does the full shift atmosphere meet OSHA standards?

   (b) To *exactly* meet OSHA standards, how much more or less time for the phosgene exposure would be permissible provided other contaminants remain as previously stated?

7.5. A plant has two identical standby generator units for emergency use. In the area of the generators, the normal noise level registers 81 dBA on the sound level meter with the generators turned off. When one generator switches on, the SLM needle jumps to 83.6 dBA.

   (a) What will the dBA reading be when the second generator also turns on (so that both generators are on)?

   (b) If both generators are on for a full 8-hour shift, will OSHA's PEL be exceeded? Will the AL be exceeded?

   (c) If one generator is on for half the shift and both are on for the other half, will the PEL be exceeded? Will the AL be exceeded?

(d) In the absence of any plant background noise, what would be the sound level contributed by a single generator? By both generators?

7.6. Carbon disulfide has the following physical properties:

Flashpoint: $-22\,°F$
Boiling point: $46.5\,°F$
Density: 1.261
Vapor density: 2.64
Lower flammable limit: 1.3%
Upper flammable limit: 50%
8-hour TWA PEL: 20 ppm

An industrial process liberates 3 cubic feet of carbon disulfide per hour into a room that measures 10 feet by 20 feet and has a ceiling height of 8 feet.

(a) What minimum general exhaust ventilation (ft^3/hr) is necessary to prevent a general *safety* hazard for the process?

(b) What minimum general exhaust ventilation (ft^3/hr) is necessary to prevent a general *health* hazard for the process?

(c) Carbon disulfide is which of the following:

1. Class IA flammable liquid
2. Class IB flammable liquid
3. Class IC flammable liquid
4. Class II combustible liquid
5. Class III combustible liquid

7.7. Four machines contribute the following noise levels in dB to a worker's exposure:

Machine 1: 80 dBA
Machine 2: 86 dBA
Machine 3: 93 dBA
Machine 4: 70 dBA

(a) What is the combined noise level exposure for this worker?

(b) The offending machine is obviously machine 3. Suppose machine 3 was at a distance of 5 feet away from the worker when the 93 dBA noise level was measured. How far away would machine 3 have to be moved to bring the worker's continuous 8-hour combined exposure from *all* machines down to the OSHA PEL?

7.8. A particular gas welding process in a confined space is suspected of producing dangerous concentrations of carbon monoxide, carbon dioxide, iron oxide particulate, and manganese fumes. Atmospheric sampling produces the following exposure data:

Period	ppm		mg/m^3	
	CO	CO_2	Iron Oxide	Manganese
8:00 A.M.–10:00 A.M.	10	1000	1	1
10:00 A.M.–12:00 noon	20	1000	4	1
12:00 noon–1:00 P.M.	25	1000	2	0
1:00 P.M.–4:00 P.M.	30	1000	3	1

Does the combined exposure represent an OSHA violation?

7.9. Name at least five pneumoconioses. Which are most dangerous?

7.10. How do fibroses differ from other pneumoconioses?

7.11. Name the two basic classes of asphyxiants and give examples of each.

7.12. What is the difference between depressants and systemic poisons? Give examples of each.

7.13. Explain the following terms:
   (a) mutagen
   (b) carcinogen
   (c) teratogen

7.14. In what ways are the threats of poisons at work different from those at home?

7.15. What is the difference between *fumes* and *vapors*?

7.16. Compare the particle sizes of the following.
   (a) zinc oxide fumes
   (b) tobacco smoke
   (c) diameter of human hair
   (d) bacteria

7.17. Explain the following terms:
   (a) TLV
   (b) PEL
   (c) TWA
   (d) MAC
   (e) STEL
   (f) AL

7.18. Name some traditional methods of detecting the presence of dangerous air contaminants and explain advantages and disadvantages.

7.19. Name three basic approaches to measuring air contaminant exposures.

7.20. What are some desirable alternatives to industrial ventilation to remove air contaminants?

7.21. What is "makeup air"?

7.22. Ten machines all contribute equally to the noise exposure of one worker, whose exposure level is 99 dB for a full 8-hour shift. When all the machines are turned off, the noise level is 65 dB. How many of the ten machines must be turned off to achieve a full-shift noise exposure level that would meet standards if the worker wears no personal protective equipment?

7.23. A worker stands on a factory floor, and a sound-level meter shows a reading of 55 dB at that point. A machine 3 feet away is turned on, and the meter jumps to 90 dB. What will the SLM read if the machine is moved to a point 12 feet away?

7.24. Suppose that an industrial process produces the following air contaminant concentration for the periods shown:

Period	Methanol (ppm)	Nitric Oxide (ppm)	Sulfur Dioxide (ppm)
8:00 A.M.–10:00 A.M.	50	5	0
10:00 A.M.–11:00 A.M.	150	10	1
11:00 A.M.–1:00 P.M.	100	5	1
1:00 P.M.–4:00 P.M.	200	10	1

Taken collectively, would these concentrations exceed permissible exposure levels?

7.25. In Question 7.24, suppose that the solvent ethanol could serve to replace the methanol in this process, but at the expense of double the concentrations of the solvent in the atmosphere. Would this help or hinder matters? Explain.

7.26. A paper mill uses liquid chlorine, delivered in 90-ton railroad tank cars, as a pulp bleaching agent. One volume of liquid chlorine produces approximately 450 volumes of vapor under normal atmospheric temperature and pressure. The density of liquid chlorine is 103 pounds per cubic foot. In the event of rupture and vapor release of 20% of the tank car contents, how much vapor by volume would be released? If the release were in a closed building with a 30-foot ceiling height but no ventilation, how large would the building have to be (in square *miles* of floor space) to contain the thoroughly mixed vapor/air ratio within the OSHA PEL? The logical conclusion to this exercise is that, with or without ventilation, it is more practical to unload chlorine tank cars outdoors.

7.27. Two solvents, benzene and chlorobenzene, are under consideration by process engineers for use in a plant for which you have responsibility for Safety and Health Management. What information can you provide to the process engineers regarding the comparative hazards of these two solvents?

7.28. A woman spills a bottle of 150 proof rum on the kitchen floor in her small one-bedroom apartment. The total apartment area is 600 square feet and has standard 8-foot ceilings. By the time she cleans up the spill, approximately 5 cubic feet of alcohol vapor has entered the air through evaporation. Noting the strong smell of alcohol in the air, she opens the window and, feeling drowsy, goes to bed and sleeps all night (8 hours). The open window permits a gradual dilution of the alcohol in the air, and by morning the concentration of alcohol in the air is down to 500 parts per million. Assuming a constant rate of decline in alcohol content of the air all night, was the alcohol concentration a hazard? If the exposure had been occupational, would PELs have been exceeded?

7.29. In Question 7.28, the woman's original intent was to bake rum cake. Had she gone ahead and made the cake, suppose that the hot oven had caused an additional 25 cubic feet of alcohol vapor to be liberated into the apartment. How would this concentration compare with the PEL?

7.30. A worker exposure to noise in a given plant is measured, resulting in the following readings for various time periods during the 8-hour shift:

8:00 A.M.– 9:00 A.M.	86 dBA
9:00 A.M.–11:00 A.M.	84 dBA
11:00 A.M.–noon	81 dBA
noon–1:00 P.M.	101 dBA
1:00 P.M.– 4:00 P.M.	75 dBA

(a) Perform computations to determine whether maximum PELs have been exceeded.
(b) Have ALs been exceeded?
(c) Given the noise exposure just described, would the employer be required to furnish hearing protectors?
(d) Would employees be required to use the hearing protectors?
(e) Suppose that an engineering control could be devised that would cut the noise level (sound pressure level) in half either in the morning or in the afternoon, but not both. Which would you select? Why?

7.31. A direct-reading gas detector tube is available for sampling concentrations of the toxic gas sulfur dioxide. Tube 5H is specified for concentrations in the range of 0.05% to 8.0% and Tube 5M is for concentrations in the range of 20 ppm to 3,600 ppm. Which tube is the more sensitive of the two tubes?

7.32. The following gas detector tubes are available for direct reading of concentrations of the toxic gas hydrogen sulfide:

Tube	Concentrations
4HT	1%–40%
4HH	0.1%–4.0%
4H	10 ppm–3,200 ppm
4M	12.5 ppm–500 ppm
4L	1 ppm–240 ppm
4LL	0.25 ppm–60 ppm

Which of these tubes would be satisfactory for detecting the OSHA-specified ceiling concentration for hydrogen sulfide? Of the satisfactory tubes, which one tests the narrowest range of concentrations? (The narrowest-range tube in the list is the only tube that was ever certified by NIOSH; the NIOSH certification program for detector tubes has been discontinued.)

7.33. A detector tube is available for testing concentrations of isopropyl acetate in the range of 0.05% to 0.75%. Would this tube be capable of detecting concentrations at the OSHA PEL? Would it be a satisfactory instrument for testing whether the AL was exceeded?

7.34. Measurements taken in a particular rayon manufacturing process reveal exposures to airborne contaminants in concentrations as shown in Table 7.5. Determine which of the substances are listed by OSHA as having exposure limits for a standard 8-hour shift and perform calculations to determine whether the given exposures, taken separately and together, exceed OSHA PELs.

TABLE 7.5 Exposure Levels

	Morning Exposures (4 hours)	Afternoon Exposures (4 hours)
Acetic anhydride	0.5 ppm (p/m)	1 ppm
Sodium hydroxide	0.2 mg/m³	0.3 mg/m³
Ammonium sulfide	3 ppm	4 ppm
Calcium bisulfide	5 ppm	8 ppm
Carbon disulfide	4 ppm	6 ppm
Sodium sulfide	0.7 mg/m³	0.8 mg/m³
Sodium sulfite	0.5 mg/m³	0.5 mg/m³

7.35. Suppose you are the Safety and Health Manager in the rayon plant described in Question 7.34. Plant engineers suggest introducing a process that uses the solvent formaldehyde. Preliminary evaluations suggest that the new process will add a slight amount of formaldehyde vapors to the plant atmosphere, perhaps one part per million by volume, in addition to current levels of the other air contaminants listed in Table 7.5. Perform calculations and evaluate the potential impact of the new process proposal upon the safety and health of the workers in the plant. What recommendations would you make to management?

8

Hazardous Materials

Percent of OSHA
general industry citations
addressing this subject

8%

A simple listing of all known hazardous materials together with a concise summary of the characteristics of each would require a book larger than this one. Furthermore, additional hazardous materials are being developed or recognized daily. The focus of this book is on only a few materials, those for which individual national consensus standards already exist. Also, the concept of the term *hazardous* in this usage tends to refer more to hazards of fire and explosion, rather than to health hazards. Materials for which the hazard is primarily one of health were discussed in Chapter 7.

FLAMMABLE LIQUIDS

The most familiar hazardous substances are flammable liquids, and the new Safety and Health Manager might expect applicable standards for flammable liquids to be the easiest to learn and apply. But unfortunately, the standards are quite complicated as a natural result of the fact that flammable liquids occur so frequently in industry and in such widely varying quantities and applications. To illustrate this point, procedures for handling gasoline in a petroleum refinery where gasoline is *manufactured* are vastly different from procedures for storing and handling flammable liquids in an ordinary factory or office. Therefore, no simple set of rules for flammable liquids is appropriate.

Familiar as flammable liquids are, most people do not really understand many of the commonly used terms, such as *flashpoint, Class I liquid, flammable, com-*

bustible, and *volatile.* Much confusion also surrounds the sources of ignition for flammable liquids and the circumstances under which gasoline, for example, will burn, explode, or not burn at all. This chapter will therefore focus first upon definitions and principles of flammable liquids ignition and then will turn to discussion of some of the problems in complying with appropriate standards.

Perhaps the most basic term should be defined first, and this is the term *liquid.* Almost everyone knows what a liquid is, but on the other hand, virtually every flammable substance exists as both a liquid and a gas, depending on temperature or pressure. A good practical rule to follow is that if the substance is normally a liquid, it is defined as a liquid. Where one can get into trouble is in classifying propane and butane, which are gases and were not intended to be considered flammable liquids, although they can be liquefied. The NFPA standard definition of flammable liquid excludes propane and butane by excluding all "liquids" having a vapor pressure in excess of 40 pounds.

The term *flashpoint* is very important to the Safety and Health Manager because it is the principal basis for classification of flammable and combustible liquids. Therefore, it is primarily flashpoint that determines the amounts of the liquid that are permitted to be stored in various types of containers. Flashpoint is the point to which a flammable liquid must be heated so that it will give off sufficient vapor to create a flash at the surface of the liquid when there has been a spark or flame applied to it. Flashpoint is not the same as *firepoint;* firepoint is a higher temperature and is the temperature at which a fire on top of the liquid is sustained.

Three principal testing methods are in use for determining flashpoint. The Cleveland open cup test is simple but is not often used because it is intended for heavy oils. The most frequently used method is the Tag closed tester method. The term "Tag" is simply an abbreviation for a French name *Tagliabue.* The third method is the Pensky-Martens closed tester method. This method uses a small stirring rod and is used for viscous liquids and liquids that form a film on the surface. This method is also more rare than the Tag test. The open cup method best simulates the in-plant situation where open vats are being utilized. The closed cup situation best simulates the situation for flammable liquids in storage.

Classification of flammable liquids also depends on *boiling point,* but even this can be confusing because liquid does not normally boil at only a single point of temperature. This range is acknowledged in the standards by designating the "10% point" as the key. The 10% point is the temperature at which 10% of the liquid has become gas. *IBP* refers to initial boiling point and is the temperature at which the first drop of liquid falls from the end of the distillation tube in a standard ASTM[1] test distillation.

Volatility refers to how readily a liquid will evaporate; it is closely related to boiling point. *Light* and *heavy* refer to high volatility and low volatility, respectively.

Flammable liquid is a term that means the same thing as *Class I liquid.* However, such liquids are further divided into IA, IB, and IC. *Combustible liquid* is the general

---

[1]American Society for Testing Materials.

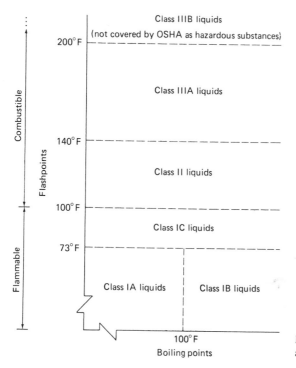

Figure 8.1. Classification of flammable and combustible liquids.

term for both Class II and Class III liquids. The entire classification scheme, which is based on flashpoint and boiling point, is explained in Figure 8.1.

Although it has become increasingly scarce, gasoline is the most widely used and plentiful flammable liquid. Because of its widespread use and because of horrible explosions and fires in its history, gasoline is blamed for a great deal of the fire hazard in the workplace and elsewhere. Some of the fears of gasoline hazards arise from ignorance rather than knowledge-based prudence, and this is detrimental to the cause of safety. In Chapter 3 it was explained that overzealous safety rules contribute to worker apathy for the rules and in turn work against the cause of safety rather than helping it. Operational and safety rules for working around gasoline and other flammable liquids are sometimes in this unfortunate category. Admittedly, gasoline can be extremely dangerous, but there is no substitute for knowledge of the mechanism of its hazards so that the worker can intelligently take sensible precautions.

Knowledge begins with exposing the falsehood of the many myths surrounding the subject. Gasoline and other flammable liquids have their share of such myths, and this book will attempt here to dispel some of them.

Perhaps the wildest myth about gasoline is as follows:

Flammable Liquid Myth 1

A lighted cigarette when brought into contact with the surface of a container of gasoline is sure to ignite it.

To the contrary, it is almost impossible to ignite a tank of gasoline *at the surface* with a lighted cigarette. As with any ordinary fire, three ingredients are essential to support combustion:

1. Fuel
2. Oxygen (usually from the air)
3. Sufficient heat

There is plenty of fuel at the surface of a container of gasoline, but both of the other two ingredients are generally insufficient to support combustion. Concentration of gasoline vapors in excess of 7.6% are too rich and will not burn, and at the surface of gasoline standing in still air the concentration is much higher than 7.6%. Also, a glowing cigarette is not hot enough in most cases to permit ignition.[2] In fact, dramatic demonstrations have been performed in which a lighted cigarette is extinguished by drowning it in a cup of gasoline! Incidentally, there are hazards involved in such demonstrations, and experimentation is not recommended. Things can go wrong, such as tiny flame on the cigarette paper being hot enough for ignition. Also, there is the problem of getting the cigarette *through* the region in which the vapors are not too rich and causing ignition before the rich area near the surface can be reached. Also, small quantities of gasoline in the surrounding area can cause vapor/air mixtures just right for combustion. These are the reasons for "no smoking" rules around gasoline.

Gasoline has a burnability range of 1.4% to 7.6% gasoline vapors in dry air. Some other flammable liquids have wider burnability ranges and are more easily ignited than gasoline. Burnabilities for some commonly used highly flammable liquids are shown in Figure 8.2. Note that though gasoline is easier to ignite at "lean" concentrations than alcohol is, alcohol will ignite at much richer concentrations. Also note the extremely broad and dangerous burnability range of carbon disulfide. The upper limit above which concentrations of flammable vapors are too rich to ignite is designated as the *upper explosive limit* (UEL). The corresponding lower limit below which concentrations of flammable vapors are too lean to ignite is the *lower explosive limit* (LEL).

Another myth regarding gasoline has to do with service station fires and other fires around underground tanks.

Flammable Liquid Myth 2

Fires in the underground gasoline tanks burn or explode with such intensity as to destroy much life and property around service stations.

Actually, fires do not burn in underground tanks of gasoline, even when a serious fire occurs above the ground. John A. Ainlay[3] states that in a study of 45 years of

---

[2]The ignition temperature for gasoline (536 to 853 °F) is higher than for wood (approximately 400 °F)!

[3]John A. Ainlay, Evanston, Illinois, is a nationally recognized authority on the chemistry of petroleum fires.

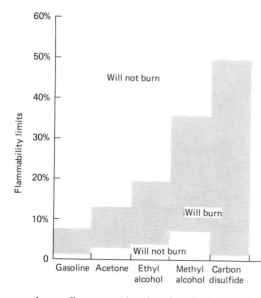

Figure 8.2. Burnability ranges of some popular flammable liquids.

petroleum fire reporting by the National Fire Protection Association (NFPA) and the American Petroleum Institute (API), there has never been a fire reported in an underground tank of gasoline currently in use. The vapor mixture in the tank is too rich to support combustion.

An abandoned empty tank is, curiously, more dangerous than a full or nearly full tank in use. The abandoned empty tank has had a chance to dry out, and vapors have gradually dispersed until the mixture may have become lean enough to result in an explosive mixture. The same is true of any *empty* gasoline drum. But a drum that is full or partially filled with gasoline will normally have a vapor/air ratio that is too rich to support combustion. Figure 8.2 shows that fire has much greater hazard potential inside drums of alcohol or carbon disulfide.

With aboveground tanks the hazard takes on a different dimension. The aboveground tank is exposed to intense heat and possible rupture during a service station fire. When a tank ruptures or explodes, tremendous quantities of fuel are suddenly added to the fire in the presence of abundant supplies of oxygen and heat.

The preceding paragraph describes ample reason for the standards that prohibit aboveground tanks for service stations, except when special conditions are met. There is, however, still another hazard for aboveground tanks. The vapor density[4] of gasoline is higher than 3 to 1. This means that unlike natural gas or other lighter-than-air materials, gasoline vapors will settle in low places. An aboveground tank is an invitation for gasoline vapors to settle in low areas of the service station, such as in service pits. The vapor density of gasoline is the reason that basements in service stations are now illegal.

Despite the hazards just described, a large number of violations of the standard prohibiting aboveground tanks have been found. Small independent service sta-

[4]"Vapor density" is the ratio of the weight of the vapor to the weight of the same volume of air.

tions and private, in-plant service stations are the most frequent violators. Many of these installations were made prior to the writing of the standard, and the owners are asking, "Why didn't you tell me the rule before I built my station?" Some feel that such a standard was intended to act as a "building code" and apply to all future installations but not require a remodeling of all existing noncomplying stations.

Misconceptions about "octane rating" are at the root of the third myth.

Flammable Liquid Myth 3

High-octane "aviation gasoline" or "premium gasoline" is much more hazardous than regular gasoline.

Octane rating has to do with the preignition characteristics of gasoline within internal combustion engines and has nothing to do with fire safety. High-octane gasoline should be given the same fire precautions as regular gasoline—no more, no less.

SOURCES OF IGNITION

Dispelling myths about flammable liquids should make personnel more cautious. The chemistry of a petroleum fire explains some of the seemingly peculiar instances of nonoccurrence of petroleum fires. But at the same time, this understanding can highlight the extreme hazard involved when conditions are right for a fire. The seeming innocence of a broken light bulb can result in a disastrous fire. In the instant before a hot light bulb filament burns out after the bulb glass breaks, that filament is hot enough to ignite gasoline vapors. This is why it is important to protect light bulbs in the presence of flammable vapors as required in both the *National Electrical Code®* and further references in federal standards.

Welding sparks are another important ignition hazard for flammable vapors. The usual temptation is to speed up a repair operation, and welding is often commenced before sources of flammable vapors are removed from the area and existing vapors purged. Welding around flammable vapors has cost the lives of many inexperienced personnel who were not aware of the hazards involved.

Associated with welding is the grinding of the finished weld. Sparks generated are not to be trusted. It is true that many grinding sparks do not reach the required ignition temperature for gasoline vapors, but some sparks do reach ignition temperature, and spark-producing grinding should not be performed in the presence of flammable vapors.

The hazards of static electrical discharge around flammable vapors are well publicized. After all, it is electrical discharge that ignites gasoline vapors with precise reliability in most internal combustion engines. Electrical arcing or static electrical discharge is easily a source of ignition.

An understanding of the principles underlying the hazards with flammable liquids is useful when applying appropriate standards. Having covered some of the basic definitions and principles, we now turn to a direct analysis of the standards

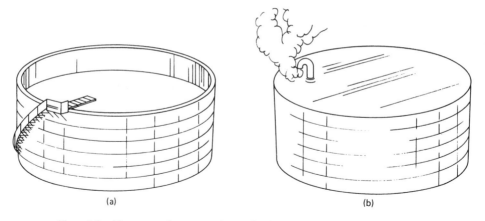

Figure 8.3. Two types of storage tanks: (a) floating roof type; (b) conventional vented type. The floating roof rises and falls with the level of the liquid inside the tank, eliminating the necessity of costly and dangerous venting.

and helpful procedures to be followed by Safety and Health Managers to bring their facilities into compliance.

Federal codes for tank storage are quite complicated and are mostly the concern of layout designers of petroleum tank farms, bulk plants, dykes and drainage schemes, refineries, and service stations. The codes also cover such elements of design as tank construction and proper venting. Most safety managers do not need to concern themselves with mastering the details of tank construction. To the contrary, it is sufficient to know where to find the requirements and to alert designers and other planners that strict codes must be followed in specifying tanks for flammable liquids.

One criterion for distance requirements between tanks is whether the tank roof is fixed or floating. The general public does not usually realize that the roofs of many petroleum tanks rise and fall with the level of the liquid inside (see Figure 8.3). A tank with a fixed roof will not fill unless it is vented, and such venting causes a costly loss of vapors. But the Safety and Health Manager should understand that the floating roof also protects against the fire hazards of releasing vapors to the atmosphere. The vapor/air space inside an empty or nearly empty tank with a fixed roof is also more hazardous than for the floating roof tank, which has little or no vapor/air space. The floating roof tank is a dramatic example of an industrial improvement that saves production cost while promoting a safer workplace.[5]

To prevent ignition hazards, electrical interconnection is required between nozzle and container when dispensing Class I liquids. But this raises questions regarding dispensing Class I liquids into containers made of plastic or other nonconductive material. It makes no sense to bond a plastic container to the nozzle because such

---

[5]The disastrous effects of improper design of flammable liquid tanks are illustrated in Exercise 8.4 at the end of this chapter.

a bond would be ineffective in equalizing the static charge (see Figure 8.4). The NFPA recognized this fact when they exempted nonconducting containers from electrical bonding requirements in NFPA standards.

Some readers may wonder at all the fuss about static electicty during dispensing operations. But filling tanks or containers *generates* static electricity due to the flow of the liquid—a little known phenomenon regarding fluid flow. The rapid flow of carbon dioxide from a fire extinguisher in use can cause static electrical shocks that make the extinguisher very uncomfortable to hold. During the loading of fuel oil into tank trucks at night, an astonishing phenomenon has been observed in which "lightning displays" are seen flashing around inside the tank!

To prevent buildup of static electricity during loading operations, the flow should be kept as quiet and smooth as possible. Filters are big static generators and are best placed as far back in the line as possible, remote from the fill spout. Another measure that can be taken to reduce static electricity is to retard the flow. Also, "splash loading" should be avoided; that is, the fill spout should extend down to a point close to the bottom of the compartment to avoid excessive splashing, which generates static electricity. When loading multicompartment trucks, the front and rear compartments are the most likely to present problems from splash loading because of the arrangement of some types of loading apparatus. The reason for this can be seen in Figure 8.5. Another static eliminator is to place a "rest area" in the delivery line, as shown in Figure 8.6. The rest area is an expansion area in the pipe which allows static charge to bleed off the liquid before rapid flow continues.

"Switch loading" is a danger from the standpoint of ignition by static electricity. With gasoline, static electricity is not as serious a problem in loading because the vapor concentration is generally far too rich to permit ignition. Even when loading gasoline into a tank that previously contained fuel oil, the vapor concentration becomes too rich as soon as loading commences. The real danger is when fuel oil is loaded into a compartment that previously contained gasoline. This is called "switch loading" and is very hazardous. The vapor concentration in such an operation is just right for ignition and a static discharge or any other source of ignition can result in an explosion that will rip the truck apart. Remedies for the problem when switch

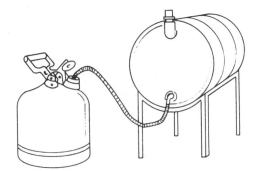

Nonmetallic safety can
needs no bonding wire
during filling operation

Figure 8.4. Bonding not necessary for nonconducting container.

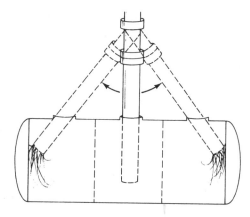

Figure 8.5. "Splash loading" occurs more often while loading front and rear compartments than while loading middle compartment.

loading is necessary are to (1) fill the tank with carbon dioxide (the cardox method), (2) use vacuum cleaners to purge the tank of gasoline, or (3) cut the loading speed down to about 30% until the tank is about one-third full.

One curious provision of appropriate safety standards requires accurate *inventory records* for Class I liquid storage tanks. Inventory records are usually for accounting and cost control, so what business has a *safety* standard in dealing with accurate inventory records? The answer is that accurate inventory records can be used to detect dangerous leaks. Unfortunately, inventory discrepancies are often attributed to "clerical error" or "unexplained loss" and are simply ignored. After a serious fire has occurred, an investigating team sometimes probes the past inventory records only to discover that the evidence of a dangerous leak had been in the records for many days.

With respect to the hazards of leaking tanks, the Safety and Health Manager must follow the regulations of two federal agencies: OSHA and EPA. The EPA cracked down on underground storage tanks in late 1988 (ref. 19) and required systems for monitoring tank and pipe leaks, automatic shut-offs for pressurized systems, specified tank and pipe construction, spill protection, and systems to prevent overfill. In addition, there are reporting requirements to notify local or state authorities whenever a new tank is installed, an old one is closed permanently, or a leak is discovered. A notable exemption to EPA's tank rule is *above*ground tanks, provided that less than 10% of the product is stored in subsystem piping.

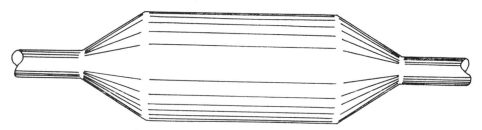

Figure 8.6. Static eliminator "rest area."

In summary, the Safety and Health Manager should approach the flammable liquids problem with a knowledge of the principles of ignition. Any new facilities should be designed and constructed with recognition of the hazards of flammable liquids. After construction and installation, commonsense rules for eliminating sources of ignition and preventing dangerous leaks should be established and enforced by the Safety and Health Manager. Training of personnel to understand the underlying principles of the hazards will help a great deal with the problem of safety around flammable liquids.

SPRAY FINISHING

A direct concern of Safety and Health Managers, especially in manufacturing plants, is the installation of facilities and proper procedures for spray painting areas or booths. This subject is a logical extension of the subject of flammable liquids covered previously because most spray painting materials are flammable. The subject is important not only from a compliance standpoint but also greatly affects insurance rates and basic insurability.

The construction and operation of a spray painting area that meets applicable code is a quite costly undertaking, and a weak Safety and Health Manager sometimes attempts to circumvent the rules to appease top management. The most common such circumvention of the rules is to call the paint facility a "small portable spraying apparatus not used repeatedly in the same location" and thus achieve exemption from spray painting standards. But if the small "temporary" setup becomes a more or less permanent arrangement, a continuing and serious fire hazard will result. Furthermore, neither the insurance company inspector nor an experienced government inspector will be fooled by a so-called temporary arrangement that has eventually become permanent, because spray paint residues will accumulate throughout the area in large quantities.

There are both health and safety considerations in spray finishing operations, but standards for spray finishing principally concern safety aspects, particularly fire. The most frequent violations of spray finishing standards are in the following categories:

- Improper wiring type for hazardous location
- Exhaust air filter deficiencies
- Cleaning and residue disposal
- Quantities of materials in storage
- Grounding of containers
- "No Smoking" signs

Also frequently violated but of somewhat less emphasis are the physical construction requirements for spray booths and the mechanical ventilation requirements.

Applying the rule of attacking the easiest problems first, the Safety and Health Manager should take steps to immediately install No Smoking signs in spraying areas and in paint storage rooms. This advice may sound superficial, but thousands of firms have received OSHA citations simply for failure to post No Smoking signs. The cost of complying with this rule is almost negligible.

After ensuring that No Smoking signs are installed, the Safety and Health Manager should investigate the wiring in the spray area to see if it conforms with *National Electrical Code®* specifications for hazardous areas. A competent electrician, knowledgeable in the provisions of the *National Electrical Code®* , is useful for this phase of the problem.

Appropriate wiring and electrical equipment classification in and around spray areas may be simplified somewhat in the decision diagram of Figure 8.7. A great deal of controversy has arisen regarding enforcement of electrical requirements around

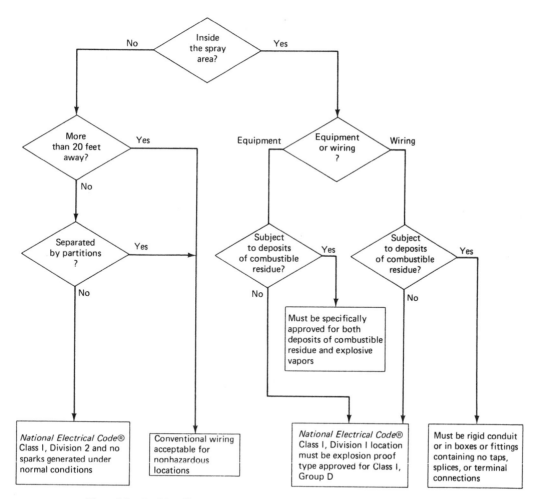

Figure 8.7. Decision diagram: How to follow OSHA standard 1910.107(c)(5) and (c)(6) specifications for spray area wiring and electrical equipment.

spray areas, particularly regarding interpretation of the legendary "20-foot distance" from the spray area. It is interesting to note that the *National Electrical Code®*, 1975 edition, reduced the required distance from 20 feet to 5 feet from the spray area.

Besides distance of travel, another issue is the *direction* of travel of the flammable vapors after they leave the paint spray booth. Since flammable vapors can travel in any direction, it is recommended that the Safety and Health Manager take the cautious rule of using Class I, Division 2, wiring in all directions from the open face of the spray area or booth, including vertically and around the corners alongside the booth. Additional guidance on the various classifications for electrical wiring is contained in Chapter 14.

Some people construe the standards to require automatic sprinkler systems for all paint spray areas, but the standards do not really specify this. If an automatic sprinkler system is used, however, it must meet NFPA requirements. The vendor who installs the sprinkler system should be required to ensure that the system conforms to all applicable codes for the installation to which it will be applied. If the system is installed inside the ducts, sprinklers are needed on both sides of the filter system.

Combustible residues contribute to the largest proportion of spray booth fires. Reactions between different materials can add to this hazard, especially when peroxides are used. The control of overspray residues involves both engineering and administrative controls and is an item that deserves the attention of the Safety and Health Manager. Residue accumulation is easy to recognize and is an embarrassing testament to poor maintenance and lack of hazard control. Once removed, the residue and debris must be properly disposed to prevent spontaneous combustion and other fire hazards from oily rags and residue debris.

It was stated earlier that automatic sprinkler systems are not required for *all* areas. However, for fixed electrostatic systems automatic sprinklers are required "where this protection is available." Automatic sprinkler systems are preferred, and if such a system is already available close by (within approximately 50 feet), it should be extended to the electrostatic spraying area. In the absence of "available" automatic sprinkler systems, the standard requires "other approved automatic extinguishing equipment" for electrostatic spray areas. Such alternate systems would include fixed carbon dioxide or dry chemical systems, and these systems will be discussed further in Chapter 10.

The use of heating devices for drying in the spray area increases the hazard by raising the temperature of overspray residues and also by increasing the vapor level in the air. In addition, the standards are not very clear concerning the use of the spray area for a drying area. The use of the spray area as a drying area is prohibited unless the arrangement does not "cause a material increase in the surface temperature of the spray booth, room, or enclosure."

DIP TANKS

Dip tanks often contain hazardous materials and are treated separately in federal standards. Care must be exercised when consulting the standard, however, because it applies only to those dip tanks containing flammable and combustible liquids.

Plating dip tanks containing hazardous acids are *not* covered by the dip tank standard unless the acid is flammable or combustible.

The following are the principal problems with dip tanks:

- Automatic extinguishing facilities
- No Smoking signs
- Dip tank covers

The lack of dip tank covers is the most frequent violation of the three. One problem with covers is that the covers must be "kept closed when tanks are not in use." It is unreasonable to expect dip tank covers to be closed during short intervals of nonuse, such as coffee breaks and other short interruptions. However, an idle period of as long as one-half shift would be considered a "not in use" period. Also, if the dip tank is discovered to be idle for any period in which the work crew and supervisor have left the area, the dip tank should be considered "not in use." Automatic closure devices for actuation in event of fire are desirable but are not specifically required. Such closure devices "shall be actuated by approved automatic devices and shall also be arranged for manual operation." Automatically closing dip tank covers are considered among the most appropriate means of automatic extinguishing facilities specified under conditions described in Figure 8.8

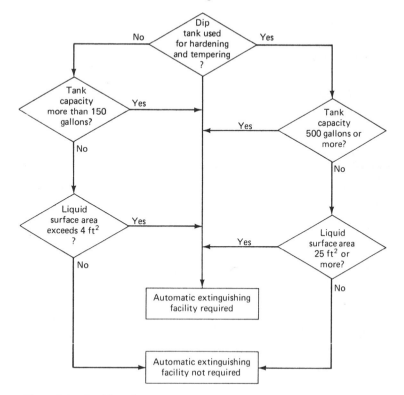

Figure 8.8. Decision chart: Automatic extinguishing facilities for dip tanks.

EXPLOSIVES

Everyone knows that explosives are dangerous, and the general public completely avoids contact with them. Only the well-trained professional knows which procedures are safe and what to do in each situation. The body of existing codes governing explosives pertains almost entirely to storage or to the construction of the magazines in which the explosives are stored.

As in flammable liquids, explosives are classified according to degree of hazard. Class A is the most hazardous, and most of the materials that the general public considers "explosives," such as nitroglycerin, black powder, and dynamite, fall into the Class A category. Class B explosives include propellants, photographic flash powders, and some special fireworks. Class C explosives are manufactured articles which contain explosives in restricted quantities. The Safety and Health Manager should take advantage of manufacturers' labeling to determine explosive classes where a questionable item is concerned.

The storage magazines for explosives are divided into two groups, also called "classes," but in the case of magazines the class designation is a Roman numeral instead of an alphabetic letter. The magazine class depends chiefly on quantity (weight) of explosives stored, not the explosives class. Class I magazines are for quantities more than 50 pounds, and Class II magazines are for 50 pounds or less.

Most facilities do not use explosives, and most Safety and Health Managers can ignore the standard. But those who do have explosives within their facilities should take the precaution of assuring that handling and especially storage procedures are in compliance with applicable code.

LIQUEFIED PETROLEUM GAS

Liquefied petroleum gas (LPG) is a commonly used fuel gas, especially in areas remote from piped natural gas utilities. All petroleum gases can be liquefied if the temperature is lowered sufficiently, but natural gas, which is chiefly methane, is very difficult to liquefy, although it is otherwise cheaper than LPG. LPG is principally a mixture of propane and butane, both of which can be liquefied more easily than methane and therefore can be transported more compactly. The expansion ratio is approximately 1:270; that is, 1 gallon of liquid converts to 270 gallons of gas at normal temperature and pressure.

The exact mix between propane and butane is a matter of climate and economics. Butane has a higher boiling point and is unsuitable for cold climates because it will not convert to a gas at low temperatures. Butane has historically been cheaper, however, so it has been used in climates such as those in the southern United States. Recently, propane has been used almost exclusively because butane prices have increased due to butane's application in the manufacture of artificial fabrics.

Propane is a product of the refinery cracking[6] process and is odorless in the natural state. For safety's sake, the odorant ethyl mercaptan is added as a stenching agent before delivery to the customer. This stenching agent facilitates leak detection.

[6]Large molecules are "cracked" into simpler molecules of more economically useful products.

However, as was emphasized for hydrogen sulfide in Chapter 7, a heavy exposure to a strong odor can overcome the olfactory system, and consequently the victim simply no longer can smell it.

One of the safety hazards of propane is that it is heavier than air (approximately 1½ times the density of air). This contrasts with natural gas (methane), which is lighter than air. This difference in properties can lead to problems when one attempts to convert a piece of equipment from natural-gas powered to propane powered.

Another hazard with propane is that the extreme cold of its liquid state can burn the flesh. In fact, the treatment is the same as for third-degree burns. One way such a burn can occur is by opening the valve too quickly and by placing the hand over the valve to feel the flow. Injury from liquid propane while opening the valve does not indicate a defective valve.

Unlike gasoline and other atmospheric-vented tanks, LPG tanks are closed, and there is no opportunity for water vapor to accumulate inside the tank. Therefore, no "bleed-off" valve for moisture is used. If water should happen to be introduced into the system, it would probably freeze the valve during expansion of the gas. The closed tank of LPG then contains no air and is a mixture of liquid and gaseous propane. The vapor pressure inside the tank depends on temperature. At 0 °F the pressure is 28 pounds per square inch (psi), and at 100 °F the pressure is almost 200 psi. The pressure relief valve for tank trucks is set at 250 psi and for cylinders at 375 psi.

Although flesh burn from the extreme cold is a hazard to be considered, the principal hazard with LPG is fire, and when an LPG fire occurs, it is usually a disaster. The fire expands quickly, and portable fire extinguishers are generally useless, although they may be useful in extinguishing other burning materials that may be threatening LPG facilities. Once a tank itself ignites, however, it is strictly a matter for professional firefighters and special techniques using large volumes of high-pressure water spray to protect firefighters who must approach the tank to close valves or otherwise control the fire. Large tanks, such as railroad cars, have resulted in spectacular fires, including the phenomenon called "BLEVE" (rhymes with "heavy"), which means "boiling liquid expanding vapor explosion."

The federal standard requiring the use of laboratory-approved equipment ("listed" by an approved testing laboratory) seems like so much red tape. But without this requirement people would try all kinds of makeshift arrangements. One individual decided to use an old hot water tank for storage of LPG. The tank exploded, killing one person and injuring another. Another temptation is to use ordinary water hose instead of approved piping. The insidious nature of this hazard is that the water hose will often withstand the LPG pressures and will seemingly work, but the LPG will attack the rubber in the hose and will eventually result in a rupture. Similar problems result from using ordinary plumbing fittings and valves which have rubber seals.

Another frequent misuse of equipment is to interchange tanks for storage of anhydrous ammonia and LPG. Anhydrous ammonia attacks brass and copper fittings in LPG tanks, making them unsafe. Particularly dangerous is damage to the tank relief valve.

Fire, welding operations, and other sources of intense heat can weaken LPG cylinders and make them no longer capable of meeting laboratory tests. The Safety and Health Manager should be on the alert for this danger and have LPG equipment recertified after a plant fire or other exposure to heat. As for welding, no welding should be permitted directly to the tank shell; it is permissible, however, to weld to existing brackets, plates, and lugs, which in turn were already welded to the tank in its original manufacture in the configuration which has laboratory approval.

Fire control for LPG tanks is quite different from fire control for flammable liquid tanks. Dikes are built around flammable liquid tanks to contain the burning liquid in event of tank rupture. But such dikes are hazardous to LPG tanks because they can cause burning near or beneath the tank which might result in explosive rupture.

In any high-pressure cylinder with a valve on the end, there is danger should the valve be accidentally ruptured or even broken off. Like a torpedo in size and shape, the cylinder becomes a dangerous missile. The most dangerous are the very high pressure oxygen cylinders used in welding (see Chapter 13). However, LPG cylinders at 200 psi can also be quite dangerous. This danger is in addition to the fact that the gas liberated in the rupture can ignite explosively. Therefore, valves must be protected, and there are two acceptable methods for this: by recessing the valve into the tank or by a ventilated cap or collar.

Safety and Health Managers often misinterpret the provision of the federal standard for LPG, which reads as follows:

> Engines on vehicles shall be shut down while fueling if the fueling operation involves venting to the atmosphere.

In-plant refueling of forklift trucks that operate on LPG normally does not involve venting to the atmosphere. It is not required that the engines be shut down in such refueling operations.

CONCLUSION

A final suggestion for Safety and Health Managers is to take advantage of community resources, public and private, for advice and assistance in dealing with hazardous materials. Local fire departments and state fire marshals can be of assistance, particularly with regard to flammable liquids, spray finishing, and codes. Some fire or police departments have explosives experts on hand. Compressed gases, LPG, and anhydrous ammonia problems can sometimes be alleviated by consulting the local distributor for these materials. Some of these distributors are backstopped by extensive training resource centers at their company headquarters and can supply booklets, guides, warning labels, films, and slide-tape programs for in-house training to combat hazards.

EXERCISES AND STUDY QUESTIONS

8.1. What is a floating roof?

8.2. Would it be proper to use a Class II magazine for storage of Class A explosives? Explain.

8.3. What is a "BLEVE"?

8.4. Figure 8.9 shows a storage tank arrangement for storing tetrahydrofuran on the second-story level inside a building located in Chicago. The first day the tanks were loaded, the following events took place. A tanker delivery truck hooked up at the receiving valve at ground level and proceeded to deliver 500 gallons. About halfway through the filling operation a company employee inside the building yelled out the window that the tank was overflowing. Soon thereafter a tremendous explosion and fire killed both the company employee and the tank truck driver. Study the circumstances and identify potential causes of this terrible accident. What is the flashpoint of tetrahydrofuran?

8.5. Compare the flammability hazards of gasoline with ethyl alcohol.

8.6. A manufacturing process uses the powerful solvent acetone. One phase of the process is to dry out the acetone in a "drying area." The drying area evaporates 2 gallons of acetone per hour. Every gallon of liquid acetone produces 41 cubic feet of vapor. How much ventilation (ft³/hr) would be needed to maintain vapor concentration below the burnable level? How many times per hour would the ventilation system exhaust a room 9 feet by 12 feet with a ceiling height of 10 feet?

8.7. In a one-time inspection of a facility, how will an insurance representative be able to determine that a temporary spraying area or a so-called touch-up area is really a more permanent arrangement?

8.8. Under what circumstances are automatic sprinkler systems required for paint spraying areas?

8.9. When are dip tank covers required to be closed? Must they close automatically in event of fire?

8.10. Compare liquefied petroleum gas (LPG) with natural gas in terms of safety hazards.

8.11. Are fire extinguishers appropriate for LPG facilities? Why or why not?

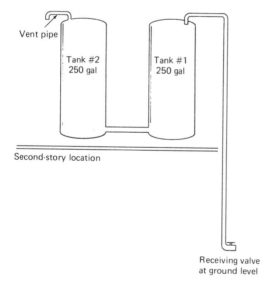

Receiving valve
at ground level   **Figure 8.9.** Tank configuration.

9

Personal Protection
and First Aid

Percent of OSHA
general industry citations
addressing this subject

10%

In a way it is unfortunate that this chapter is even necessary. The need for personal protection implies that the hazard has not been eliminated or controlled. And the need for first aid implies something even worse! When feasible, engineering control of the hazard is preferred over the use of personal protective equipment. As discussed in Chapter 1, we realize that there will always be some risks remaining, but our goal is to eliminate *unreasonable* risks, not all risks. The job of improving safety and health in the workplace will never be completely finished, so we must concern ourselves with the need to provide personal protection against hazards that have not been completely eliminated and to provide first aid when an accident does occur.

The problem of providing personal protective equipment seems straightforward and easy enough to understand. But the simplicity of the problem is an illusion, and many industrial Safety and Health Managers fall into its trap. For instance, it would seem that if the noise level in the production area is too high, the solution to the problem would be to provide ear plugs for the workers. But anyone who has actually confronted this problem knows that the solution is not that simple. For a variety of reasons, most people do not want to wear ear protection. They may be shy about the appearance of ear protection equipment, they may feel discomfort or perhaps even pain, they may feel it interferes with their necessary hearing or efficiency, or they may feel that use of personal protective equipment is their own business, not their employer's. The bases for these complaints and what to do about them will be taken up later in this chapter, but first to be covered are the hazards for which

192

personal protection may be needed and the types of equipment available to meet these needs.

The matter of personal protective equipment becomes very delicate when employees bring their own equipment to work. If the equipment is not properly maintained, who is responsible—employer or employee? As Safety and Health Manager, consider the following logic. If employees bring their own personal protective equipment to work, is it not possible that the equipment itself could represent a hazard? Personal protective equipment must be properly selected to match the hazard, and employees bringing their own equipment might falsely think they are safe when their equipment could actually be malfunctioning or inappropriate.

Even if the employee works in a job that does not require personal protective equipment, the use of faulty or inadequate equipment might tempt the worker to take chances. For instance, suppose that a maintenance worker who has no need to even approach the edge of a rooftop while maintaining air-conditioning equipment wears his or her own personal protective rope attached to a leather belt as added protection, not required by company policy. Then suppose that some unusual situation tempts the employee to approach the edge of the building and the false sense of security from the improvised "safety belt and lanyard" causes the worker to proceed without caution. An accidental fall could cost the worker's life when the leather belt breaks under a 2000-pound shock load due to the fall. Alternatively, even if the leather belt withstands the shock load, the worker could be killed by the shock itself if the lanyard is too long and is nonelastic. These are facts workers might not know, and when they improvise or bring their own personal protective equipment to work, the employer should assure that the equipment is adequate for the situation and properly maintained.

HEARING PROTECTION

As might be expected, the greatest emphasis in providing personal protective equipment coincides with the principal problem in environmental control, covered in Chapter 7—the problem of noise. If engineering or administrative control measures are unsuccessful in eliminating the hazard of noise in the workplace, management must turn to personal protective equipment to shield the worker from exposure.

The most important factor in selection of the type of noise protection is probably effectiveness in reducing decibel level of noise exposure. However, this is by no means the only important factor, and selection can be somewhat complicated. Economics is always a factor, and if limited effectiveness is all that is necessary in a given situation, cheaper devices can be selected. Employee comfort is probably at least as important a factor as economics. The worker comfort factor goes beyond the simple goal of promoting worker satisfaction; it affects the amount of protection the worker will receive. If workers find a type of ear protection uncomfortable or awkward to wear, they will use every excuse not to wear it, with a resulting loss in protection. The following paragraphs will consider the merits of various types of ear protection.

Cotton Balls

Ordinary cotton balls, without addition of a sealing material, are virtually worthless as a means of personal protection from noise.

Swedish Wool

Similar in feel to cotton, "Swedish wool" is a mineral fiber which has much better attenuation values than cotton. Swedish wool is somewhat effective as is, but is much more effective when impregnated with wax to make a better seal. One problem with Swedish wool is that it can tear when it is pulled out. To alleviate this problem, Swedish wool sometimes comes in a small plastic wrapper which is inserted with the wool. Swedish wool can be considered only fairly reusable; reuse will depend on personal hygiene, quantity of earwax, and worker preference.

Ear Plugs

The most popular type of personal protection for hearing is the inexpensive rubber, plastic, or foam ear plug. Ear plugs are practical from the standpoint of being easily cleaned and reusable. Workers often prefer ear plugs because they are not as visible as muffs or other devices worn external to the ear. But within this advantage lies a pitfall: Workers may be more complacent about using the ear plugs when it is not immediately obvious to the supervisor whether the ear plugs are being worn. The noise attenuation for properly fitted ear plugs is fairly good, falling somewhere between that of Swedish wool and the more effective earmuffs.

Molded Ear Caps

Some ear protectors form the seal on the external portion of the ear by means of a mold to conform to the external ear and a small plug. Since human ear shapes vary so widely, fit is a problem. Molded ear caps are more visible than ear plugs, which has both advantages and disadvantages, as discussed earlier. Molded ear caps may be more comfortable to the wearer, but are more expensive than ear plugs.

Earmuffs

Earmuffs are larger, generally more expensive, and more conspicuous than Swedish wool, plugs, or caps, but they can have considerably better attenuation properties. The attenuation capability depends on design, and more variety in design is possible with earmuffs. Although some workers object to wearing conspicuous earmuffs, it is interesting to note that some workers prefer them, stating that they are more comfortable than ear plugs.

Helmets

The most severe noise exposure problems may force the Safety and Health Manager to consider helmets for personal protection against noise. Helmets are capable not only of sealing the ear from noise, but also of shielding the skull bone structure from sound vibrations which can be transmitted to the sense of hearing. Helmets are the most expensive form of hearing protection but have the potential of offering protection from a combination of hazards. Properly designed, the helmet can act as a hard-hat and a hearing protector at the same time.

It must be remembered that fit is very important for all types of hearing protectors. As in noise enclosures or sound barriers, the material itself might have excellent sound attenuation properties, but if there is a leak or crack, most of the effectiveness of the device is lost.

EYE AND FACE PROTECTION

The use of safety glasses has become so widespread and so many different styles are now available that many Safety and Health Managers establish a rule that they must be worn throughout the plant. A general custom in industry is to require visitors to wear safety glasses during plant tours.

There is a difference between "street safety" glasses and "industrial safety" glasses. Visitors or employees who claim their prescription glasses are "safety glasses" probably mean that they have "street safety" lenses. Industrial safety lenses must pass much more severe tests to meet ANSI standards. This is not to say that street safety lenses are not adequate for some industrial environments. Eye protection standards are not specific about which jobs do and which do not require safety lenses—street or industrial. It is good to have exacting standards to be sure that safety glasses will meet uniform performance standards. But responsibility for deciding when eye protection equipment is necessary usually falls on the Safety and Health Manager—not the optics industry, not the standards, and not enforcement agencies.

A word of caution is in order for the Safety and Health Manager who is setting forth a policy on the wearing of safety glasses. It can be as bad a mistake to require safety glasses in those areas of the plant where there is no eye hazard as it is to not require safety glasses in areas where they are needed. The peril is that workers will not respect the safety glasses policy, and use of the glasses will not be uniform. Eye injuries can be the result—in addition to violations of code. It is easy for inspectors to establish violations when they observe workers not wearing eye protection in the presence of an established company rule requiring eye protection. But some Safety and Health Managers say that a simple plantwide rule is easiest to enforce. The costs on both sides of the question are great, and the decision to require eye protection must be made with care.

There are some jobs for which industry and enforcement agencies alike have seemed to come to a consensus on the need for eye protection. Machining operations that produce chips or sparks are almost universally conceded to necessitate the use of eye protection. Notable among these operations are those of grinding machines, drill presses, and lathes. Both metal and wood materials can produce dangerous eye hazards when machined. Corrosive liquids or other dangerous chemicals also represent eye hazards when poured, brushed, or otherwise handled in the open. When working with such materials, face protection may be needed in addition to eye protection.

More important than when to *require* eye protection is how to *educate* workers to be alert to eye hazards and the long-term consequences of eye injury. The National Safety Council has some films that bring this point home. One effective film shows emergency eye surgery in progress. Another is a poignant personal testimony of a man who was blinded on the job and the hardships he and his family have suffered since.

RESPIRATORY PROTECTION

Of even more vital importance (in the literal sense of the word *vital*) than eye and hearing protection is the need for respiratory protection from airborne contaminants. Chapter 7 discussed the different types of problems with industrial atmospheres, and determination of the type of atmospheric problem is essential in selecting the correct respiratory equipment. A well-designed and expensive gas mask is useless and might be more properly designated a "death mask" if the atmospheric problem to be tackled is oxygen deficiency, for example.

Particularly hazardous atmospheres may be referred to as IDL or IDLH, which stand for "immediately dangerous to life" and "immediately dangerous to life or health," respectively. Recently, the acronym *IDLH* has become more widely used. If a single acute exposure is expected to result in death, the atmosphere is said to be IDL. If a single acute exposure is expected to result in *irreversible* damage to health, the atmosphere is said to be IDLH.

It should be apparent to the reader at this point that there is more to respiratory protection than simply handing out respirators to workers who might be exposed to hazards. Effective protection demands that a well-planned program be implemented, including proper selection of the respirators, fit testing, regular maintenance, and employee training.

Some firms supply respirators to employees without bothering with a comprehensive respirator program, relying on the excuse that the respirators are not really required anyway because contaminants in the plant atmosphere do not exceed maximum permissible exposure limits (PELs). The Safety and Health Manager is really asking for trouble, though, by toying with a partial program. The atmospheres are no doubt marginal, or the question of a partial program would never have arisen. The marginal atmospheres might later deteriorate without warning. Employees would be lulled into a false sense of security by the superficial respirator program. Bad habits, such as negligent maintenance, inadequate fit testing, or improper equipment usage, could be developed. A whole feeling of complacency toward the use of respirators can be engendered by the use of such equipment when it is not really needed.

As with general personal protective equipment discussed earlier, the Safety and Health Manager is often placed on the spot when employees bring their own respiratory protection equipment onto the job site. In this situation the employer should take a responsible position and ensure that these employees use their respiratory equipment properly. If the employee resents what he or she perceives as the employer's interference, he or she should be reminded of the employer's continuing responsibility to eliminate hazards in the workplace, including misuse of personal protective equipment. If the equipment or its improper use can be hazardous, the employer can forbid the employee from bringing the equipment onto the premises. The Safety and Health Manager may feel some hesitation in exercising such authority over the employee's own personal property, but instances have already occurred in which employers have exercised such authority when necessary to prevent hazards.

Before going further into the subject of respiratory protection, a classification of the various devices is in order. The two major classifications are *air-purifying* devices versus *atmosphere-supplying* devices. The air-purifying devices are generally cheaper, less cumbersome to operate, and the best alternative if they are capable of handling the particular agent to which the user will be exposed. But some materials simply cannot be reduced to safe levels by air-purifying devices, and a supplied-air system is required. Another important consideration is oxygen deficiency. No amount of filtering or purifying is going to make an oxygen-deficient atmosphere safe. The only way to go in this situation is with atmosphere-supplying respirators. A summary classification of respiratory protection devices is as follows:

1. Air-purifying devices
   a. Dust mask
   b. Quarter mask
   c. Half mask
   d. Full-face mask
   e. Gas mask
   f. Mouthpiece respirator
2. Atmosphere-supplying respirators
   a. Air line respirator
   b. Hose mask
   c. Self-contained breathing apparatus

Further description of each type of device follows.

Dust Mask

The most popular respirator of all is also the most misused. Approved only for particulates (suspended solids), the dust mask (Figure 9.1) is not approved for most painting and welding hazards, although it is frequently used improperly in this way. Some dust masks are approved for mild[1] systemic poisons, but generally these masks are limited to "irritant dusts"—those that produce pneumoconiosis or fibrosis (see Chapter 7). One of the principal limitations of the dust mask is fit. Even the best-fitting models have approximately 20% leakage. A rule of thumb is that approval is valid for particulates no more toxic than lead.

[1] Dusts that have a TLV of not less than 0.05 mg/m³.

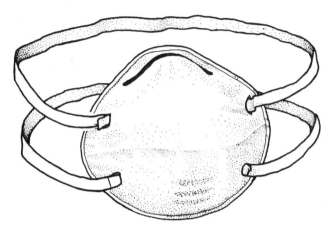

Figure 9.1. Disposable "dust" mask.

Despite its disadvantages, the dust mask is popular because it is inexpensive, sanitary, and can be thrown away after each use. Its low cost and general availability make the dust mask attractive for purchase at the local drug store and for personal use. Therefore, this is the type of respirator that the Safety and Health Manager is most likely to find employees bringing from home for use on the job. Care should be taken to educate employees about the limitations of the dust mask.

Quarter Mask

The quarter mask, sometimes called the Type B half mask, is shown in Figure 9.2. The quarter mask looks very much like a half mask except that the chin does not go inside the mask. The quarter mask is better than the dust mask, but it, too, is generally approved for toxic dusts no more toxic than lead.

Half Mask

The half mask, shown in Figure 9.3, fits underneath the chin and to the bridge of the nose. This mask must have four suspension points, two on each side of the mask, connected to rubber or elastic about the head.

Full-Face Mask

Actually, the gas mask is also "full face," but the name "full-face mask" generally refers to a mask in which the filtering chamber attaches directly to the chin area of the mask. The filters may be either dual "cartridges" or single "canisters." Both types are shown in Figure 9.4. Canisters contain granular "sorbents" which filter the air by adsorption, absorption, or chemical reaction.

Gas Mask

The gas mask is designed for filtering canisters that are too large or too heavy to hang directly from the chin. In the gas mask, the canister is suspended by its own

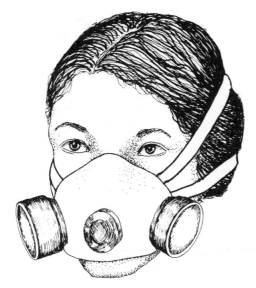

Figure 9.2.   Quarter mask.

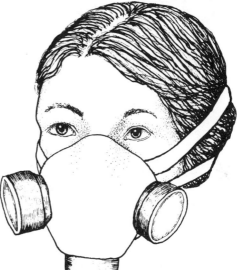

Figure 9.3.   Half mask.

harness and is typically connected to the face mask by a corrugated flexible breathing tube. The gas mask is shown in Figure 9.5.

Mouthpiece Respirator

Perhaps the mouthpiece respirator should be omitted from this discussion because this type of device is not intended for normal use. But emergencies will occur from time to time, and enabling the user to be prepared to escape when emergency does occur is the purpose of the mouthpiece respirator. Breathing is accomplished through

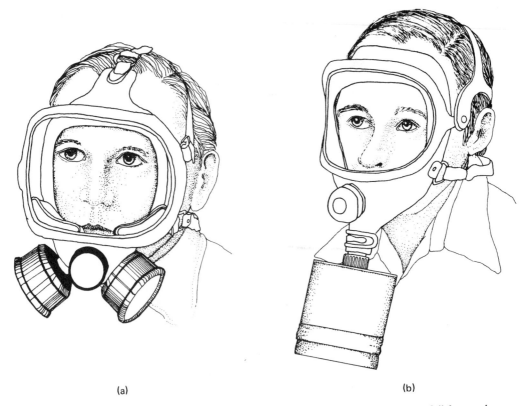

(a)   (b)

Figure 9.4. Full-face masks: (a) dual-cartridge full-face mask; (b) canister-type full-face mask.

the mouth by means of a stem held inside the teeth. A nose clip must be used to prevent inhaling through the nose. It is possible to form a good seal with the mouth and lips, but the effectiveness of the mouthpiece respirator is greatly dependent on the knowledge and skill of the user.

Air Line Respirator

The air line respirator is an atmosphere-supplying respirator and derives its name from the way in which air is supplied to the respirator mask. The air is supplied to the mask by a small-diameter hose (not over 300 feet long), which is approved together with the mask. Ordinary garden hose is not acceptable. The air is supplied by either cylinders or compressors. The method of delivery of air to the user results in three different modes of air line respirator: continuous flow, demand flow, and pressure demand.

In the *continuous-flow* mode, the air line respirator receives fresh air without any action on the part of the user; that is, the flow is forced by the apparatus. The airflow must be at least 6 cubic feet per minute to qualify a hood for use with this mode, but the flow should not be much greater because it can be responsible for a

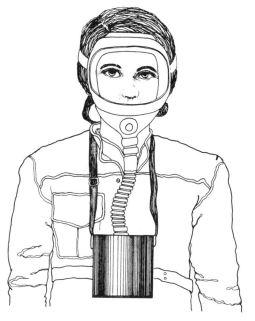

Figure 9.5. Gas mask.

high noise level inside the hood. One of the advantages of the continuous-flow mode, however, is that it permits use of a somewhat leaky, loose-fitting hood. The positive pressure differential between the inside and outside of the hood keeps the flow outward, preventing toxic agent entry. The continuous-flow mode needs an unlimited supply of air, so a compressor is used instead of tanks.

In the *demand-flow* mode, air does not flow until a valve opens, caused by a negative pressure created when the user inhales. Exhalation, in turn, closes the valve. This mode has the advantage of using less air, so it is feasible with cylinders. But the disadvantage is the need for a tight-fitting facepiece. Since inhalation causes a negative pressure differential to develop, a leaky facepiece will readily draw in the toxic ambient atmosphere. In fact, if the facepiece is too leaky, the inhalation valve will fail to open, making use of the facepiece even more hazardous than use of no personal protective equipment at all. For this reason the demand-flow mode is becoming obsolete and is being replaced by the third mode: pressure demand.

The *pressure-demand* mode has features of both the continuous-flow and demand-flow modes. As in the continuous flow, a positive pressure differential is maintained. The positive pressure is maintained by a preset exhalation valve. Despite its advantages, the demand-flow mode still requires a well-fitting mask; use by a person with a beard is not acceptable.

Hose Mask

A hose mask is a somewhat crude form of an air line respirator. The diameter of the hose is larger than in the air line respirator, permitting air to be inhaled by ordinary lung power. A blower is sometimes used as an assist. The hose mask is declining in popularity.

Self-Contained Breathing Apparatus

In this type of respiratory protection, the user carries all of the apparatus with him, usually on his back. This has the advantage of increasing the distance the user can roam because there is no umbilical cord to drag along and perhaps sever or crush. But a disadvantage is that the large pack on the back may restrict the passage of the user through a manhole or other close passage. Engineers should consider this problem when designing manholes for vessel entry when toxic atmospheres may be a problem. Many fatalities have occurred when rescue breathing packs were rendered useless because the rescuer was unable to enter the vessel with a breathing pack on his back.

    Most self-contained breathing apparatus units at the present time are "open circuit"; that is, the exhaled breath is discharged to the atmosphere (see Figure 9.6). "Closed-circuit" units recycle the exhaled breath, restoring oxygen levels. The advantage of the closed-circuit types is that the pack can be much smaller and lighter per minute of maximum permissible use. Some closed-circuit types (see Figure 9.7) have a small high-pressure oxygen tank for restoring oxygen levels after carbon dioxide is removed. Other closed-circuit types use a chemical reaction to restore oxygen levels, making possible a very compact unit. The source of the oxygen is a superoxide of potassium in which oxygen is liberated by mere contact with water. The water is supplied by the moisture in the user's exhaled breath. It takes a little while for the chemical process to function and become balanced, so the user should prime the system while still in clean air before entering the hazardous zone. A hazard with the chemical oxygen-generating unit is that the potassium superoxide must be kept sealed from moisture, except for the small amount of moisture in exhaled breath. A water flooding of the potassium superoxide unit's interior is almost sure to cause an explosion. Another hazard with the chemical oxygen-generating unit is that it supplies an oxygen-rich atmosphere, and this can be a fire hazard.

Figure 9.6. Open-circuit self-contained breathing apparatus.

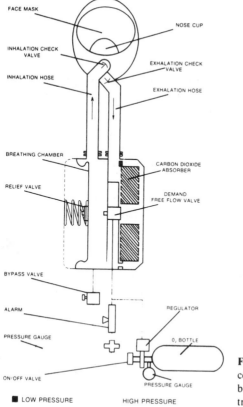

FACE MASK

NOSE CUP

INHALATION CHECK
VALVE

EXHALATION CHECK
VALVE

INHALATION HOSE

EXHALATION HOSE

BREATHING CHAMBER

CARBON DIOXIDE
ABSORBER

RELIEF VALVE

DEMAND
FREE FLOW VALVE

BYPASS VALVE

ALARM

REGULATOR

PRESSURE GAUGE

O, BOTTLE

ON/OFF VALVE

PRESSURE GAUGE

■ LOW PRESSURE   HIGH PRESSURE

Figure 9.7. Closed-circuit self-contained breathing apparatus; oxygen bottle type. (Courtesy of Rexnord Electronic Products.)

Respirator Plan

The long list of available types of respiratory protection may bewilder the Safety and Health Manager who must choose which type of protection is optimum for each application. Expert consultants can help in this situation, but one simplifying factor is that the hazard usually dictates the choice of device or at least greatly narrows the field of choice. Devices are approved for particular concentrations of particular substances, and unapproved devices must not be used.[2] Consultants can help to select between approved respirators, considering efficiency, cost, convenience, and other factors.

Sometimes the excuse is offered that no approved device exists for a particular toxic substance. But this excuse is a poor one because the worker must somehow be protected from dangerous atmospheres. There are many different kinds of

[2]For approval of respirator devices, see *NIOSH Certified Equipment,* Cumulative Supplement, (NIOSH 77-195), June 1977, and subsequent issues available periodically from NIOSH Laboratories in Morgantown, West Virginia.

respiratory equipment, just as there are many different types of respiratory hazards. It may be true that there is no approved device of a particular type for a given hazard. The employer may want to use some type of respirator, for instance, for a hazard in which there is no approved respirator as such, but another type of device *could* be used. An example would be the hazard of mercury vapor. No air-purifying device is acceptable for protection against the hazards of mercury vapor. Mercury vapor offers no warning odor or other means to detect when the canister is no longer effective. So with mercury vapor the employer must turn to another type of protection, such as air line breathing equipment. Thus for every hazardous atmosphere in which an employee must work, there must be an approved device of *some type* to protect the worker. If there is no approved device for the given hazardous contaminant, workers must be prohibited from the area.

Determining the approval status of a device for a particular respiratory hazard is no easy task. The approval authority is complicated and has frequently changed from agency to agency. Furthermore, the equipment approval lists are continually being updated. A practical solution to the problem is to seek the advice of the manufacturer of the equipment, for the obvious reason that the equipment manufacturers are likely to be most aware of the approval status of their own equipment. Not as worthy of trust is the local equipment distributor. With no reflection on the character of the local equipment distributor, such persons often carry so many lines of safety equipment that it is impossible for them to be as intimately aware of the various toxic substance approval limitations of each respiratory protection device as will be the manufacturer of the given device.

Complicating the situation is the misleading nomenclature. For instance, "organic vapor respirators" are not necessarily acceptable for commonly encountered organic vapors. Methanol, for instance, is not readily absorbed by any cartridge, and only air line equipment is effective for protection against hazardous concentrations of methanol. It is unfortunate that federal regulations require a respirator to be identified as "organic vapor respirator" because it has passed a certain prescribed test, even though the respirator may be useless for certain organic vapors. Workers are not even protected by warning labels that their "organic vapor respirators" may not be as effective for the organic vapors with which they are working. The same is true for gas masks.

Why does such a confusing and misleading system exist? Part of the problem is the complexity of organic chemistry and the myriad hydrocarbon compounds that exist. If the manufacturer attempted to label each cartridge with all the organic compounds for which it was effective, there would soon be no space on the cartridge. Any kind of shorthand coding system might confuse users in the field. Furthermore, an attempt to itemize would imply that the list was exhaustive, and the user might check no further to see whether a particular cartridge might be effective against a given substance. So the best course of action again is to check with the manufacturer for more complete and detailed lists of substances for which the organic vapor respirator cartridge is effective.

A basic principle of respirator selection is never to select a gas-absorbing respirator to be used for a gas that has no distinguishing warning properties. A moment's reflection will reveal the logic behind this principle. All respirator cartridges eventually become saturated or loaded to the extent that they are no longer effective. The users will know automatically when that point is reached through their own senses of smell, taste, or perhaps irritation or whatever the sensory warning property of the dangerous gas may be. But if the toxic gas has no warning property, the users will never know and may be subjected to dangerous exposures while wearing an approved respirator.

A good example of such misuse is the typical roofing company use of half-mask respirators when applying expandable foam. Unfortunately, the organic substances present in expandable foam do not have adequate warning properties, and there is no air-purifying device approved for such use. Therefore, air line respirators should be used.

There is one exception to the rule that air-purifying devices cannot be used with gases that do not have warning properties capable of being sensed by the user. That exception is where canisters can be equipped with an effective end-of-service-life indicator. Currently, such canisters are available only for carbon monoxide (Type N canisters).

"Paint, lacquer, and enamels" cartridges offer protection from the particulates and also from the organic solvents in the paint. Actually, the organic vapors represent the real hazard with paint, lacquer, and enamels respirators. Except for lead-based paint, the particulates in paint are not really a problem.

Respirator cartridges are not like vacuum cleaner bags; they cannot be changed from manufacturer to manufacturer without approval. The cartridge and the respirator must be approved together as a unit, although it would be possible for a respirator of one make and a cartridge of another make to be approved together if the specific combination was subjected to approval testing. Public controversy over cartridge refill approvals has charged that manufacturers have monopolistically promoted safety regulations to protect sales of their own products. On the other side of the issue is the argument that cartridge fit and proper seal are critical to the effective operation of the respirator.

For jobs requiring the use of respirators, a determination should be made to be sure that persons are "physically able to perform the work and use the equipment. The local physician shall determine what health and physical conditions are pertinent." This sounds costly and time consuming, but there is a shortcut that can expedite the personnel screening process, eliminating a large majority of the problem cases. The shortcut is to use a questionnaire that can identify obvious problems before employment.

A variety of physical conditions makes a person unfit for respirator use, whereas that person might be quite well suited to a job that does not require the use of respirators. The US Atomic Energy Commission lists some of these physical conditions that should be included in a questionnaire for screening workers from jobs requiring the use of respirators. Asthma or emphysema are pulmonary problems which may result in problems with respirators. If the environment requires self-contained breathing apparatus, the potential employee may not be fit for carrying the heavy equipment due to a back injury history or heart disease. Another physical problem is ruptured eardrums. A person with a ruptured eardrum actually breathes through his or her ears. Thus a face mask respirator that does not cover the ears may offer little in the way of protection. Epilepsy is another potential problem. Epileptics are often on medications, and these medications can present some hazards. Also, the possibility of a seizure presents the threat of the respirator being removed at the wrong time. Diabetics may also be on medications, and the use of respirators may interfere with such medications. These individuals, for instance, would probably be a poor choice for a rescue team. Even psychological factors such as claustrophobia may be important. The employee questionnaire can be a source of protection from future liability in case physical problems are present. The questionnaire should be developed with the assistance of a physician. If physical problems or variations are revealed by the preliminary questionnaire, the employee should be examined by the physician for a final determination.

An important element of an effective respirator program is fit testing or leak testing. Facial characteristics vary significantly, and fit tests are necessary to determine which particular model fits each worker best without appreciable leaking. Even the best respirators on the market will fit only a percentage of workers. Other respirators must be selected for other facial shapes.

One facial shape that is impossible to fit is the bearded one. A beard should be prohibited for any worker who must wear a fitted respirator for his safety and health. Two firemen in Los Angeles filed a discrimination suit when they were laid off because they wore beards. Their job required a possible use of self-contained breathing apparatus. The firemen lost their case. The argument against facial hair on jobs requiring respirators is well documented.

Respirator fit testing seems to imply expensive environmental chambers, but such chambers are not required. Some consultants recommend a simple plastic clothes bag suspended over the wearer's head. The subject's respirator should be equipped with an organic vapor cartridge, and the air inside the bag is contaminated with isoamyl acetate (available from the local pharmacy). If the subject smells the familiar "banana oil" odor, his or her respirator is leaking. Once fit-tested for a particular make and model of respirator, every respirator of the same make and model is acceptable for that user.

One ironic development in the field of respiratory protection is NIOSH's discovery of evidence suggesting that the substance di-2-ethylhexyl phthalate (DEHP) is a potential carcinogen. The irony is that DEHP had been used as a test agent for determining the fit of respirators!

For air supplied by compressors, care should be taken to select a "breathing air" type of compressor. What to avoid is the use of the ordinary mechanical tool air compressor for this purpose. Pneumatic tools are sometimes lubricated through a system of injection lubricants into the supplied air. This, of course, would not be satisfactory for breathing purposes.

Some confusion exists over "alarms" required for compressor failure or overheating. The reason alarms for overheating are needed is that a hot compressor can deliver deadly carbon monoxide to the air line respirator user. Some breathing-air compressors have automatic shutdowns instead of alarms, and this is entirely satisfactory for some types of industrial environments. For instance, abrasive blasting atmospheres are not IDLH (immediately dangerous to life or health), and if the compressor overheats and shuts down, workers can merely take off their respirators and immediately leave the area. But in an IDLH atmosphere a compressor shutdown could result in a fatality, and an alarm is needed.

Respirators are subject to deterioration from improper maintenance, and in addition, they can simply become unsanitary. A routine inspection before and after each use can be performed by the users themselves. These routine inspections could include a check for cleanliness, deterioration, and obvious function.

Self-contained breathing apparatus should have an effective monthly inspection. The regulator should be pressurized to check to see whether the low-pressure warning device operates. Cost for loss of pressurized air should not be an important factor because it should take less than 50 pounds of air to perform a complete checkout. Tanks should be pressurized to at least 1,800 psi. Both the cylinder and the regulator have a gauge. Instead of trusting the cylinder gauge, the regulator should be pressurized and the gauge verified.

Self-contained breathing apparatus is usually reserved for emergency use, not ordinary on-the-job use. The whole idea of emergency respirators is serious business because in an emergency there is a high probability of IDLH atmospheres. Emergency rescue equipment in IDLH atmospheres demands high-quality equipment, carefully inspected and maintained and effectively used by trained rescue teams.

Welding operations present special problems because toxic gases or fumes, or both, are often encountered, but at the same time the welder must be protected from harmful rays. Snap-on lenses are available for full-face respirators to protect against harmful rays. Another alternative is to wear the respirator beneath the hood. This requires a special welding hood, and only a half-mask respirator can be worn under this hood.

HEAD PROTECTION ✓

A primary symbol for OSHA, corporate safety departments, and just about anything related to occupational safety and health is the familiar profile of the "hardhat." So important is the symbol that many zealous safety managers have established sweeping hardhat rules throughout large general work areas. There is nothing wrong with such rules if a genuine hazard exists. But when workers sense that no hazard exists and that the hardhat rule exists as a promotional device or for window dressing, they often show their defiance by refusing to wear the hardhat.

Hardhat rules should be carefully formulated with ample consideration for the consequences both ways. Once it has been decided that a hardhat rule is necessary, the Safety and Health Manager should take steps to ensure its implementation. The evidence that was used to prove the need for the hardhat rule should be compiled into organized training packages to convince workers. After training and launching of the implementation phase, follow-up checks should be used to ensure that the rule is being followed. Corrective steps should be taken to overcome individual violations of the rule, including disciplinary actions if necessary.

Hardhats seem to have won wider acceptance than hearing protection. Besides being a symbol for occupational safety and health, the hardhat has become a symbol for rugged, physical jobs. This image has appealed to males for centuries and is becoming an increasingly appealing image for the female worker also. Management personnel have also coveted the image conveyed by the hardhat. A hardhat worn by a manager seems to imply that the manager is well founded in the operations of the enterprise and is action oriented, doing more than talking on the telephone, attending meetings, and sitting behind a desk.

MISCELLANEOUS PERSONAL PROTECTIVE EQUIPMENT ✓

Safety Shoes

Safety shoes are a more expensive undertaking than hardhats because safety shoes are worn out faster and are more expensive per item. Employees may buy their own shoes at attractive discounts in some arrangements, and this encourages actual use. Safety shoes come in a wide variety of appealing styles, and employee resistance to wearing safety shoes is largely a thing of the past.

The Safety and Health Manager is usually saddled with the decision as to which jobs require safety shoes and which do not. Although applicable national standards are explicit about the design and construction of safety shoes, the decision of where such shoes must be worn is left up to the user or to management.

One place safety shoes are clearly needed is on and around shipping and receiving docks. This should be obvious, but there has been some legal controversy over this issue. The courts have settled the question, and Safety and Health Managers should ensure that shipping and receiving dock personnel wear safety shoes.

Protective Clothing and Skin Hazards

Occupational skin disease represents approximately 70% of all occupational diseases, and over 80% of the skin disease reported is classified as contact dermatitis from irritants to the skin. The Safety and Health Manager should be alert to several skin hazard sources, such as welding, special chemicals, open surface tanks, cutting oils, and solvents.

Most welders know the value of heavy-duty protective aprons and flameproof gauntlet gloves. Leather or woolen clothing is more protective than cotton from a burnablity standpoint. Nomex™ is a treated flame-retardant fabric.

Another concern for protective clothing is chemical exposures from open-surface tanks. Gloves must be impervious to the liquid being handled and must be long enough to prevent the liquid from getting inside. If the gloves are not long enough, they can be more hazard than benefit. Many workers' hands have become more irritated than their unprotected arms simply because the gloves they were wearing permitted liquids to get inside, and then the gloves themselves became dip tanks for the hands!

Discussion of personal protective equipment for open-surface tanks is not complete without mention of "chrome holes," an ominous-sounding affliction which comes by its name honestly. Vapors and mists from chromium plating tanks can cause open ulceration, particularly in moist, tender parts of the skin. The inner septum of the nose, that part which divides the nostrils, is particularly susceptible to ulceration from chromic acid, sodium chromate, and potassium dichromate. In plating plants it is not unusual to find workers whose nasal septums have been completely destroyed by these chromium compounds. Proper ventilation and engineering controls are the most appropriate means of prevention. Worker education and regular examination can also be helpful, including periodic examination of the nostrils and other parts of the body for workers exposed to chromic acids. As with other hazards, the last line of defense would be personal protective equipment in the form of respirators with cartridges to remove the dangerous chromic acid mists.

Easily the most common hazard among skin irritants are the cutting oils used in metal machining operations. Cutting oils are useful and sometimes essential to lubricate the tool, reduce cutting temperature, remove chips, permit a higher-quality cut, and lengthen tool life. But cutting oil is not always required and is not even desirable in some situations. The manufacturing engineer usually makes the cutting oil decision for the various manufacturing processes, but there is no reason that the Safety and Health Manager should not have some input in the cutting oil decision, especially since the decision affects safety.

Cutting oils are basically of two types: natural and synthetic. The natural oils are petroleum based and are the chief culprits in a very common industrial skin disease: oil folliculitis. Oil folliculitis is basically a clogging of the hair follicles in the skin, which results in acne-like lesions. The synthetic oils are easily recognized by their familiar milky white appearance, and these oils have gained widespread use. Although the synthetic oils are not likely to cause folliculitis, they do have a nasty reputation for becoming contaminated with bacteria and presenting the hazard of skin infections. If antibacterial agents are added, these, too, can be skin irritants. Technology is working to improve cutting oils, but the job is not complete, and it appears that personal protective equipment is still in order.

Protective barrier creams for the skin are an alternative to gloves or protective clothing, but these creams are by no means a panacea. The creams must be removed and reapplied at least every break, every lunch hour, and every shift. Applying and reapplying creams cost labor time as well as the cost of the creams themselves. Also, creams are not considered as effective as gloves, even when properly applied. Creams can be used, however, when the requirements of the job make gloves infeasible.

A chief concern of the Safety and Health Manager with regard to protection of the skin is the use of various solvents within the plant. Solvents are essential for removal of grease and cutting oils, and herein lies another reason for doing without cutting oils if practical. A familiar solvent is trichloroethylene, and it is bad practice for workers to wash piece parts in trichloroethylene with their bare hands. Alternatives would include the use of wire baskets for handling the parts in the solvent, or perhaps simply substituting soap and water for trichloroethylene in some situations. For the most part, soap and water will not be effective against the oils and greases encountered, although in some situations soap and water washing is effective. The Safety and Health Manager is not doing the job unless he or she seeks out these situations and calls them to the attention of management and engineering. When alternative solvents, wire baskets, and other engineering controls are infeasible, personal protective equipment such as gloves is in order.

It seems appropriate to conclude a discussion of skin hazards, gloves, and protective clothing by mentioning one of the simplest personal protective measures of all—personal cleanliness and hygiene. Workers who are fastidious about washing their hands and bodies have frequently been found to enjoy a lower incidence of skin disease. It is easy to understand why. With skin irritants, the amount of injury is generally directly related to the duration of exposure, other factors being equal. It is easy to forget how effective soap and water are in removing injurious elements of all kinds.

What can the Safety and Health Manager do to motivate workers to adopt good habits of washing and personal hygiene? The obvious answer is training and motivational reminders in the form of signs and posters around the plant. But this obvious answer is not the only answer and perhaps is not even the best answer. The Safety and Health Manager can attempt to influence the selection and layout of convenient, well-maintained, pleasant restrooms which will boost employee morale while encouraging workers to wash regularly. There is nothing more discouraging to employees than an encounter with a washroom with no warm water, no soap, and no towels.

FIRST AID √

The Safety and Health Manager will frequently be responsible for the first-aid station and may supervise a plant nurse. The first-aid station may satisfy several additional functions besides providing immediate care for the injured. The first-aid station is often used for medical tests, screening examinations, and monitoring of acute and chronic effects of health hazards. Also, the plant nurse or other first-aid personnel may be responsible for performing some of the recordkeeping and reporting functions discussed in Chapter 2.

One adequately trained first-aid person is required in the absence of an infirmary, clinic, or hospital "in near proximity" to the workplace. No one seems to be able to determine authoritatively what constitutes "near proximity" in this context. Various interpretations around the country have been compared on this point, and the opinion has for the most part varied from 5 minutes' to 15 minutes' driving time. The interpretation has sometimes depended on whether the route to the hospital crosses a railroad track. If the workplace is not itself a hospital or clinic or is not directly adjacent to one, the Safety and Health Manager is advised to be sure that at least one, preferably more than one, employee is adequately trained in first aid.

A first-aid kit or first-aid supplies should be on hand, and the Safety and Health Manager should seek a physician's advice regarding selection of these materials. Unfortunately, medical doctors are hesitant to give such advice, probably because they fear subsequent involvement in litigation should an accident occur for which adequate materials are not available. Safety and Health Managers should do their best to obtain this advice and then document what was done.

Another first-aid consideration is the provision of emergency showers and emergency eyewash stations where injurious corrosive material exposure is a possibility. Almost everyone has seen the deluge-type shower, which is activated by grabbing and pulling a large ring attached to a chain which activates the valve. Eyewash facilities are similar to a drinking water fountain in which two jets are provided, one for each eye.

Noting the cost and permanence of a conventional eyewash facility, some enterprising and innovative individuals have marketed a simple plastic water bottle and plastic tube hanging on a frame which identifies its purpose as an emergency eyewash. Such a solution to the problem is probably not what the drafters of the standard had in mind, but the little water bottle is not without merit. For one thing water bottles can more easily be deployed to the best spots to permit instant use in case of an accident. Most workplace layouts change rather frequently, making the permanent eyewash installation cumbersome, expensive, and often in the wrong place. Furthermore, the little-water-bottle approach permits convenient introduction of antidotes or neutralizing agents for specific corrosive materials. However, woe be to the injured who uses an antidote for acid when the exposure was caustic, or vice versa. Also, the volume of water in the bottle is usually limited to a quart, or perhaps 2 quarts. This may not be enough water if a first-aid manual specifies flushing the eyes "with copious amounts of water for 15 minutes."

CONCLUSION✓

Reflection on this chapter leaves the impression that the proper provision and maintenance of personal protective equipment and first aid is not an easy task. Even after the proper equipment has been selected and procedures set up for maintenance, employees must be trained and disciplined to use the equipment properly. Inspectors can easily recognize simple violations, such as failure of individual employees to wear prescribed personal protective equipment. One pitfall for the Safety and Health Manager is the blanket specification of personal protective equipment when its necessity is marginal. Such blanket specification is a trap that will lead to employee apathy and subsequent violation of the rule.

Returning to the principle that personal protective equipment is a last resort method of protecting the workplace, engineering the hazard out of the workplace is definitely preferred. Personal protective equipment seems to be the easy and less costly way out, but an examination of the principles and pitfalls of personal protective equipment is a strong motivation toward engineering controls.

EXERCISES AND STUDY QUESTIONS

9.1. Roofing workers often apply expandable foam materials using half-mask, canister respirators. Why is this basically an incorrect and unsafe practice?

9.2. What is a chemical oxygen-generating unit? Under what conditions will it explode?

9.3. As a safety engineer, under what circumstances would you recommend a "closed-circuit" respirator?

9.4. In an IDLH situation, would you favor demand-flow mode or pressure-demand mode if both are feasible choices? Why?

9.5. In a not-so-unusual accident, two workers were killed when an employee was cleaning out a tank accessed by a manhole. A second employee saw the first employee collapse. While trying to rescue the first employee, the second employee was also overcome, and both died. What preventive measures can you suggest to prevent accidents of this type?

9.6. What are the two basic types of "safety lenses?" Which is more durable?

9.7. What is the argument against requiring safety glasses and hardhats throughout all areas of the plant, including areas where they are not needed?

9.8. Name some jobs for which eye protection should be provided.

9.9. Why are "organic vapor respirators" not to be trusted to protect against all organic vapors?

9.10. In an industrial plant a prominent sign reads:

PREVENT ACCIDENTS: WEAR YOUR HARDHAT

What principles of personal protective equipment does this sign call to mind?

9.11. Why is the provision of personal protective equipment not a very satisfactory solution to the problem of protecting workers?

9.12. What are the hazards of employee-owned personal protective equipment being brought onto the premises at work?

9.13. In what way have undersized manholes for vessel entry resulted in a large number of multiple fatalities in the United States?

9.14. Why is there no air-purifying device approved for roofing workers who apply expandable foam?

9.15. Name several alternatives to washing piece parts in trichloroethylene by workers placing their bare hands in the solvent.

10

Fire Protection

Percent of OSHA
general industry citations
addressing this subject

2%

This chapter deals with perhaps the oldest topic in occupational safety and health. But modern developments in the field of fire safety place it in a very dynamic phase. Unlike other categories of safety and health, fire safety presents the Safety and Health Manager with a wide variety of alternatives for dealing with hazards.

The field of industrial fire protection formerly consisted of little more than dealing with fire extinguishers, their selection, placement, marking, inspection, and maintenance. True, there were a few obscure standards about such topics as "standpipe and hose systems," but almost all of the activity centered around the fire extinguishers. Today the field of industrial fire protection is much more sophisticated, recognizing such alternatives as emergency action plans, fire prevention, fire brigades, fire alarm signaling systems, fixed extinguishing systems, and automatic sprinkler systems. Rather than blindly following old specific standards for fire extinguishers, Safety and Health Managers now have the opportunity to explore alternative strategies or combinations of strategies to accomplish the most cost-effective method of fire protection for their own circumstances. Even the federal standards have been changed to delegate such decision-making authority to industrial managers.

It is easy to oversimplify fire protection as meaning only fire extinguishment, but it really encompasses three fields: fire prevention, fire suppression, and personal protection (escape). Much of the criticism of old fire safety standards has been that they emphasized fire suppression and extinguishers, which may not be the safest alternative in some fire situations. Some firms wanted no extinguishers at all and desired to instruct employees to escape without attempting to extinguish fires. Their argument was that fire extinguishers are mainly for property protection, whereas the safest thing for workers to do is escape. The current standards recognize this rationale together with several shades of combination strategies. Before discussing these strategies in detail, some facts about fires should put the fire protection problem in perspective.

INDUSTRIAL FIRES

Fire is the third-ranking cause of accidental death in the United States, according to statistics reported by the National Fire Protection Association (NFPA) (ref. 57). One might think that the high-technology fire detection, protection, and suppression systems used in the United States would make possible one of the lowest fire death rates in the world. Exactly the opposite is true! The United States has one of the worst fire death rates in the industrial world, about 40 deaths per million people.

The US fire death rate is a bad one, but anyone who attempts to pin the responsibility on industry, using NFPA statistics, is in for a disappointment. The vast majority of fire deaths in the United States are in residential occupancies. The relative insignificance of industrial fires is made obvious by Table 10.1. Table 10.1 does not tell the entire story because it reports only "multiple-death fires," but other data corroborate the low percentages of injury, death, and damage that arise from the "industrial" category of fires. It should be acknowledged that some industrial fires are not reported to the NFPA because they are extinguished by automatic systems or by internal fire brigades. But this fact only strengthens the conclusion that industry, more than most other elements of our society, has done a great deal to control fire hazards. Considering the incredible exposure to flammable liquids in refineries and chemical plants and the billions of labor hours spent in industrial plants every year, it is amazing that the number of fire deaths in all industrial plants is no more than the number who die in fires in taverns and prisons, as shown in Table 10.1. In fact, more people died in a single fire in a supper club in Kentucky in 1977[1] than in all the multiple-death industrial fires in the nation in 1977, 1978, and 1979.

TABLE 10.1   Deaths in Multiple-Death Fires in 1979

Occupancy Category	Deaths	Percent of Total
Residential	939	86.6
Stores, service, and repair outlets	15	1.4
Taverns	7	0.6
Industry, mines	18	1.7
Prisons	11	1.0
Vehicles	81	7.5
Miscellaneous	13	1.2
Total	1084	100.0

Source: Compiled statistics published by NFPA (ref. 57).

[1]Beverly Hills Supper Club fire, Southgate, Kentucky, 165 lives lost.

The most dangerous industries from a fire hazard standpoint are mines, grain elevators, grain mills, refineries, and chemical plants. The fire fatalities from these four industries dwarf the total for all remaining industries combined. For general manufacturing industries the number of fire fatalities is extremely low. The last major tragedy in an ordinary manufacturing plant was the Triangle Shirtwaist Company fire which claimed 145 lives in New York in 1911. Perhaps it is the magnitude of this tragedy that has led to superior control of industrial fires and has virtually eliminated the hazard today.

FIRE PREVENTION

The best way to deal with fires is to prevent their occurrence, just as engineering controls were found in Chapter 9 to be preferable to personal protective equipment. Effective fire prevention requires anticipation of fire sources. Each facility is different and requires an individual analysis of potential fire sources. Once the hazards are identified, decisions must be made as to who has responsibility for controlling the hazards. These decisions should be documented in a fire prevention plan.

A principal cause of industrial fires is overheated bearings or hot machinery and processes. Some of these causes can be averted by adopting an effective preventive maintenance program. Such a program, while decreasing the likelihood of fire, may also extend the life of equipment. The Safety and Health Manager may see an opportunity in this strategy to save production costs while furthering the cause of fire safety.

Another ingredient in a fire prevention plan is a strategy for housekeeping. Accumulation of combustible dusts in grain elevators and paint residues in spray painting operations are good examples of how poor housekeeping can contribute to fire hazards. Even ordinary combustible paper and material waste can be a fire hazard.

EMERGENCY EVACUATION

Using the "escape strategy" for dealing with fires or other emergencies, the employer must prepare a written "emergency action plan." The emergency action plan concept has been around for many years for hospitals, schools, and institutions, and in the 1980s this concept was extended to industries in general.

Alarm Systems

Crucial to an emergency action plan is an employee alarm system. But alarm systems are not as simple as they may seem. Searching questions that must be asked are: Will persons recognize the signal as a fire alarm? and What about deaf or blind employees? Audible, visual, and tactile systems must be considered, or perhaps combinations of these systems. In small workplaces, even direct voice communication may be the best fire alarm medium. Public address systems may be used in larger facilities, but the system should provide that emergency messages take priority.

System reliability is important to fire alarms because a failure within the system may not be immediately obvious. Some sophisticated systems have built-in monitor circuits to supervise reliability. Such systems do not need testing as often as do simple alarm systems which have no such monitoring circuits. When repairs are being made, some type of backup system is needed to provide continuous protection. The backup system might even employ "runners" or telephones or other informal systems, but the Safety and Health Manager should document what the backup system is.

Fire Detection Systems

Smoke alarms and other detection devices may be used to trigger the alarm system. However, it should be noted that automatic smoke alarms are not mandatory in American industries in general. Even manual or visual systems can be considered alarm systems.

If automatic detection systems are employed, care must be taken to maintain and protect the equipment. Most detection systems are delicate instruments and will not withstand the rigors of the industrial environment. Conditions to be considered are dust, corrosive atmospheres, weather exposure, heat from processes, and mechanical damage.

People are sometimes reluctant to sound fire alarms, with tragic consequences. Hotel managers are particularly unwilling to alarm occupants. The typical response from a hotel front desk when an occupant calls in a fire alarm is to send a bellman to investigate. It is rational to consider the hazards of panic when a fire alarm is sounded. However, this rationale is sometimes permitted to be an excuse for failing to take action.

FIRE BRIGADES

Some firms may elect a strategy in which employees are organized into brigades to fight fires themselves. Such strategies should be carefully scrutinized because in the scramble to protect property, these fire brigades can be a danger to employees.

Employee Fitness

Volunteering to join the fire brigade is not sufficient to qualify the worker to fight fires. Conditions that may be hazardous include heart disease, epilepsy, or emphysema. Other conditions, such as ruptured eardrums or the wearing of a beard, may make the use of respiratory equipment ill advised. The Safety and Health Manager should be sure that the fire brigade volunteers are screened, and a physician's certificate may be necessary for questionable cases. Volunteers unfit for interior structural firefighting may be used in other tasks.

Firefighter Training

Many states have fire training academies, and Safety and Health Managers should find out what schools and academies are available for their fire brigade members.

Interior structural firefighting is more demanding, and fire brigade members assigned to such tasks should be trained at least quarterly. Other fire brigade members should be trained at least annually. Also, firefighting equipment to be used by the fire brigade should be inspected annually, and fire extinguishers should be inspected monthly.

Protective Clothing and Apparatus

If the firm elects to have fire brigade members fight interior structural fires, protective clothing and respirators must be provided. This includes protective boots or shoes, fire-resistant coats, gloves, and head, eye, and face protection.

One concept stressed for self-contained breathing apparatus is the mode for airflow into the mask. At this point the reader may want to review the three modes of respirator airflow discussed in Chapter 9. The preferred mode for firefighting is one of the positive pressure types: pressure demand or continuous flow. The only valid argument for using simple demand flow is that this mode permits longer duration exposure for a given charge. If the employer believes that the demand-flow mode is essential, quantitative fit testing is necessary for each firefighter.

FIRE EXTINGUISHERS

Fire extinguishers are still the most effective method of immediately controlling a very local fire before disastrous consequences ensue. The Safety and Health Manager needs to understand the various fire classes and the type of extinguishers appropriate for each.

Fire Classes

The fire protection field classifies fires into four categories. Application of the wrong extinguishment media to a fire can do more harm than good.

Table 10.2 describes the four classes of fires, sample appropriate extinguisher media, and the maximum travel distance specified for extinguishers for each type. Liquefied petroleum gas (LPG) fires, although technically Class B, are really not adequately addressed by any of the four classifications. Such fires are extremely dangerous, and fire extinguishers are not appropriate for their control. LPG fires should be extinguished by professional firefighters using powerful water spray systems.

The key to determining whether an extinguisher is appropriate for a given class of fire hazard is to check the approval marking on the extinguisher itself. Some types of extinguishers have been found to be hazardous and are forbidden regardless of prior approval markings. These types are listed in Table 10.3. Some extinguishers may be approved for more than one classification of fire. These multipurpose fire extinguishers typically employ a dry chemical medium. Although dry chemical extinguishers are growing in popularity, they are not a panacea. Expensive equipment such as computers can be fouled or even ruined by application of a dry chemical extinguisher when a CO_2 extinguisher would have been satisfactory. Foam or water extinguishers may also be cheaper for the more ubiquitous Class A fires.

TABLE 10.2 Four Classes of Fires and Appropriate Extinguishing Media

Fire Class	Description	Example Extinguishing Media	Maximum OSHA-Authorized Travel Distance to Nearest Extinguisher
A	Paper, wood, cloth, and some rubber and plastic materials	Foam, loaded stream, dry chemical, water	75 feet
B	Flammable or combustible liquids, flammable gases, greases, and similar materials, and some rubber and plastic materials	Bromotrifluoromethane, carbon dioxide, dry chemical, foam, loaded stream	50 feet
C	Energized electrical equipment	Bromotrifluoromethane, carbon dioxide, dry chemical	No specific maximum; distribute "on the basis of the appropriate pattern for existing Class A or B hazards"
D	Combustible metals such as magnesium, titanium, zirconium, sodium, lithium, and potassium	Special powders, sand	75 feet

TABLE 10.3 Forbidden Fire Extinguishers

1. Carbon tetrachloride
2. Chlorobromomethane
3. Soldered or riveted shell self-generating soda acid or self-generating foam or gas cartridge water-type portable fire extinguishers which are operated by inverting the extinguisher to rupture the cartridge or to initate an uncontrollable pressure-generating chemical reaction to expel the agent.

Source: Code of Federal Regulations 29CFR 1910.157

Inspection, Testing, and Mounting

In the 1970s OSHA found thousands of violations in which fire extinguishers did not have appropriate tags indicating their inspection status. OSHA has since abandoned the rule requiring such tags but still requires the employer to maintain records of annual maintenance for each fire extinguisher for up to one year after the last entry or the life of the shell, whichever is less. In addition, a visual inspection is required monthly. Many employers have found it expedient to retain the tag system even though it is no longer required. Thus when anyone requests to see the inspection record for a given fire extinguisher, it is immediately available on the tag attached to the extinguisher.

In addition to the monthly and annual inspections, fire extinguishers must receive a hydrostatic test according to a prescribed test schedule for various types of fire extinguishers. Fire extinguisher shells deteriorate from mechanical damage or corrosion and may be unsafe for containing pressures inside. The hydrostatic test places the extinguisher under a test pressure to determine whether it can safely contain the pressures to which it will be subjected in use. The test has technical specifications and must be done by a trained person using suitable equipment and facilities. The Safety and Health Manager invariably leaves the hydrostatic tests for the professional fire extinguisher service to perform.

Another big OSHA enforcement issue of the 1970s was the mounting height and identification of mounting of fire extinguishers. OSHA citations for mounting, identification, and inspection tagging of fire extinguishers amounted to more than half of all OSHA violations named in fire protection citations throughout the 1970s. These requirements have all been eliminated or drastically changed by rewording them in performance language. The current standard permits the employer the latitude of mounting extinguishers high on the wall, out of the way of forklift truck traffic, and accessed by rope and pulley. The employer may select any convenient mounting scheme provided that the extinguishers are readily accessible without subjecting employees to possible injury.

Training and Education

A facility looks well equipped and protected when fire extinguishers are placed about the workplace, readily accessible for use in an emergency. But the appalling reality is that few employees know how to use a fire extinguisher effectively, and some would even be afraid to use the extinguishers if they knew how. This is especially true of fire extinguishers or hose systems located behind glass doors. For most people there is a great reluctance to break glass even in an emergency. A study conducted among nursing staff in an Ohio hospital revealed that most knew nothing about and were afraid to use fire extinguishers. Accordingly, the Safety Manager in charge set out to train the nurses to use fire extinguishers in an emergency. A hospital bed was taken outdoors and set afire to provide a simulated fire emergency for the nurses to extinguish. The field of industrial safety has now recognized the need for fire extinguisher training, and such training has been accepted as standard for general industry. Training is required upon initial employment and at least annually thereafter.

STANDPIPE AND HOSE SYSTEMS

Some employers choose to install standpipe and hose systems for firefighting, and these systems can be employed in lieu of general distribution of fire extinguishers in most cases. Standpipe systems come in various ratings or classes and are summarized in Table 10.4.

TABLE 10.4 Standpipe and Hose System Classes

Class	Hose size	Description
I	2½ inch	For professional fire departments and persons trained to handle heavy systems
II	1½ inch	For incipient-stage fires
III	1½ and 2½ inch	For trained employee use on all levels of fires, including the more advanced stages
	"Small ⅜-inch to 1½-inch hose"	For incipient-stage fires

Equipment

Some standpipe and hose systems can be quite ancient. These old systems can usually be retained as long as they are serviceable and meet annual test requirements. But when replacements are made, new equipment should conform to all current standards. Examples of modern system changes are as follows:

1. *Shut-off*-type nozzles
2. *Lining* for hose
3. Dynamic pressure *minimums at the nozzle*
4. Hydrostatic *testing* upon installation

The water supply for standpipe and hose systems can be provided either by elevated water supply tanks or by pressure tanks. The supply must be sufficient to provide 100 gallons per minute flow for at least 30 minutes. This calculates to 3,000 gallons per use period, but remember that it takes considerably more than 3,000 gallons to provide the head pressure to maintain the 100-gallon-per-minute flow throughout the 30-minute period. One seemingly clever idea is to forget about standpipes or pressure tanks and connect hose systems to city water supplies. But the flaw in the theory is the adequacy (pressure and flow) of the supply, which usually cannot meet the 100-gallon-per-minute flow requirement.

Maintenance

The ancient systems described earlier can create some unpleasant surprises when the hose is deployed for use. Hemp and linen systems are especially subject to deterioration. After hanging on the rack for many years without use, the hose may break apart or disintegrate when taken off the rack for use in an emergency. Hose systems should be checked annually and after each use. In the case of hemp or linen hose systems, the hose must be reracked using a different fold pattern.

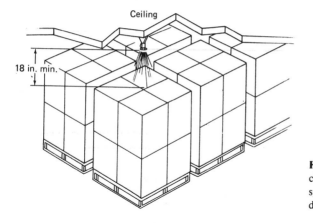

Ceiling

18 in. min.

Figure 10.1. Minimum vertical clearance between stacked material and sprinkler head is 18 inches to permit distribution of spray.

AUTOMATIC SPRINKLER SYSTEMS

Automatic sprinkler systems present a paradox since they affect employee safety but are usually installed principally to protect property and to lower insurance rates. If such a system is *voluntarily* installed by the employer to protect property, should it be required to meet personal safety standards? And if an existing system does not meet current standards, should it be dismantled and withdrawn from use? This would hardly be in the interest of safety.

Sometimes a good sprinkler system is installed only to be made ineffective by incorrect usage of the space protected. One error is to allow the sprinkler heads to become fouled by materials such as paint spray residues. If a spray area is protected by an automatic sprinkler system, a good way to protect the sprinkler heads is to cover them with paper bags. If a fire does occur, the paper bags will either burn away or be washed away by the water spray so that they do not interfere with the fire suppression action of the sprinklers.

Another error with automatic sprinklers is to stack material too close to the ceiling. This interferes with the distribution of spray from the sprinkler head. At least an 18-inch clearance must be used, as shown in Figure 10.1, to permit the sprinkler spray to be well distributed.

FIXED EXTINGUISHING SYSTEMS

Technically speaking, automatic sprinkler systems are fixed extinguishing systems, but really what is usually meant by a fixed extinguishing system is a more local system for controlling special fire hazards such as kitchen grills or tempering tanks. Again, the principal objective may be property protection and insurance rate reduction, but steps must be taken to prevent the system, which may discharge dangerous gases or other agents, from becoming a hazard to employees. Thus if the discharge of the system is not obvious, it is necessary to warn employees, perhaps with a discharge alarm, that dangerous agents are being expelled into the atmosphere. If a strategy of "total flooding" is used with a dangerous agent, an emergency action plan is necessary to assure that personnel escape. Some agents are so dangerous as to be prohibited altogether as extinguishing media; examples are chlorobromomethane and carbon tetrachloride.

Fixed extinguishing systems are somewhat like giant, fixed, automatic fire extinguishers. Many of the maintenance procedures appropriate for portable fire extinguishers are also appropriate for fixed extinguishing systems. Like portable extinguishers, fixed systems are required to be inspected annually. Corresponding to the monthly visual inspections of portable extinguishers is a somewhat more comprehensive semiannual inspection for fixed systems to determine whether containers are charged and ready for operation. If the containers are factory charged and have no gauges or indicators, they must be weighed to determine charge. A weight decrease of 5% or a pressure decrease of 10% is considered within tolerance. Discussion of specific systems follows with requirements for each.

Dry Chemical Systems

The word *chemical* in dry chemical systems should be remembered, and the question to be asked is whether the extinguishing chemical will produce any undesirable reaction with process reagents or perhaps foams and wetting agents also employed. There is more than one type of dry chemical available, and generally these chemicals are not to be mixed when filling the cylinders or containers. Mixing is permitted if the chemical to be added is "compatible" with the chemical stated on the approval nameplate of the system.

Dry chemicals used for extinguishing agents are usually not dangerous to the health or safety of personnel. However, the actual distribution of the chemical powder during an emergency may obscure vision, preventing escape. A possibility such as this calls for a predischarge employee alarm system as described earlier.

The biggest problem with dry chemical systems is the caking or lumping of the agent. Humid climates or moisture-producing processes subject the system to a greater risk of caking. Caking can render a dry chemical useless, so the chemical should be sampled annually to be certain moisture is not causing caking.

Other Fixed Systems Agents

Many fixed extinguishing systems employ carbon dioxide, Halon 1211, or Halon 1301 gases. These systems have the advantage of not requiring as much cleanup after the emergency as other systems, but all three of these gases can be dangerous to unsuspecting employees, especially if a total flooding strategy is employed. Standards must be followed in planning predischarge warning systems, employee egress routes, and maximum concentrations of gases to be released.

Water spray and foam agents are less dangerous to employees, but necessary volumes required to be effective may introduce egress hazards. Drainage must be directed away from work areas and so as not to obstruct egress paths.

SUMMARY

The many strategies for dealing with industrial fire hazards can be grouped under the general categories of prevention, suppression, and escape, or combinations of these categories. Current applicable industrial standards encompass all of these strategies.

It is necessary to keep the hazards of industrial fires in perspective. Industrial fires cause very little loss of life and injury now, compared with fire deaths elsewhere and occupational fatalities and injuries from other causes. In light of this perspective it may seem that emphasis on equipment specifications, regular inspection of extinguishers, personnel training, and written plans is somewhat misplaced. However, the excellent record of industry in controlling fire hazards on the job should not be allowed to induce complacency now that success has been achieved. There is no doubt that the adherence to strict fire codes is what has helped industry achieve such a superior control of fire hazards compared to residential and other fire exposures.

The beginning of this chapter identified fire protection as the oldest of topics in occupational safety and health. There is perhaps a correlation between the facts that industrial fire protection is an old endeavor, and it is also a very successful endeavor. Perhaps with time even the newer fields of occupational safety and health covered in other chapters of this book will achieve the same high degree of safety and health as has already been achieved for fire protection.

EXERCISES AND STUDY QUESTIONS

10.1. Should fire extinguishers be required in all industrial plants? Why or why not?

10.2. What are the three major fields of fire safety?

10.3. What argument does industry offer against providing fire extinguishers?

10.4. How does fire rank among causes of accidental death in the United States?

10.5. How does the United States rank among nations of the world in numbers of fire deaths per million people?

10.6. What is the "occupancy category" for most fire deaths in the United States?

10.7. What percentage of the total number of fire deaths from multiple-death fires is attributed to industry?

10.8. Name some items to be included in fire prevention plans.

10.9. How is preventive maintenance related to fire hazards?

10.10. How can automatic audible alarm systems fail to warn employees of a fire emergency?

10.11. Are automatic smoke alarms required in industrial plants?

10.12. Why does an employee sometimes notice a fire in the plant but fail to sound the alarm?

10.13. Are industrial plants required to have fire brigades?

10.14. How often must fire brigade members be trained?

10.15. Name some conditions that would make an employee unfit to serve in a fire brigade.

10.16. Identify the four classes of fires. Give example extinguishing media for each.

10.17. What extinguishment medium is used for LPG fires?

10.18. Name some forbidden fire extinguishers.

10.19. What are the advantages and disadvantages of dry chemical as a fire extinguishment medium?

10.20. How often must fire extinguishers be inspected?

10.21. Are fire extinguishers required to be mounted at a certain distance from the floor?

10.22. How often are employees required to be trained to operate fire extinguishers?

10.23. Name some modern requirements for standpipe and hose systems.

10.24. Why are city water supplies usually unacceptable for directly supplying hose systems for fire protection?

10.25. Are industries required to have automatic sprinkler systems?

10.26. Why are paper bags placed over sprinkler spray heads?

10.27. How close to the automatic sprinkler head in a warehouse is it permissible to stack material?

10.28. If a fixed extinguishing system has no gauges or indicators, how can its charge condition be determined?

10.29. Name some gases employed in fixed extinguishing systems.

10.30. Explain why ruptured eardrums would represent a hazard to firefighters.

10.31. Name three acceptable media for use in fire alarm systems.

10.32. What are the two principal ingredients of a fire prevention plan?

11

Materials Handling and Storage

Percent of OSHA
general industry citations
addressing this subject

6%

The usual concept of a factory is a place where things are made or materials are processed, but often the major activity in a factory consists of moving things and materials around. Lifting, a most basic material-handling activity, accounts for most back injuries, the largest workplace injury category of all. Industrial trucks, tractors, cranes, and conveyors all have the simple mission of moving materials, and they all cause injuries and fatalities every year.

The National Safety Council (NSC) charges materials handling with 20% to 25% of all occupational injuries. The size of the problem is emphasized in NSC's *Accident Prevention Manual for Industrial Operations* as follows:

> As an average, industry moves about 50 tons of material for each ton of product produced. Some industries move 180 tons for each ton of product.

Before addressing specific hazards, it would be wise to examine the general nature of hazards created by materials handling.

In materials handling, masses are usually measured in tons or pallet loads instead of ounces, pounds, or kilograms. The human body is light and frail by comparison, so bulk materials' masses can easily pinch, fracture, sever or crush. Contributing to the hazards of large masses is the reality that materials handling includes *motion* for these masses.

To illustrate the general "mass/motion" hazards of material-handling equipment, consider the following comparison of processing versus material-handling equipment. To be struck by a moving part of a processing machine may or may not cause injury, depending on the size of the machine, the motion of the moving part, and the shape or surface characteristics of the part. But being struck by an industrial truck or conveyor is almost certain to cause injury. More indirectly, the mass/motion hazards of materials handling can affect safety by impacting facilities such as gas lines or electric lines or by overloading structural components of buildings.

Another general hazard of materials handling is its automatic or remote-control nature. Materials pumps and conveyors are often started automatically on demand or from a manual switch that is located far away. Conveyor accidents are often caused by this remoteness characteristic. In another example, railroad cars often move about a yard, far from the eye of the engineer, or worse yet, they may roll almost silently, coasting independently into position from the momentum of a locomotive's momentary push, with no local control at all.

An indirect general hazard from materials handling is fire. This hazard emphasizes the storage aspect of materials handling. Warehouse fires are costly in terms of property loss, but they can also be dangerous to workers.

MATERIALS STORAGE

Materials-handling standards say that bags, containers, or bundles stored in tiers shall be "stacked, blocked, interlocked, and limited in height so that they are stable and secure against sliding or collapse." For general materials the standard does not explain what constitutes "blocking" or "interlocking" and does not specify height limits for stacks. Some industry practices, however, seem to defy the standard. Since the standard for materials storage is not specific in its requirements but does address results (e.g., "so that they are stable and secure against sliding or collapse"), the standard should be recognized as a performance standard.

Housekeeping is another consideration for materials in storage. Sloppy warehouse practices can lead to trip hazards or fire. Pest harborage can result from certain accumulations of materials and can constitute a hazard. Outside storage can become grown over in weeds and grass, which can be a fire hazard in dry weather.

The Safety and Health Manager should monitor the ups and downs of the company's sales and production fortunes. If there is a recent buildup in production in anticipation or in response to booming sales, there is likely to be a warehousing problem. Additional warehouse space is likely to be added later, after the upturn appears more permanent. Even after an expansion decision is made, a certain amount of lead time is necessary to plan and build add-on warehouse space. Meanwhile, the existing warehouse space becomes crowded and unsafe. Judgmental limits for material stacking are cast aside as creative ideas are explored to squeeze more material into a small space. Some of these new ideas will be good ones; it is not basically unsafe to try to conserve space. But during a period of warehouse crowding, special attention should be given to new hazards which may result from new procedures in the warehouse. Aisle and exit blockage are likely to occur during such periods of warehouse stress, as is the incidence of storage stack collapse. Any accidents that occur during such periods will point to some of the new hazards being generated, and these accidents should be analyzed to eliminate cause.

INDUSTRIAL TRUCKS

This category of material-handling equipment is typified by the forklift truck. Forklifts are typically either electric motor powered or have internal combustion engines. Besides forklifts there are tractors, platform lift trucks, motorized hand trucks, and other specialized industrial trucks. Not included are trucks powered by means other than electric motors or internal combustion engines. Also not included are farm tractors or vehicles primarily for earth moving or over-the-road hauling.

Truck Selection

"Ignorance is bliss" for most industrial managers who set out to buy a forklift truck. Little known is the fact that there are eleven different design classifications by type of power and by degree of hazard for which approved. These eleven design classifications are distributed over at least sixteen different classifications of hazardous locations to which the forklift truck may be exposed.

One can easily wonder why it must be so complicated to select a forklift truck. The basis for this complication is that engines and motors can be dangerous sources of ignition for flammable vapors, dusts, and fibers. To design these engines and motors to effectively prevent ignition hazards is a costly matter, and the marketplace will not support the purchase of an explosion-safe forklift truck for use in an ordinary factory location. Thus a complicated array of classifications and regulations is set up to permit the specification of the right industrial truck for the right job—no more and no less.

Industry standards for industrial truck classifications and their corresponding hazardous location codes are a maze of abbreviations and definitions. To understand these abbreviations, keep in mind that the objective is safety from *fires and explosions*. Whether the industrial truck is diesel, gasoline, electric, or LP gas powered, the more fire-safe models are designed to prevent ignition of accidental fires, and thus they are more expensive. A simple summary is contained in Table 11.1. More details are contained in the standards, but most Safety and Health Managers will need only the general idea. Another thing to keep in mind is that it is legal to use a safer, higher-classified industrial truck than the minimum required, but of course it usually will not be economical to do so. If a firm already has an EE-approved electric forklift, however, it may be expedient to go ahead and use it instead of buying a new ES- or E-approved unit which would suffice for the given application.

Summarizing these principles and several pages of applicable regulations, Table 11.2 gives the Safety and Health Manager a perspective into the approval classes of various industrial truck designs. The "classes," "groups," and "divisions" represent varieties of hazardous locations in which the definitions correspond *roughly* to those of the *National Electrical Code®* and are covered in more detail in Chapter 14. *Class* and *group* refer to the type of hazardous material present, and *division* refers to the extent or degree to which the hazardous material is likely to be present in dangerous quantities.

There are so many categories of "approvals" for industrial trucks that it is easy to lose sight of the overall objective in the approval process: to prevent fires and

TABLE 11.1   Summary of Industrial Truck Design Classifications

Diesel	Electric	Gasoline	LP Gas

D
  Standard model
  Cheapest
	E
	  Standard model
	  Cheapest
		G
		  Standard model
		  Cheapest
			LP
			  Standard model
			  Cheapest

DS
  Safer model
  More expensive
  Exhaust, fuel,
    electrical
    systems safeguards
	ES
	  Safer model
	  More expensive
	  Spark prevention
	  Surface-temperature
	    limitation
		GS
		  Safer model
		  More expensive
		  Exhaust, fuel, and
		    electrical systems
		    safeguards
			LPS
			  Safer model
			  More expensive
			  Exhaust, fuel,
			    and electrical
			    systems safeguards

DY
  Safest diesel
  Most expensive
    diesel
  No electrical
    equipment
  Temperature-
    limitation
    feature
	EE
	  Safer still
	  More expensive
	  All motors and
	    electrical enclosed
	EX
	  Safest of all
	  Most expensive
	    electrical truck
	  Even electrical
	    fittings designed
	    for hazardous
	    atmospheres		

Source: Summarized from Code of Federal Regulations 29 CFR 1910.178.

TABLE 11.2   Allowable Categories for Industrial Trucks for Various Hazardous Locations

Class	Group	Division 1	Division 2
I	A	No industrial truck permitted	DY, EE, EX
	B	No industrial truck permitted	DY, EE, EX
	C	No industrial truck permitted	DY, EE, EX
	D	EX	DS, DY, ES, EE, EX, GS, LPS
II	E	EX	EX
	F	EX	EX
	G	EX	DY, EE, EX
III		DY, EE, EX	DS, DY, E[a], ES, EE, EX, GS, LPS

[a]Permitted to continue if previously in use

Source: Summarized from Code of Federal Regulations 29 CFR 1910.178.

explosions from improper use of the wrong truck in a hazardous atmosphere. The authority for approval of industrial trucks is delegated to nationally recognized testing laboratories, such as Underwriters' Laboratories, Inc., and Factory Mutual Engineering Corporation. The prudent Safety and Health Manager will leave the application-for-approval process to the manufacturer of the equipment and simply look for the UL or FM approval classification, such as DY, EX, GS, and so on. For nonhazardous locations, even an *unapproved* truck may be used if the truck *conforms* to the requirements of Type D, E, G, or LP. This meets the spirit if not the letter of the approvals process.

In these days of high energy costs and the search for alternatives, some company managements may desire to convert an industrial truck from one energy source to another. But tampering with the design or altering the truck may invalidate the approval. Industrial truck conversions can be made, but the process is a bit tricky. The enlightened Safety and Health Manager will be alert to warn the company management not to waltz blithely into a lift truck conversion scheme without doing it properly. The conversion equipment *itself* must be approved, and there is a right way and many wrong ways to carry out the conversion. The "easy way out" is often to buy a new lift truck, a characteristic of the standards more appreciated by lift truck dealers than by lift truck users.

The hazard of fires and explosions may be the most complicated factor in the selection of forklift trucks, but it certainly is not the only factor. Far more important than the fire hazard rating of the design of the truck are the operations, fueling, guarding, training of drivers, and maintenance—subjects discussed in the next section.

Operations

One of the first items of interest to the Safety and Health Manager should be the refueling or recharging area of forklifts. Smoking is prohibited in these areas, and this fault is found more than any other. Also a problem is the charging of forklifts in a "nondesignated area." Hazards in the area include spilled battery acid, fires, lifting of heavy batteries, damage to the equipment by the forklifts, and battery gases or fumes. All of these hazards need to be addressed by the Safety and Health Manager in some way or another.

An old chemistry laboratory adage says:

> Here he lies all calm and placid;
> He poured water in his acid!

Another says:

> Do what you oughter:
> Pour acid into water!

Federal standards require the foregoing, but the language of the standards is not so colorful. Perhaps a sign in the area would achieve compliance with the rule. A better way to promote safety, though, would be to include in an employee training program an explanation of the violent and exothermic reaction that occurs when water is poured into a concentrated strong acid.

Everyone knows that arcs and sparks frequently fly when battery connections are made. What most do not know is that gases liberated during charging processes can reach ignitable concentrations. Fire is little or no hazard when a battery is simply "jumped" to another. But a forklift charging area is a different matter: Large volumes of gases are liberated. Adequate ventilation to remove gases is essential.

With hazards of gases and acids in battery-charging areas, the alternative of the internal combustion engine seems attractive. But each of the internal combustion engine choices—diesel, gasoline, and LP gas—emit another dangerous gas: carbon monoxide. Since forklifts usually operate indoors, carbon monoxide gas levels can be a problem. The 8-hour time-weighted average exposure limit for carbon monoxide is 50 ppm.

If the Safety and Health Manager determines that a carbon monoxide problem exists in the plant and that the forklift trucks are the culprits, several alternatives are possible. The obvious one is to switch to electric forklifts. Another solution might be to alter the building or to install adequate ventilation systems. Perhaps the cheapest solution of all would be to review procedures and operations to determine whether sources of emissions can be reduced or perhaps eliminated entirely. The following are key questions:

1. Are operators leaving engines running unnecessarily?
2. Can the layout of warehouses or plant facilities be revised to reduce concentrations?
3. Are faulty or worn-out lift trucks creating more emissions than necessary?

Although there are no universal general minimum lighting requirements for industrial plants, where industrial trucks are operated, safety demands that the trucks themselves have directional lights if the plant area is too dark. Truck lights are required if the general lighting is less than 2 lumens per square foot. This is really a quite low level of light since an ordinary 100-watt incandescent bulb can produce 1,700 lumens. Even in an all-black room with nonreflective walls, a 100-watt bulb could produce more than 2 lumens per square foot in an 8 foot by 12 foot by 16 foot room. Wall or other surface reflections help the overall situation, so the requirement for 2 lumens per square foot is not difficult to meet. A lighting consultant can help in this determination.

One of the greatest hazards with forklifts and other industrial trucks is the transition between dock and cargo vehicle. Figure 11.1 shows precautions to be taken. Although a highway truck is shown in the figure, the hazard also exists for loading railroad cars.

Many workers feel that because they know how to drive an automobile, they also basically know how to operate a lift truck. Unfortunately, many employers are inclined to take their word for it. But the operation of a lift truck takes a great deal

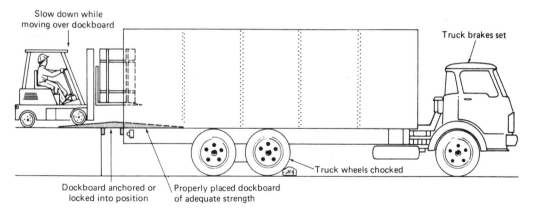

Slow down while
moving over dockboard

Truck brakes set

Truck wheels chocked

Dockboard anchored or
locked into position

Properly placed dockboard
of adequate strength

Figure 11.1. Preventing forklift hazards: transition from dock to cargo vehicle.

more skill than the operation of an automobile. Compared to an automobile, a lift truck has a much shorter wheelbase, and when the load is lifted the center of gravity is very high. This creates stability problems to which the operator may be unaccustomed. Compounding the stability problem are the small-diameter wheels found on lift trucks, making chuckholes and obstructions more hazardous. When loaded, the center of gravity of the lift truck and load together can be shifted dangerously forward. Picking up and depositing loads requires skill in proper manipulation and safe positioning. An off-center load presents a special hazard in which the load might tip in transit even though the truck itself is in a stable position.

Besides stability problems, visibility can be a problem with lift trucks. The load itself can block the view and necessitate driving with the load trailing. Driving in worker aisles presents problems of pedestrian traffic, especially at corners where visibility is limited. Although lift trucks are not silent, they may seem so in a noisy factory environment. This increases the hazard to pedestrians and the need for greater visibility for lift truck operators.

Passengers on a lift truck can be a hazard in more than one way. For one thing, the truck is often equipped to seat only the driver, and there may be no safe place for a passenger to ride. Hitchhikers also can distract the driver, whose attention is even more important on a lift truck than in an automobile. One practice particularly frowned upon is riding on the lift *forks*. The temptation is great to use a forklift truck as a personnel elevator. Actually this practice can be safe *if*

1. A safety platform is firmly secured to the lifting fork.
2. The person on the platform can shut off power to the truck.
3. Protection from falling objects is provided if needed.

An ordinary wood pallet is not generally considered to be a "safety platform," although it is often used for lifting personnel.

Some readers may consider unreasonable the rule that the person on the platform be able to shut off power to the truck. This rule also stands in the way of use of an ordinary pallet as a lifting platform. Workers, too, may resist this rule, and the Safety and Health Manager needs to be in a position to counter this resistance with training programs that effectively explain the reasons for the rule. As a rationale for the rule, ask the workers to consider the hazard caused by unexpected obstructions. A small obstruction could

1. Damage the platform
2. Tip the platform, causing the worker to lose balance
3. Injure the worker on the platform
4. Knock the worker off balance

The forklift driver is really in a poor position, due to distance or angle, to see all obstructions and to judge their distance from the lifted platform. One might argue that the lift may be entirely clear of any obstructions at all, but such lifts would be unusual. There is usually no reason for employees to be lifted unless the lift is adjacent to equipment, stacks of material, or building structure. Any of these items can present obstruction hazards.

A variation of the "forklift rider" is the "carpet pole rider." In carpet warehouses the lift truck is equipped with a single pole, which is guided into the spool of a roll of carpet to lift the roll and transport it around the plant. Workers have been known to ride these poles rodeo style to gain access to the top of a stack of carpet rolls. Normally, there should be no reason for a worker to ride the pole because the pole can be raised and guided into position by the driver without assistance.

When manipulating loads, the forklift is often operating close to observers, supervisors, or assistants giving directions to the operator. A dangerous place to stand is beneath the elevated fork, whether it is loaded or not. Another dangerous position is between an approaching lift truck and a fixed object or bench.

The foregoing examination of operational hazards should make it clear that a measure of special training for industrial truck operators is appropriate. Equipment dealer representatives can be of assistance, and most equipment manufacturers have prepared training programs for purchasers and operators of their equipment. The Safety and Health Manager should reflect on the following question: If an inspector should visit my facility and flag down a forklift driver and ask the driver to describe the training program for forklift operators, what would the driver say? Too often the response has been disappointing, perhaps even embarrassing.

One thing to check is parked, unattended forklift trucks. The first thing to do is determine whether the lift truck is really unattended or not. If the operator cannot see the truck, it should be considered unattended. Even if the operator can see the truck, if the truck is more than 25 feet away, the truck is unattended. If the truck is unattended, the motor should be turned off. Even if the operator *is* close by, if the operator is *dismounted,* the fork must be fully lowered and the controls neutralized. The next thing to check is whether the brake is set, and if the truck is on an incline, whether the wheels are blocked. Finally, the horn should be tested.

Federal standards take very seriously the question of maintenance, inspection, and service for industrial trucks. No latitude is permitted to allow defective industrial trucks to continue to be operated until the next regular overhaul. Any condition, such as an inoperative horn, defective brake, or broken headlight in a forklift is cause to remove it from use until repaired.

Most people are amazed to learn that federal standards require industrial trucks in use to be inspected for safety *daily*. Compare this rule with procedures for automobile safety inspections, which most states require *annually*. If the industrial truck is used on a round-the-clock basis, safety inspections are required after every *shift*. It would be wise for the Safety and Health Manager to institute some type of procedure or record to ensure that this job gets done and that proof of performance will be kept on file.

A final note on the subject of industrial trucks is the provision of an overhead guard to protect the operator from objects falling from the elevated load. More and more forklifts are being equipped with these overhead guards required for protection against falling objects such as small packages, boxes, and bagged material, not for protection against the impact of a full capacity load. These overhead guards are not to be confused with the much sturdier rollover protective structures (ROPS) described in Chapter 15, although the same structure can be used for both purposes if it is constructed properly. Some loads are unitized, and objects within the load are secured from falling back on the operator. For such applications the hazard to the operator from falling objects disappears, and the overhead guard becomes unnecessary.

CRANES

An industrial truck is convenient for general material handling of palletized loads. But some material-handling jobs cannot be handled by an industrial truck. Larger, heavier, more awkward loads require the versatility of a crane, especially if the path of travel is complicated.

A crane is a construction industry device that is typically seen lifting heavy steel beams to lofty places. Although this image is accurate, it is incomplete. Cranes are also used extensively in general industry, although they usually take a different form. Cranes in industrial plants are generally limited in travel by a track or overhead runway structure, typified by the *overhead traveling crane* shown in Figure 11.2. Such cranes are popularly called *overhead bridge cranes* or simply *bridge cranes* by the workers who use them. Some models, such as those in Figure 11.2, are operated from a cab mounted on the crane itself. Others are operated from the floor by means of a hanging cord control called a *pendant,* or from a fixed remote station called a *pulpit*. *Gantry* cranes have legs that support the bridge above the railway. *Cantilever* gantry cranes have extensions on one or both ends of the bridge; these extensions extend the reach of the crane outside the area between rails on which the crane travels. One characteristic common to all overhead and gantry cranes is that the trolley, which

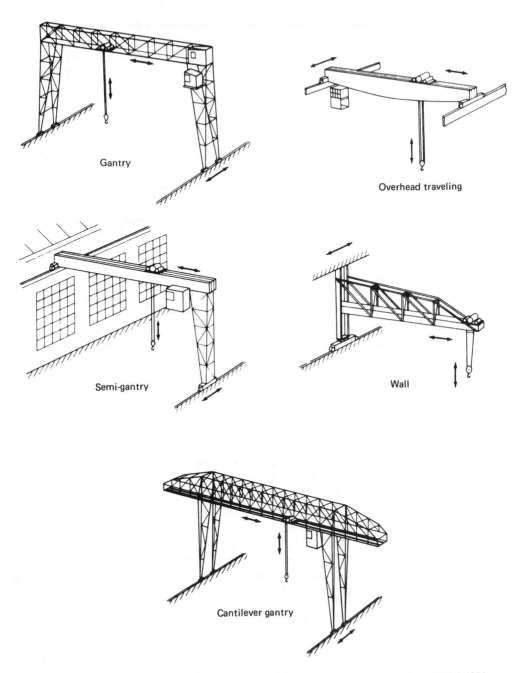

Figure 11.2. Various configurations for overhead cranes. (ANSI standard B30.2.0-1976, reprinted with permission of ANSI and ASME.)

carries the hoisting mechanism, rides *on top of* the rail on which it travels. Overhead cranes whose trolleys are not so mounted are called *underhung cranes* or *monorails,* depending on type or application. Safety standards for overhead and gantry cranes are different from standards for monorails.

The Safety and Health Manager's chief concern with regard to overhead cranes should be that workers will overload the crane. The rated load should be plainly marked on each side of the crane, and if the crane has more than one hoisting unit, each hoist must have its rated load clearly marked. So easily recognized are crane load markings that their absence is conspicuous.

Even if the rated load marking *is* on the crane, workers will sometimes be tempted to exceed the rated capacity. Everyone knows that engineers provide for a safety factor in their designs, so virtually every crane will support more than its rated load, even without damage to the crane. But uncertainties in the actual weight of the load, dynamic loads during transport, shock loads during lifting, variations in crane components, and unavoidable design variabilities can combine to result in a very dangerous situation even when rated load capacities are exceeded "slightly" or only "occasionally." As with running a stop sign, an alert driver can almost always avoid an accident, but once in a great while another driver will be approaching the intersection at a high rate of speed or the visibility will be obstructed, or the stop sign violator will be complacent or distracted, and a serious accident will occur. It becomes very difficult for the Safety and Health Manager or anyone else in the plant to police the use of cranes properly within their rated load limits at all times. This is where training becomes important so that workers will understand the risks and consequences of their actions. Recalling the principles set forth in Chapters 2 and 3, another way to control this problem is to take the trouble to make a big example out of any accident or near accident which occurs as a result of overloading a crane, even if no injuries occur.

Many overhead cranes operate outdoors, where wind can be a hazard. Wind alone is usually not dangerous, but a wind load in combination with a working load can result in dangerous structural damage to the crane. Automatic rail clamps are required for outdoor storage bridge cranes. The idea with these clamps is to lock the bridge to the rail if the wind exceeds a certain velocity. This idea sounds good because it protects against wind hazards just described and also protects the bridge from unintentional and uncontrolled rolling in a high wind or brake failure. But like some other devices, the safety device carries with it a hazard of its own. Think about it for a moment: In what mode of crane operation would the tripping device most likely engage the automatic rail clamp? The answer is when the bridge is traveling *at top speed* against the wind. The sudden engagement of the rail clamp when the bridge is traveling at top speed is likely to injure the operator in the cab, dangerously swing the load, damage the crane, or all three. Therefore, the crane needs either a visible or an audible alarm or both to warn the bridge operator *before* the rail clamp engages.

If the crane operator rides in a cab mounted on the bridge or the trolley, the operator must somehow have a way to gain access to his or her station. One idea that comes to mind is to use a portable ladder, but this is not such a good idea. Since the bridge travels, the operator, cab, trolley, and bridge may roam far from the point at which the operator mounted. The idle ladder standing far away would be an invitation for someone to remove it for some other use or perhaps just to get it out of the way. A ladder used for crane cab or bridge access should be of the fixed type. Stairs or a platform, or both, can also be used but must require no step over a gap exceeding 12 inches.

Overhead cranes designed or constructed in-house sometimes do not adhere to accepted principles for crane design and overlook some of the necessary safety features specified by applicable standards. Besides cab access, discussed earlier, there are specifications for footwalks for safe maintenance of the trolley and bridge. These footwalks need standard toeboards and handrails, as described in Chapter 6.

Ideally, footwalks will have at least 78 inches headroom. Sometimes, however, this is not practical because the crane may be near the ceiling of the building. The standards recognize this difficulty and permit less than 78 inches headroom. However, a footwalk becomes rather ridiculous if the headroom is less than 48 inches and would more properly be called a "hands-and-knees" walk. In these situations, footwalks should be omitted, and a stationary platform or landing stage for crane maintenance workers should be installed.

A critical hazard to the wire rope of a crane or hoist results from drawing the load hook or hook block up too far—to the point at which the load block makes contact with the boom point of the crane or other mechanical assembly for reeving the wire rope. This event is known as *two-blocking,* the term being derived from the physical contact of two blocks in the reeving system. Upon the incidence of two-blocking, continued travel of the load block causes a severe tensile stress to be immediately imparted to the wire rope, and this stress usually stretches or breaks the wire rope. The blocks may also be damaged.

Two-blocking is a very serious hazard that has resulted in many fatalities. A sudden break in a load-bearing wire rope constitutes an obvious hazard, especially if personnel are under the load or are themselves being hoisted aloft by the crane. As a matter of ergonomics, it can be shown that it is very difficult for a crane operator to be sufficiently vigilant to prevent an occasional dangerous brush with two-blocking, especially for construction cranes. So many fatalities have resulted from two-blocking that crane construction standards now address this hazard, and electromechanical devices are on the market for the purpose of preventing two-blocking or the damage that can result from two-blocking. These mechanisms, usually electromechanical, are commonly called *anti-two-block devices.*

An obviously serious hazard with operation of an overhead bridge crane is overtravel. Devices to control overtravel would include trolley stops, trolley bumpers, and bridge bumpers. Bumpers and stops are somewhat different in that bumpers absorb energy and reduce impact, whereas a stop simply stops travel. The stop is simpler and can consist merely of a rigid device which engages the wheel tread. For safety, such a stop needs to be at least as high as the wheel axle centerline. Bumpers soften the shock by absorbing energy but are not as positive as a true stop. For example, bridge bumpers need only be capable of stopping the bridge if it is traveling at 40% of rated load speed or less. If the crane travels only at slow speeds, shock loads are not significant, and bridge bumpers and trolley bumpers are not necessary. There may be other circumstances of operation, such as restriction of crane travel, which eliminate the need for bridge bumpers or trolley bumpers.

Related to bumpers and stops are rail sweeps. If a tool or some item of equipment happens to obstruct one of the rails on which the bridge travels, a catastrophic accident could occur. Therefore, bridge trucks are equipped with rail sweeps, projecting in front of the wheels, to eliminate this hazard. There is nothing magical about the term *sweep;* even part of the bridge truck frame itself may serve as a sweep.

One obstruction that can be encountered by an overhead crane is another overhead crane that runs adjacent to it with side rails parallel. Proper clearance should be provided between the two adjacent bridge structures. Unfortunately, such clearance may make it impossible for one crane to transfer its load to the domain of the adjacent crane. Some factories resolve this problem by using retractable extension arms on the crane. The trouble with these retractable arms is that the operator may forget and leave them extended, resulting in a collision between the two adjacent cranes.

Electric shock is a concern with overhead cranes, and the concern is principally in two areas:

- Shock from exposure to current-carrying portions of the crane's power delivery system.
- Shock from a shorted connection in a hanging control pendant box (see Figure 11.3).

A third area of concern for electrical shock is the accidental contact of live, high-voltage overhead transmission lines. This is a hazard for mobile cranes with booms (discussed in Chapter 15) because of their wide use in the construction industry. Exposed live parts in the crane's power delivery system are usually protected by their remoteness or "guarded by location." Some older models might present hazards and need modification.

A greater hazard is the possibility of shock from a pendant control. The electrical conductors can be strained if they are the sole means of support for the pendant. The control station must be supported in some satisfactory manner to protect against such strain. If a failure does occur, it is likely to be at a connection within the box. This raises the possibility of a dangerous short to ground through the operator's body. Pendant controls take much abuse and need to be of durable construction.

Not related to electrical shock but on the subject of pendant control boxes is the requirement that boxes be clearly marked for identification of functions. Some older model or homemade cranes might have pendant control boxes without function markings. Without a great deal of trouble or expense, the Safety and Health Manager can check the cranes in-house for compliance and get the control functions marked on the pendant controls.

Figure 11.3. Hand-held pendant control for overhead crane.

One hazard to think about with overhead cranes is what would happen if a temporary power failure occurred. Suppose that the crane were in the process of lifting a heavy load by means of its hoist mechanism. Obviously, no one would want the crane to drop its load to the floor upon power failure, but the hazard does not end here. Suppose that the crane bridge is traveling in a horizontal direction, whether loaded or unloaded. Upon power failure the bridge would stop, and this might not be dangerous. But when power is *restored,* dangerous lurching action might occur. In fact, upon a power failure the crane operator might even leave the crane cab! Several design alternatives can protect against these hazards. One solution is to equip the control console with spring-return controllers. Pendant boxes can be constructed with spring pushbuttons instead of toggle switches. A disconnect device can neutralize all motors and not permit a reconnect until some sort of positive "reset" action is taken. Even if the power remains on, it could be a hazard to inadvertently switch on a lever at the wrong time. Notches, latches, or detents in the "off" position can prevent such inadvertent actions.

Brakes are of obvious importance to the safe operation of a crane. But a large number of crane operators do not use brakes but instead rely on a practice called "plugging" by the industry. The operator merely reverses the control and applies power in the opposite direction, thereby stopping the load. While no OSHA standard prohibits the practice of "plugging," it should be pointed out that plugging is not as effective as applying a brake under extreme conditions, such as stopping a large, fast-moving load. Under no circumstances should a crane operator depend entirely on plugging due to the brake being inoperative. In fact, plugging would be completely useless in event of crane motor failure.

A crane has many moving parts, many of which are located far from the operator's console in the cab or from a floor operator holding a pendant. Moving machine parts are hazardous, and the remoteness characteristic amplifies the hazard. Moving parts are hazardous not only to personnel; they can be hazardous to the crane itself, which in turn can be indirectly hazardous to personnel as well. Hoisting ropes, for instance, can possibly run too close to other parts in some crane configurations and in some positions of the bridge and trolley. The result can be chafing or fouling of the hoist rope. If the configuration of the equipment will permit this situation to develop, guards must be installed to prevent chafing or fouling. Such moving parts as gears, set screws, projecting keys, chains, chain sprockets, and reciprocating components need to be checked to see if they present a hazard, and if so they must be guarded.

As with conveyors and other material-handling equipment, overhead cranes are often large and widely distributed items of equipment. Electrical power is distributed over long distances, sometimes by means of open runway conductors, and portions may be so far away as to be obscured from view from the location of the power supply switch. Picture the insecurity of the maintenance worker who is compelled to repair a crane and is in direct contact with an exposed 600-volt (but deenergized) runway conductor, but the power supply switch is so far away that it is out of sight! Therefore, such switches should be so arranged as to be *locked* in the open or "off" position.

Ropes and Sheaves

Safety standards for wire rope strength state that "Rated load divided by the number of parts of rope shall not exceed 20% of the nominal breaking strength of the rope." A safety factor of 5 is implied as follows:

$$\frac{\text{rated load}}{\text{number of parts of rope}} \leq 20\% \times (\text{nominal breaking strength}) \qquad (11.1)$$

It will later be explained that the "number of parts of rope" is a multiplying factor that enables a multiple-sheaved block-and-tackle assembly to be loaded much higher than the wire rope load. Thus

$$\text{wire rope load} = \frac{\text{rated load (including load block)}}{\text{number of parts of rope}} \qquad (11.2)$$

From equations (11.1) and (11.2),

$$\text{wire rope load} \leq 20\% \times \text{(nominal breaking strength)} \qquad (11.3)$$

Multiplying each side of the (in)equality by 5, we have

$$5 \times \text{(wire rope load)} \leq 100\% \times \text{(nominal breaking strength)} \qquad (11.4)$$

Rearranging to create a ratio yields

$$\frac{\text{nominal breaking strength}}{\text{wire rope load}} \leq 5 \qquad (11.5)$$

Since the ratio of the strength to the load is at least 5, the *safety factor is 5.*

The term *parts of rope* refers to the mechanical advantage provided by the block and tackle assembly. *Parts of rope* is computed by counting the number of lines supporting the load block. Of course, all the lines make up one continuous line which is reeved through several sheaves to achieve mechanical advantage. The concept is best explained by a picture; Figure 11.4 shows five different reeving combinations. Note that the mechanical advantage is numerically equivalent to the number of parts of line.

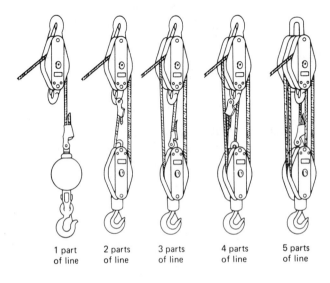

| 1 part of line | 2 parts of line | 3 parts of line | 4 parts of line | 5 parts of line |

Figure 11.4. Five different reeving combinations. The mechanical advantage is equal to the number of "parts of line" supporting the load block.

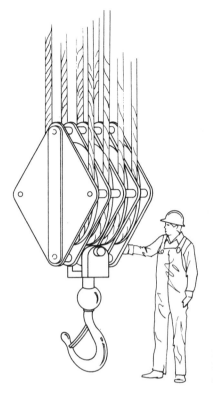

Figure 11.5. Massive load block. The load block becomes part of total load on the line.

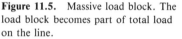

   One additional caution is in order when determining the appropriate maximum load to be applied to a given reeving setup. The weight of the load-carrying sheave must be added to the weight of the load to be picked up to arrive at the total load of the line. The weight of the load block cannot be ignored, as is emphasized by the massive load block displayed in Figure 11.5.

   One way to prevent overloading the wire rope and the crane itself is to provide a hoist motor which can develop insufficient torque to overload the wire rope. The combination of such a hoist motor and correct reeving for the crane design will result in no overloading. Under this arrangement the crane simply will be unable to lift any load that would damage the crane or exceed its safety factor. Most overhead cranes today are designed this way. How fortunate it would be if the human back had this design feature.

   Any time a rope is wound on a drum, the rope end anchor clamp bears very little of the load when several wraps are on the drum. The friction of the rope on the drum holds the load. But if the drum is unwound to less than two wraps, dangerous loading of the anchor clamp may result in failure, and the wire rope will break free from the drum. Usually, the overhead crane is set up so that, even if the load block is lying on the floor, several wraps remain on the hoist drum. The floor may not be the extreme low position for the crane, however. The Safety and Health Manager should look around for pits or floor openings into which the overhead crane might operate, resulting in dangerous unwinding of the hoisting drum.

"Don't saddle a dead horse" is a familiar safety slogan which refers to the improper mounting of wire rope clips employing U bolts. Such a clip assembly bears a resemblance to a saddle, and the U bolt represents the cinch strap. The U bolt places more stress on the wire rope and has less holding power than the clip, and therefore should not be placed on the live portion of the rope when a loop is formed. The "dead" end of the rope gets the U bolt, and the "live horse" gets the clip. Right and wrong methods are illustrated in Figure 11.6. Unfortunately, some workers in the field, unsure of the correct method, place the clips *both ways* in alternating fashion, thinking they are "playing it safe." Such an arrangement may be even more unsafe than "saddling dead horses" with every clip.

Before leaving the subject of wire rope, the hazard of rope whip action should be emphasized. Wire rope "cable" seems so heavy and inflexible that it seems unnatural that it could crack like a whip or any fiber rope. It is difficult for anyone to visualize the tremendous tension forces on a wire rope during use in a material-handling operation, until that wire rope breaks. Most workers have never seen what happens when a wire rope breaks; perhaps this explains why so many workers stand too close while the rope is drawn taut by the load. The hazard is very serious, and an injury accident is very likely to be a fatality.

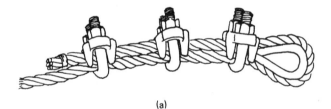

(a)

(b)

(c)

Figure 11.6. "Don't saddle a dead horse." Right and wrong ways to secure wire rope loops using U-bolt clips. (a) Incorrect—"saddle" is on dead end of rope; (b) incorrect—clips are staggered both ways; (c) correct—all clips are placed with the saddle assembly on the live portion of the rope and the U bolt on the dead end.

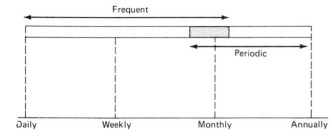

Figure 11.7. Inspection intervals for overhead cranes.

Crane Inspections

Almost everyone is aware of the long, perhaps tedious checklists for inspection of an aircraft every time it flies. An aircraft must not fail in any catastrophic way during operation, and this makes frequent inspections worthwhile, even if they are repetitious and rarely uncover any defects. In some ways a crane is like an aircraft; it, too, must not fail.

With regard to crane inspections, the standards use the terms *frequent* or *periodic* to specify when various items on the crane should be checked. Such usage is an attempt to avoid being overly specific in telling the employer *what* to do and *how often*. Some broad guidelines describing the meaning of these terms are illustrated in Figure 11.7. Note that there is some overlap, as monthly inspections can be considered either frequent or periodic.

The crane manufacturer is a good source for detailed guidance in what to look for in the frequent inspections. This type of inspection is performed by the crane operator, just as a pilot inspects his or her aircraft before a flight. This analogy between aircraft and cranes might be a starting point for a safety theme in a training program for crane operators.

The frequent-inspection routine should include a daily visual inspection of hoist chains, plus a monthly inspection with a signed report. In the field, the term *hoist chains* has been widely misinterpreted to include chain *slings* for handling the load, but there is a separate standard for slings. Figure 11.8 identifies which chain is hoist and which is sling.

Crane hooks take a great deal of abuse and are critical to the safe operation of the crane. Although they usually contain considerable overdesign, damage or wear can reduce the margin of safety. Telltale signs of an abused and dangerous hook are illustrated in Figure 11.9.

A more thorough inspection of crane components is needed for "periodic" intervals. Whereas daily inspection of crane hooks is merely visual, the periodic inspection calls for a more scientific approach, such as using magnetic particle techniques for detecting cracks. More extensive checks for wear are appropriate also, such as the use of gauges on wire rope sheaves and chain sprockets.

For most kinds of plant equipment, safety testing is customarily done by an independent testing laboratory such as Underwriters' or Factory Mutual. But the safety of an overhead crane is in part a function of the installation method and proper ad-

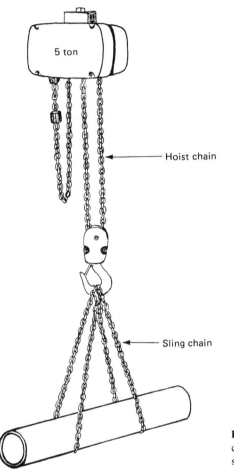

5 ton

Hoist chain

Sling chain

Figure 11.8. Hoist with sling. Hoist chain is not to be confused with the sling chain.

justments at the site. Therefore, an actual rated load test is needed prior to initial use to confirm the load rating of the crane. This is a bit tricky because if the crane is subjected to too high a load, the crane may fail, but if it is not tested with a high enough load, why conduct the test? Standards specify that the maximum load during the test should go 25% higher than the crane's load rating. This will provide some assurance that the crane will withstand its rated load. During use, however, the crane should not be loaded higher than its rated load. Any extensive repair or alteration of the crane may subject it to a retest.

Wire Rope Wear

The two principal moving parts of an overhead crane are the wire rope and the drum and sheaves upon which they travel. In addition, the wheels for the bridge and trolley move as do a few other items on the crane. Whenever parts move and contact other

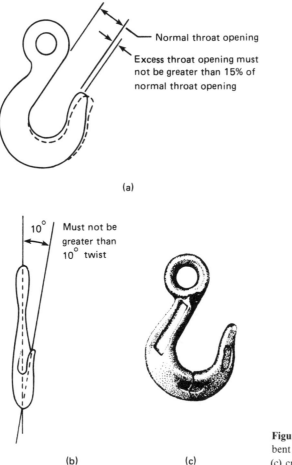

Normal throat opening

Excess throat opening must not be greater than 15% of normal throat opening

(a)

$10°$ Must not be greater than $10°$ twist

(b)   (c)

Figure 11.9. Defective crane hooks: (a) bent hook; (b) twisted hook; (c) cracked hook.

parts, they are subjected to wear. Wear on the bridge and trolley wheels may eventually lead to problems but is not likely to cause crane failure. Wear on the drum and sheaves is more dangerous because of the harm this can bring to the wire rope. But worn sheaves and drum *alone* are generally not the *direct* cause of crane failure. The wire rope remains as the most critical moving part. With continued use every wire rope will eventually wear out and fail, a hazard that generally is not tolerable. Some way has to be devised to predict the failure of wire rope and withdraw it from use before a catastrophe.

It is not a simple matter to determine when a wire rope needs replacing. A wire rope has many individual wires, and it is easy to cut or break away any one of these small wires. Almost everyone has seen broken wires in an old wire rope (or "cable," as laypersons often say), which raises the question, Is such a wire rope dangerous? Safety and Health Managers do not survive long in their companies if they go around ordering wire rope replaced every time they find a broken wire. And if they did survive, their companies would not survive. Before delving into this issue, some basics about wire rope need to be examined.

A wire rope should really be considered a machine because the individual wires move upon each other while the rope flexes, resulting in friction and wear. Furthermore, unless the individual strands are able to move properly during flexure, tremendous tensile stresses may be placed on some of the wires, causing them to break. Rust, kinks, and other types of abuse can interfere with the movement of the wires and lead to the stresses that cause wire breakage. As some wires break, the tensile stresses on other wires increase, leading to their breakage as well. Eventually, the force of the crane load will be sufficient to overcome the tensile strength of all remaining wires combined, and the wire rope fails.

Even a well-maintained wire rope will be subject to *wear* on the individual wires, especially the outside wires. As the wear causes the diameter of the wire to become smaller, the tensile force increases the tensile stress on that single wire due to its reduced cross-sectional area. Thus the worn wire, too, may break due to stress concentra-

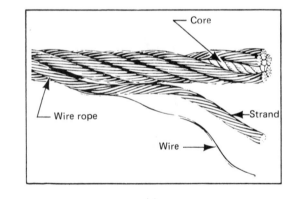

(a)

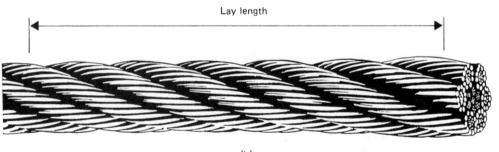

(b)

Figure 11.10. Wire rope components. (a) This wire rope has six strands plus an inner core. Do not confuse "strands" with "wires"; (b) Wire rope lay. "Lay" is one complete wrap of a single strand around the rope. Since this rope has six strands, lay is the length from the first hump to the seventh hump, as shown in the diagram. (*Source:* Courtesy of the Construction Safety Association of Ontario, ref. 77.)

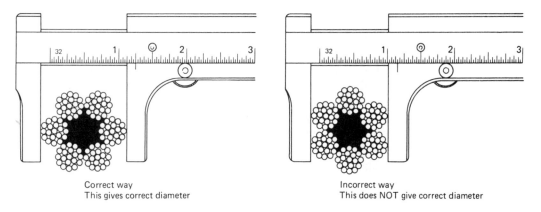

Correct way
This gives correct diameter

Incorrect way
This does NOT give correct diameter

Figure 11.11. Gauging wire rope. (Rotate rope to select largest diameter.) (*Source:* Courtesy of Construction Safety Association of Ontario.)

tion, even if there are no kinks or rust, and even if the wires move properly to distribute the load among all wires.

For obvious purposes, a wire rope is overdesigned and will withstand more than its rated load. Equally obvious from the previous discussion is that all wire rope will eventually develop broken wires with continued use. Broken wires are permissible to an extent, but beyond a certain point the rope becomes dangerous. A means of evaluating the *degree* of wire rope deterioration is awkward and difficult but nevertheless *necessary* to prevent catastrophic failure.

The ANSI standard recommends[1] a procedure for counting the broken wires. Figure 11.10 diagrams the components of wire rope, defining the terms *strand* and *lay*. If there are more than 12 randomly distributed broken wires in a single strand in a single lay, there should be cause for questioning the continued use of the rope. A good place to look for broken wires is around end connections. Sometimes an expert can evaluate the remaining strength in a deteriorated rope after inspection. This possibility is acknowledged by the ANSI standard, which advocates the use of good judgment.

Another measure of rope condition is the amount of reduction of rope diameter below nominal. Figure 11.11 shows that when gauging a wire rope, the caliper can be placed upon a small diameter or a larger diameter. The convention for wire rope terminology is to use the larger diameter for designation as the nominal diameter of the rope. Most people pay little attention to wire rope, but wire rope is no small matter considering that a quarter million tons of wire rope is sold annually.

Operations

The actual handling and moving of the load by the crane is a function of the skill and attitude of the crane operator and the workers who attach and secure the sling

[1]ANSI B30.2–2.4.2

or lifting device. As is the case with motor vehicles, the *operator* of the crane is probably the most important factor in preventing accidents.

A good deal of skill is required to attach the load safely, especially if a sling is used. Slings will be discussed later in this chapter. The hoist rope is not intended for wrapping around the load, as this kind of abuse may damage the rope and at the same time be an inadequate support for the load. Misplacing the attachment off the line of the center of gravity can cause dangerous swings when the load is lifted. After the load is lowered, the tendency is to consider the hazards removed. But disengaging the load attachment can also result in dangerous shifts in material that can injure the inexperienced or unwary worker on the floor.

SLINGS

Slings are used to attach the load to the crane, helicopter, or other lifting device. Slings come in a great many varieties and are very important in the safety of material-handling operations. Components of the sling assembly are often subjected to much larger forces during lifting than is the hoist rope or other material-handling equipment. Because the skill of the user is so important in the proper application of the sling, slings are often misused, resulting in much more abuse and damage to slings than to the components of the crane.

The most important point to remember for the safe use of all slings is that the stress on a sling is greatly dependent on the way it is attached to the load. Figure 11.12 shows two different ways of applying a sling to pick up identical loads. If the angle of the supporting legs of a sling is sharp, as in Figure 11.12b, the advantage of multiple legs can be lost. Using too short a sling is the most common cause for this condition. The "rated capacity" of a sling is the working load limit under *ideal* conditions; if the sling is applied at leg angles other than specified in the rated capacity table, the capacity can be drastically reduced due to the physics of the applied forces. Therefore, *rated capacity* is an incomplete term without the accompanying leg angle.

Note the following progression in capacities of ½-inch alloy steel chain as the number of legs increase:

Single (at vertical)	11,250 pounds
Double (at 60° from horizontal)	19,500 pounds
Triple (at 60° from horizontal)	29,000 pounds
Quadruple (at 60° from horizontal)	29,000 pounds

It might be assumed that the capacity of the sling increases as the number of legs or supporting members is increased. But note that no increase in capacity is shown in going from three legs to four legs. The reason for this is that, as in a four-legged chair, three legs will actually carry the load. At times the weight distribution might be equal among legs of the sling, but usually the balance is not so perfect. As the load shifts around, it is entirely possible for one of the four legs to become slack and for the other three to bear the entire load.

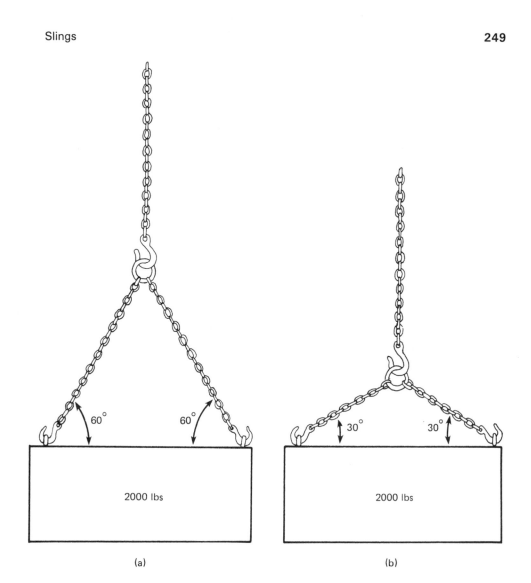

Figure 11.12. Comparison of sling tensile forces for two different methods of lifting identical loads: (a) tensile force on the sling is approximately 1150 lb; (b) tensile force on this sling is approximately 2000 lb.

Alloy steel chain, besides being very strong, is very durable and capable of withstanding the physical abuse that industrial slings routinely receive. The ordinary carbon steel chain commonly found in hardware stores should not be used for slings. Wire rope slings can be just as strong as steel chain slings, but wire rope is more subject to wear. Individual wires in the wire rope are more easily broken, with eventual incapacitation of the sling.

Wire rope slings also have a specification for the permitted number of broken wires; the rule is no more than ten randomly distributed broken wires in one rope lay, or five broken wires in one strand in one rope lay. Note from our earlier discussion that this is a bit more strict than the requirement for wire rope for overhead cranes.

Selection of the proper sling for a given application can consider several factors besides rated load. The nature of the item to be lifted, its surface finish, temperature, sling cost, and environmental factors must be considered. The Safety and Health Manager is usually not the person who makes this decision, but there is increasing reason for the Safety and Health Manager to have a voice in what is done in this area. Many factory superintendents, supervisors, and material-handling workers are not aware of the many federal standards confronting the criteria for selection of slings. In fact, many of these personnel do not even have a clear understanding of the inherent hazard mechanisms surrounding industrial slings. Therefore, it is recommended that the Safety and Health Manager provide consultation and advice in the selection and use of industrial slings in the interest of worker safety.

For some criteria, such as load markings, repair procedures, proof testing, and operating temperatures, the requirements for various types of slings are not identical, and even vary in curious ways. Some of this is due to the physical differences in sling types, and some is due to the various origins and rationales for the requirements for various slings. Table 11.3 is intended to summarize some of the more curious differences between requirements for various types of slings. It is by no means exhaustive. For instance, nylon web slings are not permitted in the presence of acid or phenolic vapors. In the case of caustic vapors, polyester and polypropylene web slings and web slings with aluminum fittings are not permitted. The Safety and Health Manager can use Table 11.3 as a first check in an in-house inspection or purchase decision; then further details should be checked out in the standards.

It must be reemphasized, however, that the skill and training of the worker who uses the sling to attach the load is more important than all of the detailed specifications and sling standards combined. This is a good place for the reader to reflect on the fatality described in the following case study:

Case Study 11.1   Two inexperienced workers had the assignment of hoisting a 40-foot bundle of channel steel. The question was where to attach the hoist hooks to the load. The solution they selected was based upon their experience in lifting loads with which they were familiar—handheld loads. To these workers the heavy, steel straps used to secure the bundle seemed to be a natural attachment point. But these straps were not intended to be used as a substitute for a sling. Their strength was insufficient, and the angle of attachment was severe. The angle of attachment will always be severe when bundle straps are used in this way, because to do their job the straps must be tight. When the load was lifted one of the wires gave way and one of the workers was killed.

TABLE 11.3 Comparison of Certain Requirements for Slings

Type of Sling	Rated Capacity Markings Required?	Repairs Permitted?	Employer Required to Maintain Records of Periodic Inspection?	Employer Required to Maintain a Certificate of Proof Test?	Safe Operating Temperature (°F) Maximum	Safe Operating Temperature (°F) Minimum	Proof Test of Sling Required?	Repairs Records Required?[a]
Alloy steel chain	Yes	Yes	Yes	Yes	1000[b]	None specified	Yes	No, except for welding or heat treating
Wire rope								
With fiber core	No	Yes	No	Yes, for welded end attachments	200	None specified	Yes, for welded end attachments	No
With nonfiber core	No	Yes	No	No	400[c]	−60[c]	No	No
Metal mesh	Yes	Yes	No	No	550[d]	20[d]	Yes	Yes
Natural fiber rope	No	No	No	No	180	20	No	N/A
Nylon fiber rope	No	No	No	No	180	20	No	N/A
Polyester fiber rope	No	No	No	No	180	20	No	N/A
Polypropylene fiber rope	No	No	No	No	180	20	No	N/A
Nylon web	Yes	Yes	No	Yes, for repaired slings	180	None specified	Yes, for repaired slings	Yes
Polyester web	Yes	Yes	No	Yes, for repaired slings	180	None specified	Yes, for repaired slings	Yes
Polypropylene web	Yes	Yes	No	Yes, for repaired slings	200	None specified	Yes, for repaired slings	Yes

[a]N/A, not applicable.
[b]If alloy steel chains are working in temperatures exceeding 600 °F, load limits are reduced.
[c]Seek manufacturer's recommendations for use outside temperature range.
[d]Impregnated metal mesh slings have more restricted temperature requirements.

CONVEYORS

Hazards with conveyors can be quite serious, and workers seem to sense these hazards more than with some other types of machines. It seems that all of us have implanted somewhere in our imaginations the vision of the beautiful maiden tied to the conveyor in the sawmill about to be sawed in half unless she submits to the villain, or in another version, signs over the deed to her ranch. The truth is that sometimes workers *are* caught in industrial conveyors and not only killed but dismembered or even pulverized beyond recognition. The horror of this truth instills a healthy respect on the part of most workers around industrial conveyors.

By contrast, some of the worst conveyor hazards are very innocent in appearance. In-running nip points which themselves might not even seriously injure a hand or arm can start an irreversible process once the employee is caught, resulting in an employee's entire body being drawn into a machine. Loose clothing in particular can get caught, and the employee, even before being injured in the slightest way, can become doomed.

Belt Conveyors

On a belt conveyor, in-running nip points are seemingly everywhere. Pulleys are required to drive the belt, change direction of the belt, support the belt, and tighten the belt. One side of every pulley is always an in-running nip point. Defense against this hazard generally consists of one of three means: isolating the nip points, installing guards, and installing emergency tripping devices.

The best method of protection is to isolate the in-running nip point so that an employee would not or could not come into the danger area. If isolation is impractical, a guard can sometimes be installed to keep out the worker's body or extremities. The design of the guard must vary with the application, and sometimes it is difficult to make the guard practical because it may interfere with the operation of the conveyor. Because of body geometry, distance from the danger zone is a design factor in building the guard, a principle that is covered in more detail in Chapter 12.

If both isolation and guarding are impossible or infeasible for the application, the workers can be protected by some type of emergency tripping mechanism. A wire or rope can be run along the length of the conveyor so that a worker, falling into the conveyor, can grab the trip wire and stop the machine. Unfortunately, this method of protection requires the overt action of an alert worker or coworker.

Overhead Conveyors

Large appliance parts or vehicle assemblies are often handled by overhead conveyors. Hooks attached to a moving chain support each item as it is moved. This type of conveyor is particularly suited for products that have delicate or finished surfaces because so little of the conveyor actually contacts the product. For the same reason, overhead conveyors are very useful also for paint spray or finishing operations.

Overhead conveyors avoid many of the risks of belt conveyors by eliminating

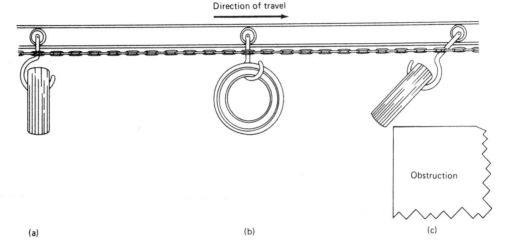

Figure 11.13.   Three different orientations for conveyor hooks: (a) hook trails the load; (b) side orientation; (c) load trails the hook (dangerous).

many of the in-running nip points and by removing the moving parts from worker access. But overhead conveyors have hazards of their own, such as dropping conveyored materials to the plant floor or upon worker stations. Screens or guards can protect against this hazard but not entirely, because the moving parts must be accessible to work stations for processing. A good rule is to place screens or shields under the conveyor whenever it passes over an aisle or other area where personnel are likely to gather. Another good place for screens is where the conveyor chain moves up or down an incline. Such movements cause the loads to shift on their hangers and increase the possibility of a falling load.

Figure 11.13 shows three different orientations for the hangers or hooks that support the work held by an overhead conveyor. Note how much safer is the orientation in which the work is held in front of the hook. If the work encounters an obstruction, it is more likely to catch and stop the conveyor if the work is in front of the hook. If the work trails the hook, an obstruction may lift and knock off the load.

Screw Conveyors

Screw conveyors can be very dangerous. The very principle of their operation is an ingoing nip point at the intake. Complicating the hazard is the fact that in order to operate at full capacity, the intake must be completely submerged in the material to be transported. Submerged usually also means hidden, so an unseen serious hazard exists at the intake. Finally, there may be a need for the worker to be fairly close to the screw conveyor for many applications, in order to shovel or distribute material into the intake.

A simple and often effective way to protect workers from the hazards of the screw intake is to box the intake area in a small screen enclosure which allows passage of the material but keeps out fingers, hands, and feet. If even a coarse screen mesh is too fine to permit passage of the material, an enclosure with larger openings may be necessary, perhaps with openings large enough to admit a finger or hand. This type of enclosure can be made safe, too, by making the box large enough that the worker's *reach* will not permit entry of hands or fingers into the danger zone even if the openings are large enough for hands or fingers. This follows principles of machine guarding which are discussed in more detail in Chapter 12.

LIFTING

Before closing this chapter on materials handling, we return to the subject of lifting. At the beginning it was stated that back injuries, mostly from lifting, are the biggest compensable injury category of all. Lifting injuries are very complex and very difficult to control. Naturally, the amount of weight lifted is important, but many other factors determine whether an injury occurs. Even a lightweight lift of 5 to 10 pounds can cause serious back injuries if conditions are just right (rather, wrong). The physical condition of the person doing the lifting is also important.

Much emphasis has been placed on technique, with the most frequently heard saying being, Lift with your legs, not with your back. Unfortunately, the rule is rather difficult to follow because almost everyone is able to lift a heavier weight with the back than with the legs. Lifting with the legs requires squatting down and then lifting both the load and the lifter's body. This requires a great deal of leg strength for heavy lifts and is especially difficult when the worker is unaccustomed to lifting with the legs. Training and exercise with light loads can help in developing the technique,

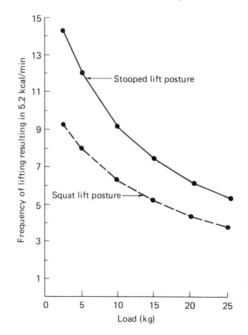

Figure 11.14. Comparison of the number of lifts that can be made using two different lift postures while consuming identical amounts of energy. (*Source:* From NIOSH, ref. 58; adapted from ref. 32.)

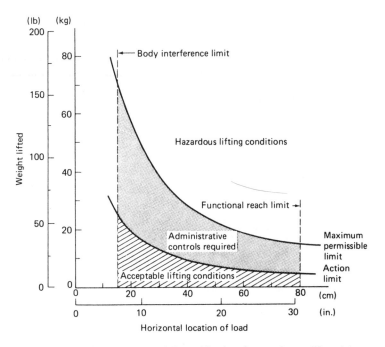

Figure 11.15. NIOSH-recommended specification for maximum lift weights at various horizontal distances for infrequent lifts from floor to knuckle height. (*Source: From NIOSH, ref. 58.*)

but there are other disadvantages of lifting with the legs. Chaffin and Park (ref. 7) have shown that if the shape of the load is such that it must be brought out in front of the knees, lifting with the legs *increases* the compression force on the lower back! Also overlooked by the often-quoted rule is the fact that lifting with the legs takes as much as 50% more energy as lifting with the back, especially when the load is light and the frequency of lifting higher, as can be concluded from Figure 11.14.

The capacity to lift varies greatly with the horizontal position of the load, which is determined largely by the shape of the object being lifted. Various independent studies of this relationship have been analyzed by NIOSH, resulting in a proposed specification for maximum weight lifted versus the horizontal distance of the load from the center of gravity of the body. This specification is summarized in Figure 11.15, but it must be remembered that the graph represents merely a NIOSH recommendation—not an established standard.

SUMMARY

This chapter has brought recognition to the fact that the handling of material in a manufacturing plant can be as hazardous as the industrial process itself. The basic nature of materials-handling hazards were examined; then hazards of specific machines and equipment were discussed.

The Safety and Health Manager needs to be aware not only of the proper safety features to seek in new equipment, but also of the inspection, service, and maintenance of equipment in-house. However, for <u>industrial trucks, cranes, slings, and perhaps *all* material-handling equipment, the operator's skill, attitude, and awareness of hazards are probably more important to the safety of the worker than are the safety features of the equipment itself.</u>

EXERCISES AND STUDY QUESTIONS

11.1. Why are four-legged slings rated no higher than three-legged slings?

11.2. An often-heard safety slogan is "Don't saddle a dead horse!" What does this slogan mean?

11.3. What *safety* design feature do most modern overhead cranes have that is unfortunately not characteristic of the human back?

11.4. Name a part of the plant the Safety and Health Manager should give special attention to when production and sales are booming.

11.5. In the diagram, which orientation of the 4,000-pound load will place less stress on the sling that handles it?

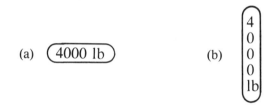

11.6. Suppose that a company was able to save some money by trading in its Type LPS forklift truck for a Type DY. Would this introduce some safety problems? What if the trade were from DY to LPS?

11.7. Why might spring pushbuttons be a better crane control than toggle switches?

11.8. In the diagram in Figure 11.16, what is the mechanical advantage? If the nominal breaking strength of the rope is 5,000 pounds and the load block weighs 200 pounds, what is the maximum rated payload of the hoist (not including the load block)?

11.9. What is a rail sweep, and why is it needed?

11.10. Name in order of preference three methods of protection for in-running nip point hazards on conveyor belts.

11.11. Why are overhead crane access ladders specified to be of the fixed type?

11.12. Because of rising gasoline costs, a company desires to convert existing gasoline-powered forklifts to LPG powered. What implications would such a decision have?

11.13. Identify a "performance" standard in the standards for materials handling. Explain why it is a performance-type standard.

11.14. Explain the following terms as applied to industrial cranes: bridge, trolley, pendant, pulpit, gantry, cantilever gantry.

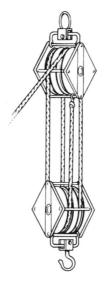

Figure 11.16. Blocks and pulleys for Question 11.8.

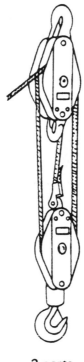

3 parts of line

Figure 11.17. Blocks and pulleys for Question 11.17.

11.15. Name at least four characteristics of forklift trucks that require more skill for safe operation than is required for operation of an automobile.

11.16. At least four general characteristics of materials handling contribute to its intrinsic hazard potential. Name and explain four such characteristics.

11.17. A 2,000-pound payload is supported by the crane block-and-tackle assembly shown in Figure 11.17. In addition to the payload, the load-carrying sheave weighs 100 pounds. What is the approximate load on the wire rope? How many parts of rope are used in the reeving as shown in Figure 11.17?

11.18. What minimum nominal breaking strength is specified by safety standards to be suitable for the application described in Exercise 11.17?

11.19. Suppose the wire rope in use in the reeving shown in Figure 11.16 is rated at 2,000 pounds and the load-carrying block weighs 150 pounds. The objective is to lift a load weighing 3,000 pounds. Does the assembly as described meet safety standards?

11.20. What maximum payload rating would you assign to the block-and-tackle assembly described in Exercise 11.19? What is the nominal breaking strength of the wire rope?

11.21. A sling with three legs is used to pick up a 1,000-pound load. The load is distributed equally among the three legs. When the load is picked up, each leg forms an angle of 30 degrees with the horizontal. What is the stress on each leg?

11.22. A three-legged sling distributes its load equally among its three legs. When the load is picked up, each leg forms an angle of 60 degrees with the horizontal. The rated load of the chain used in the sling is 6 tons. What maximum payload can this sling pick up?

11.23. From a safety standards aspect, what is the significance of whether a crane trolley rides on top of the rail or hangs from the lower flange?

12

Machine Guarding

32%

Percent of OSHA
general industry citations
addressing this subject

Most people think of a machine guard when industrial safety is mentioned, and for good reason. More efforts and resources have been expended to guard machines than for any other industrial safety and health endeavor. To modify or guard a single machine is generally not a major project when compared with installing a ventilation system or a noise abatement system. But although each machine guarding modification is usually small, the aggregate becomes a major undertaking involving plant maintenance, operations, purchasing, scheduling, and, of course, the Safety and Health Manager. The Safety and Health Manager should take a leadership role in the implementation of machine guards—enumerating problem areas, setting priorities, selecting safeguarding alternatives, and ensuring compliance with standards.

GENERAL MACHINE GUARDING

If Safety and Health Managers are to be able to "enumerate problem areas" and "set priorities" as just suggested, they need to be knowledgeable about what makes a machine dangerous. Despite wide differences between machines, some mechanical hazards seem to be shared by machines in general, and these mechanical hazards will be discussed first.

Mechanical Hazards

Following are general machine mechanical hazards listed in approximate order of importance:

1. Point of operation
2. Power transmission
3. In-running nip points
4. Rotating or reciprocating machine parts
5. Flying chips, sparks, or parts

In addition to the mechanical hazards just listed are others, such as electrical, noise, and burn hazards. These hazards are usually controlled by other methods, however, and are covered elsewhere in this book. It is predominantly the mechanical hazards that are controlled by machine guards, the subject of this chapter.

Although the sequence of priority in the foregoing list is only approximate, there is no question about which hazard should be at the top of the list. By far the largest number of injuries from machines occur at the point of operation, where the tool engages the work. This machine hazard is so important that it is discussed in a separate section later in this chapter, detailing various strategies and guarding devices in controlling the hazard.

The power transmission apparatus of the machine, typically belts and pulleys, is the second most important general machine hazard. Belts and pulleys are usually easier to guard than is the point of operation. Access to the belts and pulleys is usually necessary only for machine maintenance, whereas the point of operation must be accessible, at least to the workpiece, every time the machine is used. Although belts and pulleys are easier to guard, they are easily overlooked by the Safety and Health Manager. A section in this chaper is devoted to belts and pulleys, due to their overall importance to safety.

Machines that feed themselves from continuous stock generate a hazard where the moving material passes adjacent to or in contact with machine parts. This hazard is called an "in-running nip point" or an "ingoing nip point." Even on machines not equipped with automatic feed, in-running nip points occur where belts contact pulleys and gears mesh. Figure 12.1 shows examples of in-running nip points. In-running nip points are not only direct hazards but can cause injury indirectly by catching loose clothing and drawing the worker into the machine.

Rotating or reciprocating moving parts can present hazards similar to those of belts and pulleys and in-running nip points; in truth there is overlap between these categories. But rotating or reciprocating moving parts bring to mind other parts of the machine that might need guarding. Particularly dangerous is a part of the machine that moves *intermittently*. During the motionless part of the cycle workers might forget that the machine will later move. Material-handling apparatus, clamps, and positioners are in this category. The most intermittent motion of all is *accidental* motion. It pays to consider what would happen in event of a hydraulic failure, broken cotter pin, loosened nut, or some other accidental occurrence. Would the guard on the machine protect workers in these circumstances? Is the risk of occurrence of enough significance to warrant installing a guard?

The fifth item listed earlier, flying chips, sparks, or parts, is not necessarily the least significant; it merely stands alone as a somewhat different category. Many

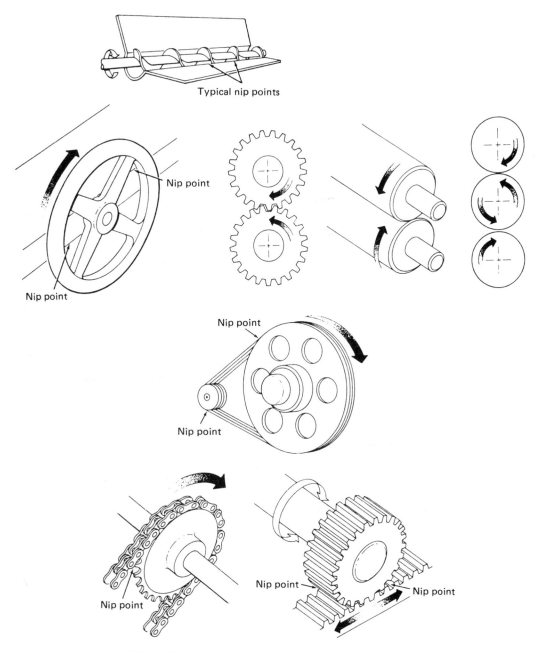

Typical nip points

Nip point

Nip point

Nip point

Nip point

Nip point

Nip point

Nip point

Nip point

Figure 12.1. In-running nip points. (*Source:* From OSHA, ref. 12.)

machines obviously throw off chips or sparks from the area of the point of operation. Flying objects should also be included because sometimes the product being manufactured actually breaks, and pieces may be hurled at the operator. It is also possible for parts of the machine to break and fall on or be thrown at the operator. One means of protecting workers from flying chips and sparks is by personal protective equipment. But this is not nearly as effective as using machine guards to protect the operator and other workers in the vicinity. Often, the term *shield* or *shield guard* is used to describe guards that protect the operator from flying chips, sparks, or parts.

Guarding by Location or Distance

The easiest and cleverest way to guard a machine is not to use any physical guard at all but rather to design the machine or operation so as to position the dangerous parts where no one will be exposed to the danger. This is usually in the realm of machine design, and more and more attention is being given to safety in modern machine designs. But even without altering a machine, the machine can be turned and backed into a corner so that its belts, pulleys, and drive motor are impossible to reach during normal operation. A good example would be a portable concrete mixer. Admittedly, this strategy makes the motor and drive difficult to reach for maintenance, but on the other hand, so do ordinary bolt-on guards.

Guarding "by distance" refers to the protection of the operator from the danger zone by setting up the operation sequence such that the operator does not need to get close to the danger. For some difficult-to-guard machines, such as press brakes, the distance method of guarding the point of operation (see Figure 12.2) is expressly permissible. Press brakes are used to bend sheet metal, and their long beds make it difficult to guard their points of operation. When the workpiece is a large piece of sheet metal, the operator of necessity stands well back from the point of operation and is thus safeguarded by "distance."

Tagouts and Lockouts

A surprising number of industrial machine accidents occur not when the machine is in operation, but when it is down for repair. A worker simply turns a machine back on, not realizing that it is down for repair and that a maintenance worker is still close to or inside the machine!

Such accidents sound like freak occurrences, but that is because most of us are more accustomed to small machines around the home, where only a few persons, usually family members, may be in the area. But factory machines may be large, and their repair status may not be obvious. Large numbers of personnel may have access to the machine, and a miscommunication between operating supervisors and maintenance crews can easily occur. Human beings have been literally mangled and digested by large industrial machines.

Two simple safety procedures for preventing accidents of this type are the tagout system and the lockout system. In the tagout system, the maintenance worker places a tag on the on–off switch or control box, so that anyone who might have occasion to turn the machine back on will be warned by the tag to leave it alone. The lockout

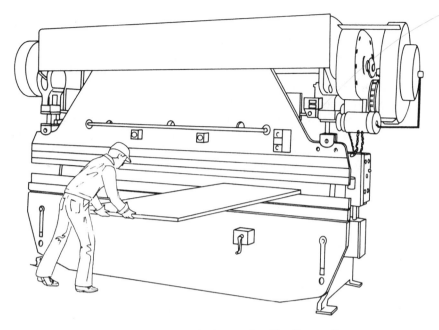

Figure 12.2. Press brake guarding "by distance."

system (see Figure 12.3) provides positive protection to the maintenance worker because he or she is the only person who has the key to the lockout. Note in Figure 12.3 that the lockout has several positions for separate padlocks for each maintenance worker exposed. Each maintenance worker therefore has an element of personal responsibility and control for his or her safety while working on the machine.

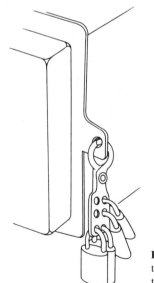

Figure 12.3. Lockout system for protection of maintenance workers while the machine is being repaired.

The tagout system is simpler, but the lockout system is gaining in favor. It would seem that a lock would not be necessary, because after all, who would turn on a machine when a maintenance worker's tag warned not to? But factories are operated by humans, and various mistakes can lead to a tagout accident. For instance, the maintenance worker can forget to remove the tag after the repair work is complete. Operating personnel may think that the maintenance personnel have forgotten to remove the tag and ignore it. The reader can no doubt think of other scenarios that would lead to an accident with tagouts but that would have been prevented with the lockout. When the maintenance worker has the only key to the lock, there is no way for an operator to turn the machine back on—that is, with the assumption that the maintenance worker was careful enough to actually use the lock.

Interlocks

Contrasted with the lockout is a safety device called an *interlock*. Modern clothes dryers stop rotation as soon as the door is opened, and thus comply with industry safety standards for revolving drums, barrels, and containers. Even if the drum itself is closed, its rotation can present a hazard unless it is guarded by an enclosure. An interlock between the enclosure and the drive mechanism is specified to prevent rotation whenever the guard enclosure is not in place.

A tumbling machine is a popular industrial machine which uses a revolving drum to rotate metal parts in the presence of an abrasive tumbling medium to improve surface characteristics of the metal parts. Many tumbling machines found in industry do not have interlocked enclosure guards, and Safety and Health Managers should check for such deficiencies.

Trip Bars

Large machinery layouts are often difficult to guard but can be provided with trip bars that stop the machine if the operator falls into or trespasses into the danger zone. The operator's hand or body deflects the bar that trips a switch. Figure 12.4 illustrates an emergency trip bar on a rubber mill, a very dangerous machine.

It is sometimes impractical to place the trip bar such that any time the worker falls into the danger zone the bar will trip automatically. An alternative in these situations is to provide a triprod or tripwire for the worker to grab to switch off the machine. Examples are shown in Figure 12.5. Such devices deserve some careful study and experimentation to be sure that workers are able to reach the triprod or tripwire if they get into trouble.

Fan Blade Guards

A very prominent safety standard is the one that requires fan blades to have guards whose openings do not exceed ½ inch. Literally millions of ventilation fans are distributed throughout industries all over the country, and many of them have guards with openings larger than ½ inch. New fan designs are generally built according to

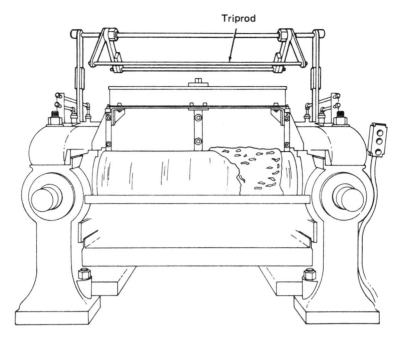

Figure 12.4. Pressure-sensitive body bar on a rubber mill. (*Source:* From OSHA, ref. 12.)

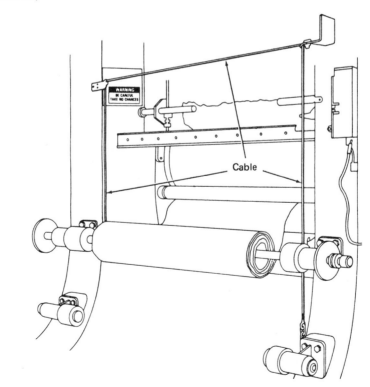

Figure 12.5. Safety triprods and tripwires. (*Source:* From OSHA, ref. 12.)

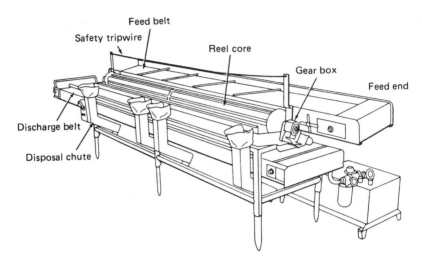

Figure 12.5.   *(Continued)*

the latest standard, but a principal problem is the retrofit of old fans to meet the ½-inch-opening-size standard.

An enterprising safety equipment manufacturer got the clever idea of marketing a nylon mesh that could be wrapped around the existing noncomplying guard and then drawn up tightly with a drawstring in the rear (see Figure 12.6). The nylon mesh was ½-inch gauge, and if it was drawn tight, it kept workers free from danger. Before rushing out to purchase these inexpensive nylon mesh guards, the Safety and Health Manager should realize that they are not a panacea. First of all, the nylon mesh, or any guard for that matter, cuts the efficiency of the fan. Furthermore, any fan will gradually accumulate oil and lint on its blade surfaces and on the guard. The nylon mesh seems even more susceptible to oil and lint buildup than does the metal guard, generating a nuisance to maintenance in keeping the guard clean or else resulting in a gradual deterioration in the effectiveness of the fan.

To deal with the fan blade guard dilemma, many innovative guard and fan blade designs are beginning to appear on the market. Some of the new guards are removable for easy washing, and some small plastic-bladed fans have no guards at all. A very lightweight fan with plastic blades, powered by a small motor, presents no hazard to personnel. With no guard at all, the increased efficiency can go a long way toward making up for the low-power motor on such an unguarded fan.

Anchoring Machines

Another troublesome machine guarding standard is the rule for anchoring fixed machinery to the floor to keep it from "walking," or moving. Such anchoring is required for all machines "designed for a fixed location." Machines that have reciprocating motions, such as presses, have a tendency to "walk" unless securely anchored. Drill presses and grinding machines can also be hazardous unless anchored.

Figure 12.6. Nylon mesh guard for ventilation fan blades.

One interpretation of the phrase "designed for a fixed location" is any machine that has mounting holes in the feet or bases of the legs. It is true that these holes are for the purpose of anchoring machines, but the mere presence of the holes is not proof that the machine must be anchored. The mounting holes might simply be a convenience feature for ease in shipping or to permit the machine to be mounted at the discretion of the user for whatever reason, such as security purposes instead of safety.

SAFEGUARDING THE POINT OF OPERATION

Injury statistics bear witness to the fact that the point of operation is the most dangerous part of machines in general. On some machines the point of operation is so dangerous that some type of safeguarding is required for every setup; mechanical power presses are one example. Taking guidance from the specific rules for mechanical power presses, the Safety and Health Manager can extend the principles to other machines because most of the safeguarding methods specified will work also for other machines.

A general classification of methods of safeguarding the point of operation is into guards and devices as follows:

1. Guards
    a. Die enclosures
    b. Fixed barriers
    c. Interlocked barriers
    d. Adjustable barriers
2. Devices
    a. Gates
    b. Presence-sensing devices
    c. Pull-outs
    d. Sweeps (no longer accepted for mechanical power presses)
    e. Hold-outs
    f. Two-hand controls
    g. Two-hand trips

The reader will note that nowhere in this list do hand-feeding tools such as tongs appear. Hand-feeding tools (see Figure 12.7) are helpful in eliminating the *need* for operators to place their hands in the danger zone, but it must be emphasized that these tools do *not* qualify as guards or devices for safeguarding the point of operation.

Guards

The function of a guard is obviously to keep the worker out of the danger area, but many guards do not succeed in this function. Some guards merely screen part of the danger area around the point of operation, but this can be hazardous. Too many workers will defeat the purpose of the guard by reaching through, over, under, or around the guard, exposing themselves to perhaps a greater hazard than if the guard were not present. Of course, every guard has to be removable between jobs for maintenance or setup purposes, but even this must be kept inconvenient, or operators will take it upon themselves to unfasten the guard to get it out of their way. Wing nuts or quick-release devices should not be used to secure guards. Nuts and bolts are better, but even better than ordinary nuts and bolts are recessed head fasteners such as Allen screws.

Most guards are metal, and popular design uses expanded metal, sheet metal, perforated metal, or wire mesh as filler material. A secure frame is needed to maintain the structural integrity of the guard. When a guard panel becomes larger than 12 square feet, the rigidity of the guard is in jeopardy, and additional frame components are needed. Many types of ordinary wire mesh are unsuitable because the wires are not secure at the cross points. Ordinary window screen would be in this category. Galvanized screen is better as are some types that are welded or soldered at the cross points.

One principle of machine guarding borrowed from the power press standard is the maximum permissible opening size. A contemplation of human anatomy results in the conclusion that the farther away from the danger zone, the larger can be the openings in the guard without creating a hazard. If the guard is at arm's length from the danger zone, an opening of several inches still might not be dangerous. But if

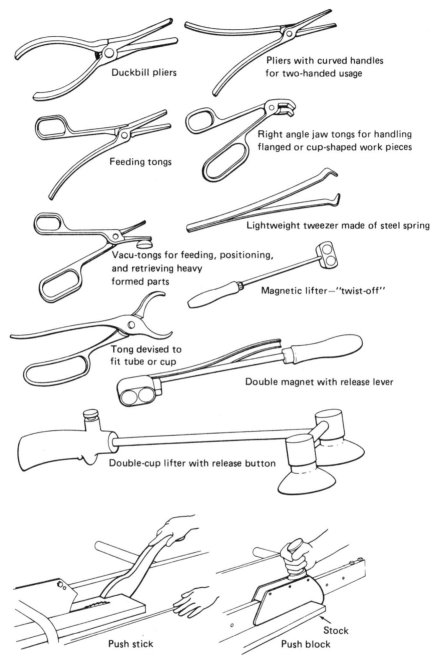

Duckbill pliers

Pliers with curved handles for two-handed usage

Feeding tongs

Right angle jaw tongs for handling flanged or cup-shaped work pieces

Lightweight tweezer made of steel spring

Vacu-tongs for feeding, positioning, and retrieving heavy formed parts

Magnetic lifter—"twist-off"

Tong devised to fit tube or cup

Double magnet with release lever

Double-cup lifter with release button

Push stick

Push block

Stock

Figure 12.7.   Hand-feeding tools help with, but do not replace, the function of a point-of-operation guard. (*Source:* From OSHA, ref. 12.)

TABLE 12.1 OSHA's Specification for Maximum Permissible Guard Opening Size Versus Distance from the Point of Operation

Distance of Opening from Point of Operation Hazard (in.)	Maximum Width of Opening (in.)
½–1½	¼
1½–2½	⅜
2½–3½	½
3½–5½	⅝
5½–6½	¾
6½–7½	⅞
7½–12½	1¼
12½–15½	1½
15½–17½	1⅞
17½–31½	2⅛

Source: OSHA Standard 1910.217, Table O–10.

the guard is immediately adjacent to the danger zone, no opening should be large enough to permit a finger to reach through. Standard guard opening sizes are specified in Table 12.1. The principle behind the standard openings is illustrated in Figure 12.8. Some companies have made available a simple go–nogo guard gauge (see Figure 12.9): Insert the point of the gauge through the guard. If the gauge reaches the danger zone, the guard opening is too large.

  Visibility through the guard is a problem with some guards. An old practice was to paint all guards orange. But the bright orange color makes it difficult to see

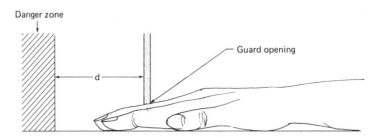

Figure 12.8. Maximum permissible guard opening should depend on distance to the danger zone.

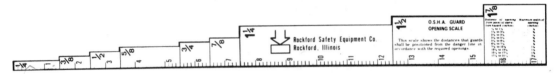

Figure 12.9. Guard opening size gauge. (*Source:* Courtesy of Rockford Safety Equipment.)

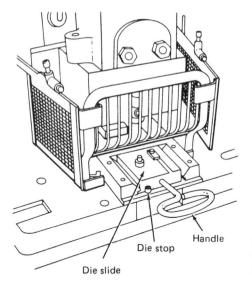

Die slide

Die stop

Handle

Figure 12.10. Die enclosure guard used with sliding die for feeding. (*Source: From OSHA, ref. 12.*)

through the guard to the point of operation of the machine; black is by far the superior color for point of operation guards.

Die Enclosures

Punch presses and similar machines have mating dies that close on each other to act as the point of operation. The space between the upper and lower dies is the danger area, and the die enclosure guard is designed to enclose only this small area. The advantage over other guard types is that the die enclosure guard is small, but still it is not the most popular guard. Since dies vary widely in size and shape, the die enclosure guard must be essentially custom-made to fit the die or at least the die "shoe" which acts as a base to hold the die. Another disadvantage is that the die enclosure guard is essentially right at the point of operation, which permits no latitude for the guard mesh size or grillwork spacing. The maximum permissible opening size at this point is ¼ inch, and this may limit visibility. Figure 12.10 illustrates a die enclosure guard.

Fixed Barriers

Fixed barrier guard is a general term for a wide variety of guards that can be attached to the frame of the machine. Figure 12.11 shows one example of a fixed barrier guard, but remember that there is no set style or shape for fixed barrier guards. Even the mesh or spacing of the bars is variable, depending on the distance of the guard to the point of operation (refer back to Table 12.1). Large fixed barrier guards can permit large distances between the guard and the point of operation and a coarser mesh for the guard material.

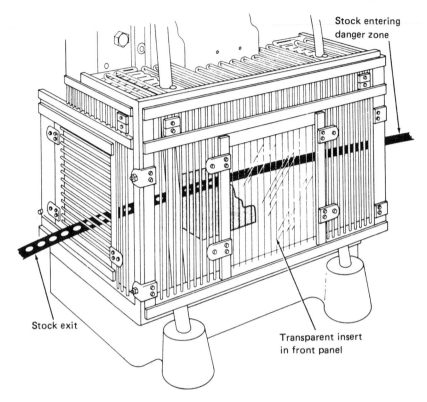

Stock entering
danger zone

Stock exit

Transparent insert
in front panel

Figure 12.11. Fixed barrier guard. (*Source:* From OSHA, ref. 12.)

Interlocked Barriers

More sophisticated is the interlocked barrier guard shown in Figure 12.12. An interlock, typically electrical, disables the actuating mechanism whenever the guard is opened. But the interlock is not required to stop the machine if it has already been tripped, and thus it usually affords inadequate protection to an operator who is attempting to hand-feed the machine. If the barrier is so easy to open and close that the operator can open it and reach in while the machine is still moving, the interlocked barrier is not doing its job. Instead of a guard, such an arrangement would be more properly labeled a gate, a device that will be discussed later.

Adjustable Barriers

Guard manufacturers have devised some clever ways for guards to be adjusted to individual applications during the setup. Unlike the fixed barrier guard, the adjustment is temporary, and the same guard can be reshaped later for a different setup. The trick with adjustable barriers is to make them easy enough to adjust to be practical but not so easy as to tempt an unauthorized person to tamper with them or gain access to the danger zone. Figure 12.13 shows one type of adjustable barrier guard.

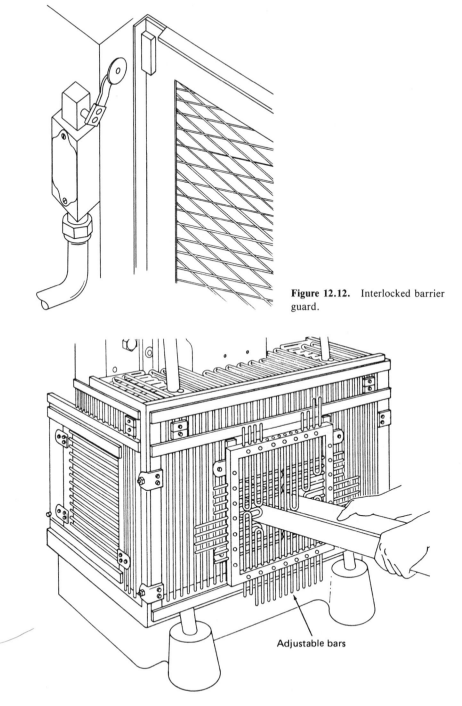

Figure 12.12. Interlocked barrier guard.

Adjustable bars

Figure 12.13. Adjustable barrier guard. (*Source:* From OSHA, ref. 12.)

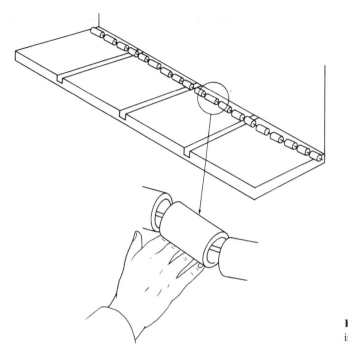

Figure 12.14. Awareness barrier
installed on a shear.

Awareness Barriers

Some people confuse the term *adjustable barrier guard* with the term *awareness barrier*. An awareness barrier (see Figure 12.14) is not recognized as a guard and does not meet the guarding criteria of keeping the operator's hands or fingers out of the danger zone. Although the awareness barrier is not a guard, it does provide a reminder that the hands are in danger. In the style pictured in the figure, metal rings or cylinders lie on the table and are lifted by the operator's fingers when the fingers are too close to danger. The operator at that point could go on to reach further into the machine, which could result in injury, but training and good judgment should inhibit him or her from taking this action. Contact with the awareness barrier should be a learned cue to immediately withdraw the hands. The effectiveness of awareness barriers remains in doubt, as some feel that a mere deterrent is not enough to protect the operator. An added complication is that the awareness barrier may obscure from view the real danger. Many operators believe that it is not only a matter of convenience, but also one of safety to be able to see the actual point of operation.

Sometimes the term *awareness barrier* is also used to describe a simple rope or chain suspended in front of a danger area with perhaps a sign hanging on it to warn personnel to keep out. An example is the back side of a metal shear, as shown in Figure 12.15. The chain will not ensure that personnel will stay out of the point of operation or danger zone, but it will warn employees of the danger.

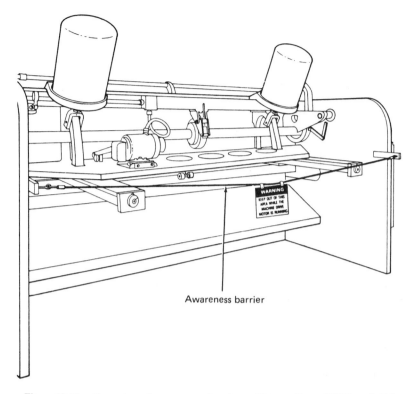

Awareness barrier

Figure 12.15.   Rear view of power squaring shear. (*Source:* From OSHA, ref. 12.)

Jig Guards

The design of a jig guard is integral to the engineering of the manufacturing operation. The guard has the function of both protecting the operator and facilitating the
operation to increase productivity. There is nothing standard about a jig guard because
it is designed to fit the individual workpiece and hold it in place while the machine
performs the cut or other operation. Jig guards usually move with the work while
the operation is performed. The jig guard depicted in Figure 12.16 is being used to
notch cross members in the manufacture of four-way pallets. This clever guard keeps
the circular table saw blade covered at all times, either by the jig guard between cuts
or by the workpiece itself during the cut.

   Point of operation guards are excellent when the machine can be fed effectively by automatic means or through a guard window without the operator's having
to reach into the danger zone. But the only feasible way to feed some machines is
manually, and some of these machines are very dangerous. This class of machine
is typified by the mechanical power press, to be discussed next. The only way to assure
protection of the operator while hand feeding these very dangerous machines is to
use some kind of *device* that keeps the operator's hands out of the danger zone while

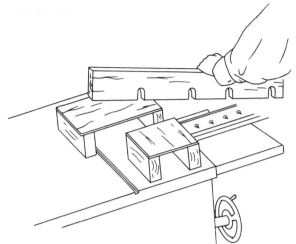

Figure 12.16. Jig guard for notching cross members of a four-way shipping pallet. (Idea credited to the Occupational Center of Central Kansas, Inc.)

the machine cycles and does its work. Some ingenious devices have been developed for this purpose and will be discussed in the next section.

POWER PRESSES

Punch presses are at the same time the most inherently dangerous and one of the most useful production machines in industry. The epitome of mass production machines, the press excels when huge volumes of identical products are required. Mass production depends on interchangeable manufacture, which in turn requires machines that successively produce parts that are essentially identical. The power press is superbly qualified for this task.

Figure 12.17 illustrates two popular model power presses. The term *power press* is really a general one that encompasses hydraulic-powered models and forging presses in addition to the popular mechanical punch press. The outstanding feature of a power press is the set of mating dies that close on each other to cut, shape, or assemble material, or to do combinations of these operations in one of more strokes. The more subtle features of presses, such as methods of power transmission and control of the stroke, are important determinants of permissible safeguarding. It is at the mechanical models, powered by flywheels, that the most restrictive press standards are aimed.

Press Hazards

There are, of course, reasons that so much importance is attached to safeguarding the point of operation of mechanical power presses. The injury record for power presses is not a good one, as indicated by estimates that even today there are about 2,000 press operator work-related amputations each year in the United States (ref. 78)! When feeding the press by hand, the operator is close to danger every time the

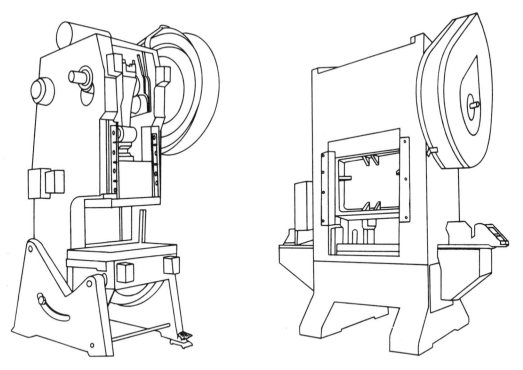

Figure 12.17. Typical power presses: (a) open-back inclinable (OBI) model; (b) straight-side model.

dies close, and this happens thousands upon thousands of times in a press operator's career. One slip and in a fraction of a second a finger or a hand is amputated. Such accidents were commonplace in the first half of the twentieth century. Shortly before World War II, there developed an awareness that even the careful operator could become a victim of the power press, and efforts were initiated to eliminate the hazard.

To understand the nature and significance of the power press hazard, it is necessary to study the interaction between human being and machine. In a hand-feed setup the press and the operator are alternating actions in a rhythm that cycles every few seconds and in some bench press operations may cycle in a fraction of a second. Figure 12.18 depicts the sequence of actions in a typical press cycle employing hand feeding without safeguarding. Production incentives motivate the operator to higher and higher speeds as skill develops. The operator learns to develop rhythm with the press motion. The sound of the press tripping mechanism, the closing dies, and other press motions can become cues to the operator to make a hand or foot motion. The process involves eye–hand–foot coordination every cycle. Operating a power press is not as difficult as simultaneously patting your head and rubbing your stomach, but it is easy to visualize the hazards involved in hundreds of thousands of repetitive cycles.

One of the biggest causes of accidents with power presses is an attempt on the

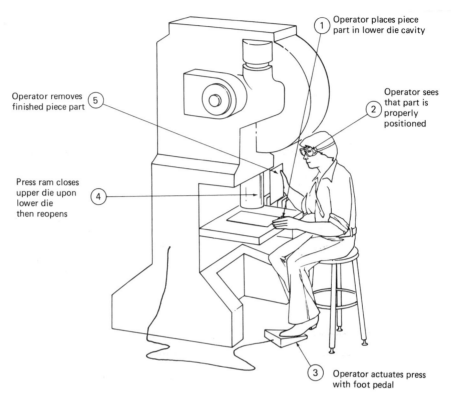

Figure 12.18. Press cycle operation sequence (no safeguarding).

part of the operator to readjust a misaligned workpiece in the die. The motivation
is a very powerful one to reach back in to correct the error even after the ram has
been tripped. If the operator lets the error go, the misaligned workpiece can wreak
havoc when the dies close. At the very least, the workpiece will be ruined. More likely is that the expensive dies will be broken or ruined, and it is very possible that the
press frame itself will be damaged or broken. One misaligned workpiece can result
in many thousands of dollars damage to the dies and the press itself. But even worse,
the misaligned workpiece or the dies can be fragmented when the dies close, subjecting the operator to injuries from flying metal pieces. It is no wonder, then, that the
operator has a powerful urge to reach back in to correct a misaligned workpiece.
The operator's eye will see the error, and the hand will reach in—even though the
foot may have already depressed the pedal to actuate the press.

      The somewhat insidious nature of power press hazards motivated the standards
writers to initiate a policy of "no hands in the dies." The theory was that tongs or
other tools and devices could be used to feed workpieces, eliminating the need for
the worker to put his or her hands into the danger zone. In addition to the tongs
or other feeding tools, press guards or safety devices would prevent operators from
placing their hands or fingers into the danger zone, even if they tried, while the dies
were closing.

The theory worked for most jobs, but some applications presented awkward feeding problems that defied solution with the no-hands-in-the-dies approach. It became clear that the metal-stamping industry would not be able to comply with a rigid no-hands-in-the-dies rule in all situations, and support for the theory began to crumble. In its place new rules were established for assuring the reliability of safeguarding devices for protecting operators at the point of operation.

Press Designs

To understand the somewhat complicated rules for safeguarding a power press, it is necessary to examine the basics of how power presses work. Most power presses are mechanically powered, although there are a large number of hydraulically powered models, generally recognizable by the presence of a large hydraulic cylinder above the ram. The cylinder usually resembles the large cylinder in an automobile service station lift. Hydraulic power presses are expressly excluded from coverage by standards for mechanical power presses, as are pneumatic presses, fastener applicators, and presses working with hot metal, *even if they are mechanically powered*. Also excluded is the press brake, a power press with a long bed (refer back to Figure 12.2) which is used to bend sheet metal. Shears, because they employ blades instead of dies, are not considered within the definition of mechanical power presses.

The easiest way to distinguish a mechanical power press from other types is by the presence of a large heavy flywheel which, by its rotation, carries energy which is imparted to the ram when the press is tripped. The flywheel is *usually* mounted on one side of the press near the top, as was shown in the typical press in Figure 12.17.

One of the most important features is whether the press is *full revolution* or *part revolution*. This refers to the method of engaging and disengaging the flywheel to deliver power to the ram. Full-revolution types make a positive engagement that cannot be broken until the crankshaft and flywheel make one complete revolution together. During this revolution the press ram goes down, the dies close and then reopen, and the ram returns to its full-up position to await another cycle. At the end of the revolution, the flywheel is disengaged and rotates freely under the power of the motor.

The part-revolution machine typically has a friction clutch which can be disengaged at any time during the press cycle. Compressed air is used to engage or disengage the clutch instantly at the discretion of the operator. Upon disengagement of the clutch, a brake is applied which immediately or almost immediately stops the ram. It is easy to see the advantage of being able to interrupt a press stroke at any point in the cycle, but the part-revolution advantage does not stop there. The instantaneous engagement is also valuable in causing the press to cycle quickly once engaged, giving less time for the operator to get into trouble by making an afterthought reach into the point of operation.

The Safety and Health Manager must know which presses are full revolution and which are part revolution in order to know how to equip the press with the proper safety equipment. Thousands of dollars have been foolishly spent by purchasing the wrong safety equipment for a power press. One way to get a rough idea whether

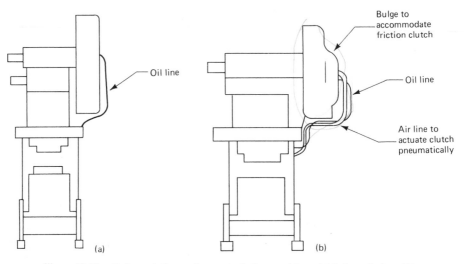

Figure 12.19. Full-revolution and part-revolution machines: (a) full revolution; (b) part revolution.

the press is full revolution is by the age of the press. Most presses are full revolution, and invariably the older presses are full revolution unless they have gone through a renovation process. Not incidentally, the average age of mechanical power presses in the United States is steadily increasing, and at the time of publication of this book statistics suggest that about half of all presses in the United States are more than 20 years old!

Since part-revolution presses employ a friction clutch, the housing on the flywheel often has a bulge to accommodate the clutch, as shown in Figure 12.19. The clutch is actuated pneumatically, and this generally means that an extra line can be seen on the outside of the flywheel cover running to the center of the clutch, as shown in the figure. Full-revolution clutch machines may also have a small line running outside the flywheel housing, but this is an oil line for lubricating purposes. None of these criteria for distinguishing machines is 100% reliable, so they should be used only as a means of preliminary screening for possible troublespots. Press engineers or the equipment manufacturer's representative can be consulted for an authoritative determination.

Point-of-Operation Safeguarding

Once having determined whether a press is full or part revolution, the Safety and Health Manager or engineer can proceed to a determination of the most effective means of safeguarding its most dangerous zone, the point of operation. There are at least ten recognized methods of power press safeguarding, but acceptability of each depends on the press configuration and method of feeding. The various safeguarding methods can be divided into the following four categories, ranked according to degree of security:

1. Methods that prohibit the operator from reaching into the danger zone altogether
2. Methods that prohibit the operator from reaching into the danger zone any time the ram is in motion
3. Methods that prohibit the operator from reaching into the danger zone only while the dies are closing
4. Methods that do not prohibit the operator from reaching into the danger zone but that stop the ram before the operator can reach in

Only categories 1 and 2 are to be trusted for a full-revolution press. Some category 3 methods are permissible for full-revolution presses, but with category 3 there is exposure to the hazard that the press will "repeat." So far, no antirepeat mechanism has been found to ensure absolutely that a full-revolution press will not repeat with an extra unwanted stroke. These repeat strokes are a terrifying possibility, but fortunately that possibility has been significantly reduced in recent years. Recalling that it is likely that about half of all presses are more than 20 years old, the threat of repeats is still something to consider.

Category 4 methods of safeguarding are permitted by OSHA only for part-revolution presses, and this is where many Safety and Health Managers go wrong. It should be obvious that any method that depends on stopping the ram to protect the operator must be installed only on part-revolution presses. Full-revolution presses, by definition, cannot be stopped. It is surprising, though, how many category 4 devices are seen in industry installed on full-revolution presses. It is true that these devices can afford some protection on full-revolution presses by locking out the tripping mechanism while the operator is in the danger zone. Once the press is tripped, however, these devices are powerless to stop the ram.

Press Guards

The section on guarding earlier in this chapter described types of guards for use on machines in general. Four of these types—die enclosure, fixed barriers, interlocked barriers, and adjustable barriers—are acceptable on mechanical power presses. Indeed, the die enclosure guard is almost exclusively used on mechanical power presses, although it is not as popular as some of the other types of guards. The fixed barrier guard is a very popular method of guarding power presses that employ automatic feeding of coils of strip stock and automatic ejection of the finished parts. The interlocked press barrier guard is not permitted for hand feeding, but a gate *device* (not a guard) can be used. Indeed, *none* of the four types of guards is permitted for hand feeding the press (putting the hands in the dies) because by definition of a press guard "it shall prevent the entry of hands or fingers into the point of operation by reaching through, over, under, or around the guard."

With only one exception, a guard or some type of point-of-operation safeguarding device must be installed on *every* mechanical power press. That one exception is where the full-open position of the ram results in a gap between the mating dies of less than ¼ inch, too small to permit entry of the fingers (see Table 12.1), and thus too small to be a hazard. But for all other mechanical power presses, even the automatic-feed models or the robot-fed setups, point-of-operation safeguarding is required. An accident can occur even with automatic feeding if a worker tending the automatic setup attempts to adjust a workpiece during operation.

We have already established that guards cannot be used where the operator feeds the press by putting his or her hand in the die. We have also stated that virtually every mechanical power press is required to have point-of-operation safeguarding. Then does this make hand feeding illegal? The answer is no, and the key is the difference between the terms *guarding* and *safeguarding*. *Safeguarding* is a more general term and encompasses a variety of mechanical or electromechanical devices that protect the operator even when hand feeding is used. The standards are specific about which devices can be used with which types of machines and appropriate setups. Each of these configurations for mechanical power press devices will now be considered.

Gates

Gates look somewhat like a guard (see Figure 12.20) but are different in that they open and close with every machine cycle. In contrast to interlocked barrier guards, gates *can* be used for manual feeding. Gates are used almost exclusively with mechanical power presses.

There are two types of gates: Type A and Type B. The Type A gate is the safer of the two because it closes before the press stroke is initiated, and it *stays closed*

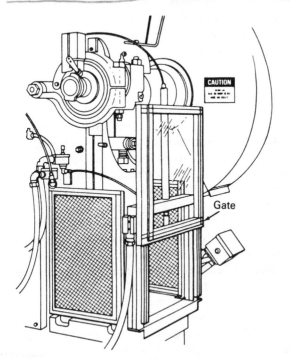

Figure 12.20. Gate devices. (*Source:* From OSHA, ref. 12.)

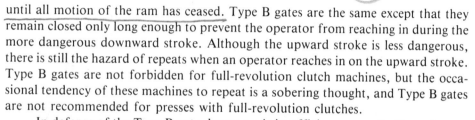
until all motion of the ram has ceased. Type B gates are the same except that they remain closed only long enough to prevent the operator from reaching in during the more dangerous downward stroke. Although the upward stroke is less dangerous, there is still the hazard of repeats when an operator reaches in on the upward stroke. Type B gates are not forbidden for full-revolution clutch machines, but the occasional tendency of these machines to repeat is a sobering thought, and Type B gates are not recommended for presses with full-revolution clutches.

In defense of the Type B gate, however, is its efficiency over the Type A gate. A substantial percentage of the press cycle time can be saved if the operator can start reaching in as soon as the ram starts back upward. The savings may be only a fraction of a second per cycle, but over hundreds of thousands of cycles, the difference can be significant. The labor and overhead cost associated with the operator's time is saved, and also conserved are productive capacity of the press and the plant floorspace in which it is housed. It is possible for these combined costs to exceed $50 an hour. This represents an incentive to modernize the power press equipment so that it qualifies for the most efficient safeguarding systems.

Presence-Sensing Devices

Modern electronic detection devices have made their way into the machine guarding industry, and there are several devices of this type available for protection of the point of operation. One type uses a bank of photoelectric cells to set up a light screen, the penetration of which will immediately stop the ram. Figure 12.21 illustrates this type of device.

It becomes a game for workers to defeat these devices intended for their protection. It is obvious that if a worker can reach over or around the light screen, the machine will not stop. Alternate points of entry not covered by the sensing device should be guarded so that the operator cannot reach into the point of operation without tripping the device.

Another way to defeat a photoelectric device is by aiming flashlights or other sources of light to maintain the sensors in an energized condition at all times even if the worker's hand has broken the field. Even when deliberate attempts to defeat the device are not a factor, ordinary ambient light in the factory may keep sensors energized so that the screening device will not function effectively. A device that cannot be trusted to function when the outcome is so critical has little or no safeguarding value.

To compensate for the ambient light problem, some presence-sensing devices function in the infrared frequencies rather than in the visible light spectrum. This makes the "light screen" invisible, and this feature may have some advantages, too.

Another way to trick a light screen is to somehow squeeze between the beams. If the sources and sensors are tightly spaced, squeezing between the beams becomes impossible for a part of a human body. The concept can be demonstrated, however, by carefully slipping a paper or cardboard edgewise between adjacent beams. Some commercially available screens have a sophisticated programmed scan that crisscrosses

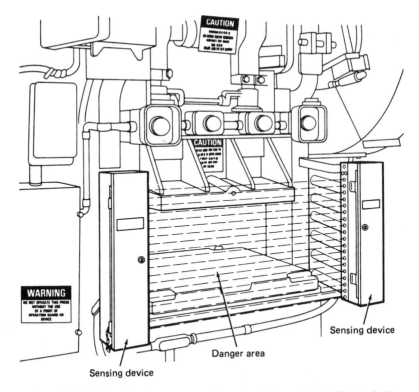

CAUTION

CAUTION

WARNING

DO NOT OPERATE THIS PRESS
WITHOUT THE USE
OF A POINT OF
OPERATION GUARD OR
DEVICE

Sensing device

Danger area

Sensing device

Figure 12.21. Photoelectric presence-sensing screen. (*Source:* From OSHA, ref. 12.)

the field in such a complicated fashion as to prevent defeating the device. Such crisscrossing may reduce the number of sensors required to protect a given area. The concept is illustrated in Figure 12.22.

Another type of presence-sensing device uses a conductor to set up an electromagnetic field in its vicinity. Many variables affect the tripping threshold of these devices, sometimes called *radio-frequency sensors,* and this has damaged their reputation. For instance, one person's body, due to its mass or conductivity characteristics, might trip the mechanism from a distance of 2 feet from the point of operation. Another person might not trip the device until he or she has actually entered the danger zone. If this type of device is used, it should be "tuned" to have the proper sensitivity for a given operator and setup. Figure 12.23 illustrates the electromagnetic field type of presence-sensing device. Late-model versions of this device have been found to be very effective.

Presence-sensing devices are quite practical for hand feeding in conjunction with a foot switch. Thus if the operator's rhythm is off and the foot switch is depressed too soon (while the operator's hand is still in the sensing field), the machine will not operate. Even more important, if the operator sees a misaligned workpiece and attempts to reach in after the ram has started its downward motion, the sensing field

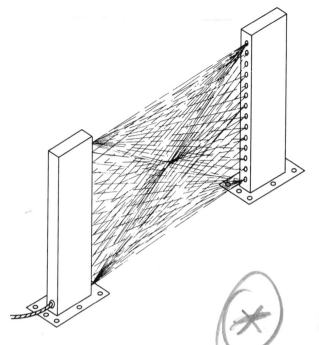

Figure 12.22. Programmed crisscross scan for sensing plane.

will detect this action and stop the ram. Thus the device not only prevents operator injury, but it also arrests a costly smash-up of the dies and possible damage to the press itself.

From what the reader has already learned, it should be obvious that this advantage of the presence-sensing device is feasible only for part-revolution presses.

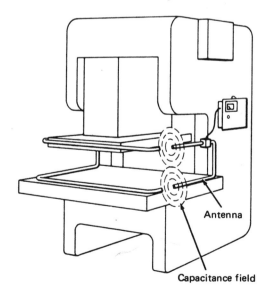

Antenna

Capacitance field

Figure 12.23. Electromagnetic field type of presence-sensing device. (*Source:* From OSHA, ref. 12.)

Indeed, the use of presence-sensing devices for safeguarding the point of operation on full-revolution mechanical power presses is prohibited.

Although the presence-sensing device is designed to stop the ram before the operator can reach the danger zone, the worker should not tempt the machine. The author knows of a case in which a new machine[1] equipped with a presence-sensing device was being demonstrated by a proud worker to his family during an open house. The worker repeatedly thrust his hand into the machine to prove that the photoelectric cell was quicker than his hand. Finally, he succeeded in beating the machine and lost the ends of his fingers. This "accident" actually happened.

Presence-sensing devices are like gates in that they are subject to the issue of when to return access to the point of operation to the operator. Since presence-sensing devices are permitted only for part-revolution presses, it seems sensible to afford them the same production efficiencies permitted the Type B gate. Therefore, it is permissible to deactivate the sensing field on the upstroke. This process of bypassing the protective system is called "muting." As soon as the ram reaches its home position, the system is reset to its protective mode. Muting permits the same production efficiencies that the Type B gate holds over the Type A gate.

While we are discussing efficiency, why not go one step further and eliminate the foot switch? It would be feasible for the press to be restrained by the control system for as long as the operator's hand or arm breaks the sensing field during hand feeding. Then, as soon as the operator withdraws his or her hand from the danger zone, the press would *automatically* cycle without a signal from the operator! Feasible? Yes. Legal? No. The presence-sensing device is prohibited as a press tripping mechanism, although many European factories do employ this highly productive mode of operation for power presses.

There is a way that a presence-sensing device *can* be used to trip the press—when the presence-sensing device is used in conjunction with another safeguarding device, such as a gate. Thus the gate represents the safety device, and the presence-sensing field acts as a tripping device. This would be a complicated and expensive system and should be considered rare.

Presence-sensing devices must be designed to adhere to the general fail-safe principle stated in Chapter 3. Thus if the presence-sensing device itself fails, the system must remain in a protective mode. A failure in the device must prevent the press from operating additional cycles until the failure is corrected. But such a failure must not deactivate the clutch and brake mechanisms, which are essential in stopping the press. If a failure in the device causes an interruption of the main power supply to the machine, the clutch must automatically disengage. Of course, the clutch and brake system should have this characteristic, regardless of the choice of safeguarding devices; the clutch and brake are simply additional examples of systems that should be designed in accordance with the general fail-safe principle.

One final note about failures in the presence-sensing system should be mentioned. It is not enough that the failure inhibits the operation of the press; the system

---

[1]A printing press, not a punch press, in this instance.

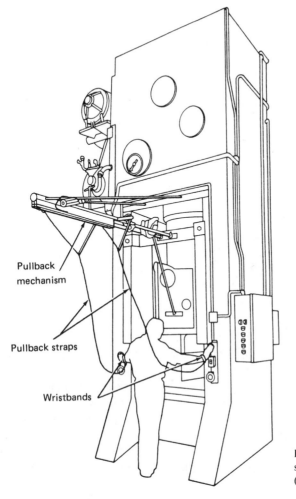

Figure 12.24. Pullback devices for safeguarding the point of operation. (*Source:* From OSHA, ref. 12.)

must also indicate that a failure has occurred. This is usually done with a trouble light on the panel.

Pullbacks

A very popular method of safeguarding a power press is by means of cables mechanically linked to the travel of the ram. These cables are attached to wristlets which pull the operator's hands out of the danger area as the ram makes its downward stroke. Figure 12.24 shows one example setup of pullbacks, or "pull-outs" as they are sometimes called.

One reason for the popularity of pullbacks is their versatility. They can be used with virtually any power press, regardless of power source or type of clutch. However, the pullbacks do have their disadvantages.

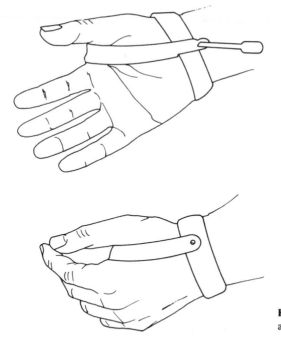

Figure 12.25. Close-ups of wristlet assembly for pullback device.

     Proper adjustment is very important to the effectiveness of pullbacks, especially with respect to close work done on bench model machines. Even the method of attachment to the wrists is important because ordinary wristlets permit too much variation in reach. Figure 12.25 is a close-up of a wristlet assembly which minimizes variations in an operator's restricted reach. Even with properly designed wristlets, proper adjustment is critical. Differences in operators' hand sizes can be a factor, but much more important are variations in die setups. A large die set will of course have a danger zone that extends closer to the operator, requiring an adjustment in the pullback limit of reach.

     Recognizing the hazard of improper adjustment of pullbacks, safety standards require an inspection for proper adjustment at the start of each operator shift, following a new die setup, and when operators are changed. This inspection requirement, and especially the *frequency* of the inspections, decidedly cools the interest of employers in the use of pullback devices. Were it not for this problem and the fact that many workers dislike wearing pullbacks, these devices would be extremely popular. These drawbacks notwithstanding, the pullback device remains one of the most popular press safeguarding devices.

     It is usually the diesetter or operating supervisor who makes the actual check of the pullbacks for proper adjustment. The concern of the Safety and Health Manager should be that the job is done effectively and that a record is made of the inspections. Simplicity is the key to the inspection record system for pullback devices. Simplicity will help to ensure that the job gets done and will also minimize the impact on production efficiency. One convenient method is to use a tag attached to the pullback device itself, with blank lines for indicating "date inspected" and "initials" by the responsible party. In multiple-shift operations, in new die setups, or in operator changes, more than one line would be initialed in one day on the tag, but the tag could be easily designed to accommodate multiple entries on the same day.

Sweeps

Very popular in the past, and still seen on some power presses, are devices that sweep away the operator's hands or arms as the dies close. These devices, illustrated in Figure 12.26, have fallen into disfavor as a means of protecting the operator. Operators even fear injury from the sweep device itself, as it comes swinging down in front of the machine. The design dilemma of these devices is that they must be powerful enough to inflict injury themselves in order to be effective in positively removing the operator's hands from the point of operation. But the overriding reason for their disfavor is that the design and construction of sweeps are inherently inadequate as a press safeguarding device. Sweeps are no longer recognized as adequate point-of-operation safeguarding devices on mechanical power presses.

Hold-Outs

A simplification of pullbacks is the hold-out (sometimes called "restraint") device, which is feasible only for setups in which it is unnecessary for the operator to reach into the danger area. Figure 12.27 shows that hold-outs look almost exactly like pullbacks, but the big difference is that the hold-out reach is fixed and does not per-

Figure 12.26. Sweep device. This device no longer qualifies as acceptable safeguarding for mechanical power presses. (*Source:* NIOSH.)

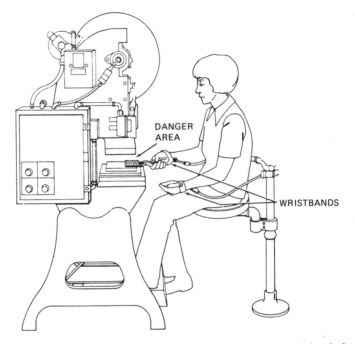

Figure 12.27. Hold-outs or restraints for restraining the operator's hands from reaching into the danger zone at *all* times (compare with pullbacks). (*Source:* From OSHA, ref. 12.)

mit the operator to reach in at all, even between strokes of the machine. If tongs, suction cups, or other gripping devices can be used to feed a machine manually, it is feasible to use hold-outs instead of pull-outs to protect the operator. Even without these gripping devices, long workpieces can be hand-fed into the machine without actually placing the hands in danger. For these applications, hold-outs are appropriate. If the operator's hands must enter the zone between the dies, however, hold-outs are infeasible as safeguarding devices.

It would seem that protection for the operator would be unnecessary in applications where tongs or other feeding devices are used instead of the hand to feed a machine. However, this idea does not recognize the strong tendency for the worker to reach in when something goes wrong. Therefore, although hand-feeding tools are somewhat useful in promoting safety, they are not recognized as point-of-operation safeguarding devices. Another means must be used to *ensure* safety, and hold-outs are a good method when hand tools are used.

Two-Hand Controls

Since people have only two hands, neither hand will be injured by the point of operation if the machine can require both of them to operate the controls, or so the theory goes. The theory is a good one, but some sophistication is necessary to ensure that the device achieves its goal. Workers take pride in "beating the system" or in trick-

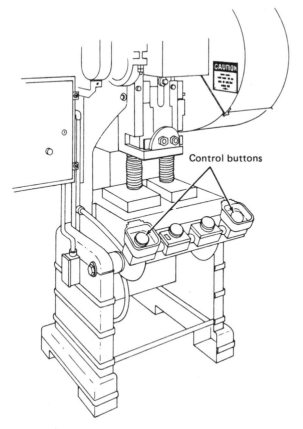

Control buttons

Figure 12.28. Two-hand control. (*Source:* From OSHA, ref. 12.)

ing the machine to make it operate without their using the controls. One trick is to use a board or rope to tie down one control so that the operator can operate the machine with one hand and feed it with the other. Another trick is for workers to use their heads, noses, or even toes to depress one of the controls. Workers have been known to try almost anything to defeat the safety features of a machine in order to achieve a production breakthrough and receive a higher production incentive payment. This points to the power of money over personal safety in motivating workers. It may also reveal some inefficiencies in safety devices as currently designed—inefficiencies that unduly compromise productivity in the name of safety. As was pointed out in Chapter 3, industry workers and managers will tolerate some slowdown of an operation for the cause of safety, but not much.

Figure 12.28 illustrates a two-hand control device and some of the features that attempt to prevent the worker from defeating the device. Note the smooth, rounded surface of the palm button, which makes it convenient for the palm but not for tying down. Also note the cups around the button, which are intended to foul attempts at tying down the button. Control circuitry can also be used to detect foul play and stop the machine if the buttons are not depressed concurrently and released between cycles.

Controls versus Trips

The term *control* implies a more sophisticated device than a mere "trip" to actuate the machine. Within the context of safeguarding the point of operation, a two-hand control means a device that not only requires both hands concurrently to actuate the machine but also stops the machine by interrupting its cycle if the controls are released prematurely. The nature of some machines does not permit such a degree of control over the machine cycle. Two-hand controls are infeasible for these machines.

For full-revolution presses and other machines that cannot be stopped once their cycle has begun, two-hand trips are used instead of two-hand controls. Two-hand trips require both hands to start the machine cycle, but once started there is no protection for the operator. If the machine cycle is fast, and if the operator is far enough away to stay out of danger, two-hand trips effectively protect the operator. But if the machine is slow or the trip station is close enough, the operator can reach into the machine *after* it has been tripped. This gets into the subject of safety distances, covered in the next section. After studying the section on safety distances, it will be easy to see that controls are superior to trips.

Safety Distances

In reviewing the safeguarding devices for protection of operators from the points of operation of machines, we see that most of them protect the operator by making it impossible to reach into the danger zone after the machine cycle has begun. However, two of these devices—the presence-sensing device and the two-hand control—rely on a capability of interrupting a machine in *midcycle*. Every mechanical machine has inertia, so some time must elapse between the signal to stop and the complete cessation of motion of the machine in the point-of-operation area. If the inertia is great, an operator might be able to reach quickly into the danger zone before the protective device is able to completely stop the machine. Therefore, the operator station must be moved back from the point of operation a sufficient distance so that a reach into the danger zone before the machine stops is impossible.

In addition to the presence-sensing device and two-hand control, the two-hand trip device must also be located at a sufficient distance, as was mentioned earlier. Although the two-hand trip is not capable of stopping the machine, protection is incomplete unless the distance to the danger zone is great enough to prevent the operator from reaching in after releasing the palm buttons.

To compute a "safe distance" for a presence-sensing device or a two-hand control, it is necessary first to compute the stopping time of the machine. Figure 12.29 shows one model of a stop-time measurement system that connects one sensor to a palm button and the other to the machine motion. Upon a signal that the palm button has been depressed, the system starts counting in fractions of a second until motion of the machine ceases. Keep in mind that we are speaking here of the motion of the machine's ram, die, cutter, or other part *at the point of operation* that could cause injury. Flywheel motion or motor rotation may continue and indeed due to their inertia would usually be impossible to stop in time to be of any benefit. The

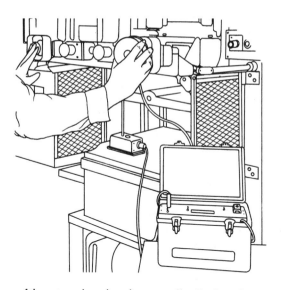

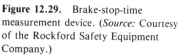

Figure 12.29. Brake-stop-time measurement device. (*Source:* Courtesy of the Rockford Safety Equipment Company.)

resulting stopping time is generally displayed as is shown in the model in Figure 12.29. The reader should take care not to confuse the brake stop-time measurement device in Figure 12.29 with a "brake monitor" to be discussed later.

Once the stopping time has been determined, this time must be multiplied by the maximum speed the hand can move toward the point of operation, as in the following formula:

$$\text{safety distance} = (\text{stopping time}) \times (\text{hand speed constant}) \tag{12.1}$$

OSHA relies on a maximum hand movement speed of 63 inches per second, sometimes referred to as the hand speed constant; this constant is credited to L. Lobl of Sweden. Consider the following example:

Example 12.1

A power press is protected by an infrared-beam-type presence-sensing device. A stop-time measurement system is used to compute the time between breaking of the infrared beam and stoppage of the press ram. This stop time is found to be 0.294 second. The safety distance then is computed to be

$$\text{safety distance} = 0.294 \text{ second} \times 63 \text{ inches/second}$$
$$= 18.5 \text{ inches}$$

Thus all points in the plane of the infrared sensing field must be at least 18½ inches away from the point of operation for this power press.

The calculation would have been identical if the device in the example had been a two-hand control (but not a two-hand trip!). The palm buttons of the two-hand control would have to be placed 18½ inches away from the point of operation.

The safety distance calculation for the two-hand *trip* uses an entirely different logic from that used for the two-hand *control*. True, the same hand speed constant of 63 inches per second is used. But with two-hand trips the idea is for the machine to complete the dangerous part of its cycle before the operator can reach in after releasing the palm buttons. Thus the *slower* the machine, the more dangerous it is when using trips. It is paradoxical that the *faster* the machine, the harder it is to stop and the greater must be the safety distance for two-hand *controls,* whereas for *trips* it is just the opposite: The *slower* the machine, the greater must be the safety distance.

With many machines, hydraulic presses for example, the commencement of the cycle is virtually instantaneous with the depressing of the palm buttons. But with machines that commence the cycle with the mechanical engagement of a flywheel, there is an inevitable delay as the engagement mechanism waits for its mating engagement point on the flywheel. The delay can be considerable for a slow-rpm machine, especially if it has only one place on the flywheel suitable for making the mechanical engagement. The process is best explained by a diagram, as in Figure 12.30. Figure 12.30a shows a very unlucky happenstance in the position of the flywheel with its engaging point just past the tripping mechanism at the moment of actuation. This follows the principle that the *worst state* of the machine should be used to determine how to make the operation safe. Since the position of the flywheel at the moment the machine is tripped is purely a matter of chance, it is equally likely that the flywheel will be in the lucky position indicated in Figure 12.30b. But since this position cannot be counted on, the safety distance is computed assuming the flywheel is in the position indicated in Figure 12.30a.

One complete rotation of the flywheel completes the machine cycle for most machines. One-half of this rotation of the flywheel is the dangerous portion of the stroke, the closure motion of the machine. Thus, adding a complete revolution for engagement plus one-half revolution for closure motion, the resultant danger period is one and one-half revolutions of the flywheel for a machine whose flywheel has only one engagement point. For machines that have several engagement points evenly spaced around the flywheel the danger period is shorter, depending on how many engagement points exist. To see this, consider an example of a machine with four engagement points. In the worst case, the farthest the engagement point could be

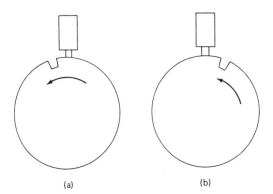

(a)   (b)

Figure 12.30. Two possible positions of the flywheel when a machine is tripped: (a) unlucky position—the engagement point has just passed the tripping mechanism; (b) lucky position—the engagement point is approaching and is very near the tripping mechanism.

from the tripping mechanism at the moment of tripping would be 90° or one-quarter revolution. Add this to the one-half revolution during machine closure, and the total danger period becomes three-quarters revolution.

Summarizing the calculation of safety distances for two-hand *trip* devices is the following formula:

$$\text{safety distance} = \frac{60}{\text{rpm}} \left(\frac{1}{2} + \frac{1}{N} \right) \times 63 \qquad (12.2)$$

where   rpm = flywheel speed in revolutions per minute while engaged
          N = number of engaging points on the flywheel

In equation (12.2) the factor 60/rpm is used to determine the time in seconds required for the flywheel to complete one revolution. The computation in parentheses is the worst-case number of flywheel rotations until closure, and the factor 63 is the hand speed constant in inches per second. The computed safety distance is in inches.

It is important to remember that equation (12.2) is to be used for two-hand *trips,* not two-hand controls. It is easy to determine that two-hand controls are superior to two-hand trips. From a production efficiency standpoint, two-hand controls can normally be placed much closer to the machine, which facilitates hand feeding. Two-hand trips can be more dangerous unless moved to a greater safety distance, which in turn impairs productivity.

Figure 12.28 was captioned "two-hand control," but the figure could also represent a two-hand trip. The device in Figure 12.28 could have been mounted on a pedestal so that it could be moved closer or farther from the point of operation, dependent on calculation of the formula for the given setup. Two-hand devices can also be mounted directly on the machine if the safety distance is short enough to permit this.

Neither equation (12.1) nor (12.2) is constant for a given machine; both are dependent on a given setup because of variations in die dimensions and die weight, which in turn affect flywheel speed, stopping time, and dimension of the danger zone on the machine. The pedestal is movable, but only a supervisor or safety engineer should be capable of moving it. This can be accomplished by locking the pedestal in position with a key or by bolting it to the floor and forbidding operators to unbolt it, or some other means of fixing the position. Failure to fix the position results in a temptation to the operator to move the pedestal closer to the machine to speed up production.

Because of the superiority of two-hand controls over two-hand trips, both in terms of safety and productivity, many industrial plants are converting their older machines equipped with trips to the more modern two-hand controls. Such a conversion is not a mere change in the two-hand device, but also generally represents a major modification to the machine itself and its power transmission apparatus, so that the machine can be classified as part revolution instead of full revolution. Brake monitoring and control reliability features are also required on part-revolution presses and will be explained next.

Brake Monitoring

It can be seen from the preceding discussion that the stopping time is very important in computation of permissible safety distance for part-revolution machines. But stopping time is dependent on the brake, and unfortunately brakes are subject to wear. Stopping time is also dependent on die setup, which can change from lot to lot. Therefore, it is naive to apply the principle of safety distances and then to walk away and trust the press to always respond in the same way as it did the day it was tested for safety distances. So for every part-revolution press whose safeguarding device depends on the brake, a brake monitoring system is needed to monitor the brake *every stroke*. Note how different this is from the brake stop-time measurement system of Figure 12.29, which would be set up only occasionally to check or set safety distances. In contrast, the brake monitor is a permanent installation on the press, monitoring the overtravel of the ram past the top stop every stroke.

Since the system is mechanical, there will be some overtravel, and a tolerance must be set that permits this overtravel. Employers can set this tolerance as high as they desire, but it will not be to their advantage to set the tolerance very high because a larger overtravel means a longer stopping time, which means a greater safety distance, which means reduced efficiency. There is no safety distance for Type B gates, but the employer is still allowed to reasonably establish a "normal limit" for ram overtravel.

The brake monitor can be engineered to measure either stopping time or overtravel distance. The most popular style is electromechanical, with a pair of limit switches triggered by a cam linked to the press crankshaft. This type is commonly known as a "top-stop" monitor (see Figure 12.31). The first switch signals the application of the brake, and the second signals overrun. The cam must not trigger the overrun switch until a new cycle is initiated. Sooner or later the brake will deteriorate, and the overrun switch will be tripped, which means that brake-stopping-time tolerance has been exceeded. At that point the monitor system must provide an indication to that effect.

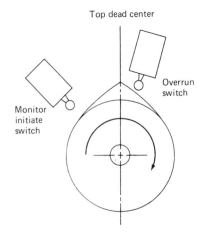

Figure 12.31. "Top-stop" brake monitor. (*Source:* From *Manufacturing Engineering*, February 1976; courtesy of the Society of Manufacturing Engineers.)

In addition to the brake monitor, a control system is required to ensure that the press will cease to operate after a failure in the point of operation safety system occurs, *but* the brake system will *not* be shut down due to the system failure. This requirement is a direct application of the general fail-safe principle stated in Chapter 3.

The reader should be reminded that brake system monitoring and control reliability is not required on all power presses. The logic is that on many press setups, the brake monitor and control system would have marginal benefit, whereas on others, they would be of critical importance due to the selection of safeguarding methods, the mode of operation, and the construction of the press itself. Table 12.2 summarizes the various options for guarding or safeguarding presses, the need for brake monitors and control systems, and alternate permissible configurations.

TABLE 12.2 Power Press Safeguarding Summary[a]

Guard Type or Device	Full Revolution		Part Revolution	
	Hands In	Hands Out	Hands In	Hands Out
Guards Barrier guards (fixed, adjustable or die enclosure)	Illegal	Inspect weekly	Illegal	Inspect weekly or Brake monitor and control system
Interlocked barrier guards	Illegal	Inspect weekly	Illegal	Inspect weekly or Brake monitor and control system
Movable barrier devices Type A gate	Inspect weekly	Inspect weekly	Inspect weekly or Brake monitor and control system	Inspect weekly or Brake monitor and control system
Type B gate	Inspect weekly	Inspect weekly	Brake monitor and control system must detect top-stop overrun beyond limits	Inspect weekly or Brake monitor and control system
Devices Presence sensing	Illegal	Illegal	Safety distance and Brake monitor and control system	Safety distance and Inspect weekly or Brake monitor and control system
Pull-out	Inspect: each shift, each die set-up, each operator	Inspect: each shift, each die set-up, each operator	Inspect: each shift, each die set-up, each operator	Inspect: each shift, each die set-up, each operator

TABLE 12.2 (Continued)

Guard Type or Device	Full Revolution		Part Revolution	
	Hands In	Hands Out	Hands In	Hands Out
Sweep	Does not qualify as safeguard	Does not qualify as safeguard	Does not qualify as safeguard	Does not qualify as safeguard
Hold-out (restraint)	Illegal	Inspect weekly	Illegal	Inspect weekly or Brake monitor and control system
Two-hand control	see Two-hand trips	see Two-hand trips	Safety distance and Fixed control position and Brake monitor and control system	Safety distance and Fixed control position
Two-hand trips	Safety distance and Fixed trip position and Inspect weekly	Safety distance and Fixed trip position and Inspect weekly	Inspect weekly or Brake monitor and control system	Inspect weekly or Brake monitor and control system

[a]This summary compares inspections requirements, brake monitoring and control system requirements, safety distances, and legal and illegal arrangements. The summary does not include detailed specifications, requirements for multiple operators, and other details too numerous to include in this summary. For details, see OSHA Standard 1910.217.

GRINDING MACHINES

Grinding machines are in almost every manufacturing plant—on the production line, or in a toolroom, or maintenance shop.

There are two or three items which create the most trouble as follows (shown in Figure 12.32):

- Failure to keep the workrest in close adjustment (within ⅛ inch) to the wheel on offhand grinding machines
- Failure to keep the tongue guard adjusted to within ¼ inch
- Failure to guard the wheel sufficiently

These rules may seem "nitpicking," but there is a grave hazard with grinding machines that most people do not know about—the breakup of the wheel while rotating at high speed. It does not happen very often, but when it does, the injuries to the operator can be fatal. It is against this hazard that all three of the items just listed are aimed, even the workrest adjustment requirement.

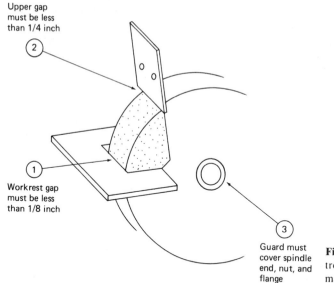

Upper gap
must be less
than 1/4 inch

②

①

Workrest gap
must be less
than 1/8 inch

③

Guard must
cover spindle
end, nut, and
flange

Figure 12.32. The three biggest troublespots on ordinary grinding machines.

A severe stress can be placed on the grinding wheel if the workpiece becomes wedged between the workrest and the wheel. A large gap invites the workpiece to become pinched and then drawn down by the wheel resulting in a severe wedging action, as illustrated in Figure 12.33. The forces of this wedging action threaten the integrity of the bonded abrasive wheel, possibly causing it to break up and hurl stone

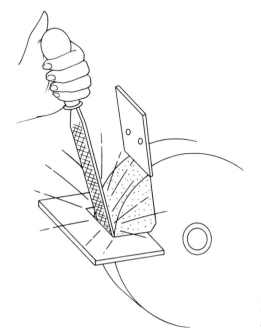

Figure 12.33. Severe wedging action due to a large gap between workrest and wheel.

fragments at the operator at near-tangential velocities. The only protection to the operator from these flying stone fragments are a good wheel guard, a properly adjusted tongue guard, and any personal protective clothing the operator might be wearing.

It is a simple matter for anyone to check the gap adjustment on the grinding machine workrest—easy to check, but not so easy to *keep* in adjustment. Since the wear of the grinding wheel causes the gap to gradually widen, constant surveillance is needed to assure that the gap is maintained less than $\frac{1}{8}$ inch. There is little tolerance for overcompensating for the wear because $\frac{1}{8}$ inch leaves little margin for setting the workrest closer than required. Since there is no way to avoid frequent adjustment, some convenient means need to be initiated to make adjustments quickly and efficiently. To this end, a workrest gauge is recommended, as illustrated in Figure 12.34. The go–nogo character of this gauge aids in a quick and sure decision as to whether to adjust the workrest every time it is checked.

Another easy check is the maximum spindle speed; it should not exceed the maximum speed marked directly on the wheel. Operation of an abrasive wheel above its design speed subjects the wheel to dangerous centrifugal forces which also could lead to wheel breakup.

Sometimes an abrasive wheel has manufacturing imperfections or transportation damage which make it dangerous. Before the wheel is mounted, it should be

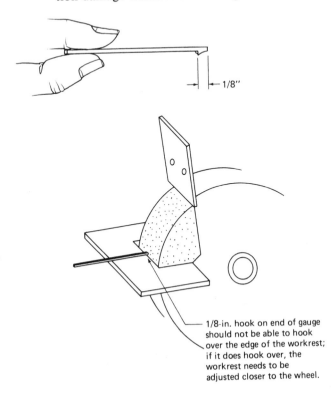

— 1/8''

1/8-in. hook on end of gauge
should not be able to hook
over the edge of the workrest;
if it does hook over, the
workrest needs to be
adjusted closer to the wheel.

Figure 12.34.   Grinding machine workrest gauge.

visually inspected for such damage or imperfections. Invisible imperfections can sometimes be detected by tapping the wheel gently with a nonmetallic implement such as the plastic handle of a screwdriver or a wooden mallet. A good wheel typically produces a ringing sound, whereas a cracked wheel will sound dead. The reason nonmetallic objects are used to make the "ring test" is that metallic objects might themselves ring, giving the impression that a defective wheel is good.

SAWS

Saws have some obvious hazards and some not-so-obvious ones. Almost everyone respects the danger of a power saw, but serious injuries continue to occur, and acceptable means of guarding both obvious and subtle hazards from saws need to be considered.

Radial Saws

If the Safety and Health Manager learns about no other saw, he or she should become familiar with this one. Radial saws or "radial arm saws" can be quite dangerous, and in addition they are difficult to guard. Figure 12.35 depicts a typical radial arm saw, but although the blade can be seen to be partially guarded on the top half, the lower portion of the blade is exposed. Figure 12.36 illustrates one type of lower blade guard for radial arm saws. The guards are very unpopular and are often removed by employees.

   Another problem with radial saws is the return-to-home position. The saw should be mounted such that "the front end of the unit will be slightly higher than the rear, so as to cause the cutting head to return gently to the starting position when released by the operator." A radial saw that creeps out while running is an obvious hazard, but one that is adjusted too much at an angle will "bounce" on the home position stop—another hazard. A radial saw also will have a stop that prevents the operator from pulling the saw off the radial arm or beyond the edge of the saw table. This stop should be adjustable for limiting travel of the saw head only as far as necessary.

   A radial saw is usually used as a cutoff saw and is mounted as shown in Figure 12.35. It is possible, though, to reorient the cutting head 90° so that the blade is parallel to the table and thus to convert the saw to a ripsaw for long pieces of stock. In this mode of operation the saw head is locked into position, and the material is pushed into the saw. There is a hazard, though, if the material is fed into the saw in the wrong direction. To be safe, the material must be fed *against* the rotation of the blade. If the material is fed *with* the rotation of the blade, especially if the feed rate is fast, the saw teeth are likely to grab the workpiece and pull it into the machine at high speeds, possibly pulling the operator's hands into the blade with the material being sawed. It is not uncommon that a radial saw fed incorrectly in this manner will hurl the workpiece through the machine and toss it across the room.

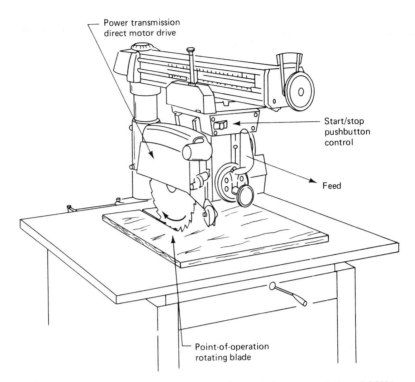

Power transmission
direct motor drive

Start/stop
pushbutton
control

Feed

Point-of-operation
rotating blade

Figure 12.35.   Typical radial saw, not properly guarded, and in violation of OSHA standard. (*Source:* From NIOSH, ref. 48.)

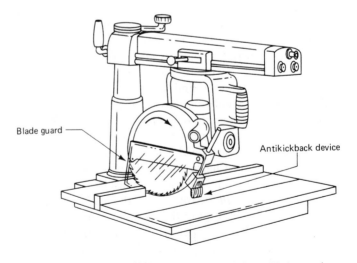

Blade guard

Antikickback device

Figure 12.36.   Radial arm saw equipped with lower blade guard.

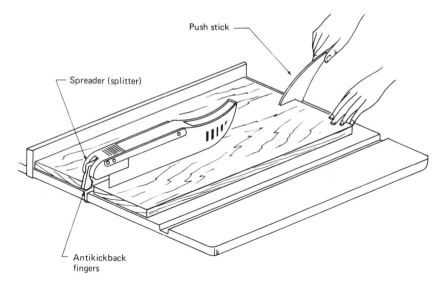

Figure 12.37. Hood guard, spreader, and antikickback fingers on a table saw.

Table Saws

Table saw is an everyday term for a hand-fed saw with a circular blade mounted in a table. Unlike the radial saw, the table saw head always remains stationary during a cut while the work is fed into it. With table saws, the three biggest problem areas are hood guards, spreaders, and nonkickback fingers (see Figure 12.37). Antikickback protection is more important for ripsaws than for crosscut saws.

The hood guards present the most problems because the obstructed view makes the saw operator's job more difficult and awkward. Although most of the hood guards in the field are metal, most new machines come with transparent plastic guards. But the rapidly rotating saw blade can cause a static charge to develop on the nonconducting plastic guard, causing it to become covered with sawdust so that the blade cannot be seen. Also, the plastic guard is easily scratched, further reducing visibility through the guard.

It is true that hood guards will not absolutely prohibit contact of the operator's hands with the saw blade. The hood guard, with its spring action, acts more as an "awareness barrier," as was discussed earlier in the section on safeguarding the point of operation on general machines. But there is another reason for using a hood guard: to protect the operator from flying objects. The saw blade rotates at 3,000 rpm, and this produces large centrifugal forces and high tangential velocities. Consider the following calculation for a popular make of a table saw:

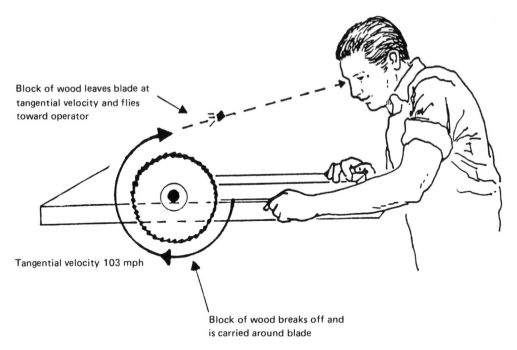

Block of wood leaves blade at
tangential velocity and flies
toward operator

Tangential velocity 103 mph

Block of wood breaks off and
is carried around blade

Figure 12.38.   Broken saw tooth or chips of wood are a hazard to the saw operator.

Blade speed: 3,450 rpm
Blade diameter: 10 inches
$$\text{Tangential velocity} = (\text{rpm}) \times (\text{blade perimeter})$$
$$= 3{,}450 \text{ rev/min} \times 10 \text{ in.} \times \pi$$
$$= 108{,}385 \text{ in./min} \times \frac{60 \text{ min}}{1 \text{ hr}} \times \frac{1 \text{ mile}}{5{,}280 \text{ ft}} \times \frac{1 \text{ ft}}{12 \text{ in.}}$$
$$= 102.63 \text{ miles per hour}$$

This means that if a tooth breaks off the saw or if a small chip or block of wood breaks off and is carried on the blade for nearly a full revolution as in Figure 12.38, it will be propelled at the face of the operator at a velocity of over 100 miles per hour. It is little wonder that eye protection is considered necessary for operating a table saw.

Kickback

Kickback is the word used to describe the situation in which the entire workpiece is picked up and thrown back at the saw operator. The energy for the kickback comes from the sawblade itself. The rotation of the blade is toward the operator. At the front of the blade where the saw first meets the work, the direction of blade motion is toward the operator and *down*. But at the back of the blade the direction of motion is toward the operator and *up*. Because the saw teeth are slightly wider than the blade thickness, a correctly aligned workpiece will contact the blade only at the point at which the cut is taking place. But if the workpiece should shift slightly, the exiting portion of the cut at the back of the blade will become misaligned, causing the edge of the material adjacent to the cut to contact the blade as it emerges from the table. This contact can result in a sudden and powerful upward thrust which causes the material to break contact with the table surface. Further misalignment becomes almost inevitable at this point, and the workpiece is grabbed firmly by the blade. If the workpiece is extremely thin or fragile, it will break up at this point, with thin portions or chips continuing with the blade down under the table. But a much more likely result is that the rigid material cannot follow the blade, and it is hurled at tangential velocity directly at the operator.

Both the spreader and the nonkickback fingers are intended to help prevent kickback. The spreader keeps the saw kerf open or spread apart in the completed portion of the cut so that the material will not contact the blade. The nonkickback fingers, or "dogs," are designed to arrest the kickback motion should it start to occur. The shape of the dog permits easy motion in the direction of the feed. A backward motion, however, causes the dog to grip the material and prevent a kickback.

Band Saws

It is essentially impossible to guard the point of operation in most band saw operations. However, the unused portion of the blade can feasibly be guarded. A sliding guard is used for this and is moved up or down to accommodate larger or smaller workpieces, respectively. This sliding guard, together with the wheel guards, permits the entire blade except for the working portion to be guarded.

Hand-Held Saws

Hand-held circular saws are subjected to a variation of the hazard of kickback, except that it is the *saw* that is kicked back instead of the material! Proper training and operator respect for the saw are important, as are a clean, sharp blade, a "dead-man" control, and a retractable guard for the lower portion of the blade. A dead-man control is simply a spring-loaded switch (button or trigger) that will immediately cut off power to the saw if the operator releases the switch.

The rectractable guard for the lower portion of the blade on a hand-held circular saw is perhaps analogous to the lower blade guard on a radial saw and the hood guard on a table saw. However, on a hand-held circular saw the retractable guard is much more important. If the operator locks the retractable guard open with a small wedge, as operators will sometimes do, the saw becomes very hazardous both before and after the cut. The blade is exposed, and direct damge or injury can result if the saw is dropped or set down on a surface.

Cutting aluminum with a hand-held circular saw can be a real problem. Hand-held circular saws are widely used to cut aluminum extrusions to length to manufacture windows, storm doors, shutters, and other architectural aluminum extrusion products. The problem is that the saw blade gets very hot, reaching the melting point of aluminum at around 1,200 °F; drops of molten aluminum start flying off the blade onto the guard, causing quite a mess. The aluminum later solidifies, causing the guard to stick and malfunction.

One prominent manufacturer of architectural aluminum extrusions wrestled with this problem for five years before turning to alternate means of protecting the operator. Reverse polarity brakes were installed on the saws, which caused the blade to come to an immediate halt as soon as the operator released the trigger. This removed most of the hazard, because it is while the blade coasts to a stop after use that the lower blade guard is most important. For added protection, the workers were provided with protective gloves and pads for the hands and wrists.

Chain Saws

Without doubt, the most hazardous hand-held saw of all is the chain saw. Binding of the blade can cause the saw to kickback and cause severe, perhaps fatal, injury to the operator. A dull chain or poorly lubricated chain can overheat and break, resulting in severe injury to the operator or to the other workers in the area. The Consumer Product Safety Commission is leading the effort to improve hand guards and minimize kickback.

Power Hacksaws

Power hacksaws are difficult to guard, more so than their cousins, the horizontal band saws. The band saw can feasibly be guarded with an adjustable guard along all portions of the blade except the portion actually performing the cut. Because of the reciprocating action of the power hacksaw, however, a much more complicated guard would be required to adjust back and forth *during* the cut every stroke. Guards, guardrails, or guarding by location would be appropriate for this hazard. Some modern power hacksaws are equipped with enclosures that contain the entire stroke of the reciprocating blade.

BELTS AND PULLEYS

Almost every industry has a wide variety of belts and pulleys and other driving systems for transmitting power from motors to machines. The hazards are well understood, and the technology is simple. Belts and pulleys represent a good target area for the Safety and Health Manager to institute a low-cost program of in-house safety improvement.

It should be recognized that not all belts and pulleys are hazardous, and the standards acknowledge this fact by excluding certain small, slow-moving belts. The exclusions are intricate and are best represented by a decision diagram, such as the one in Figure 12.39. A height of less than 7 feet off the floor or working platform is generally considered a working zone where personnel need protection from belts and other machine hazards.

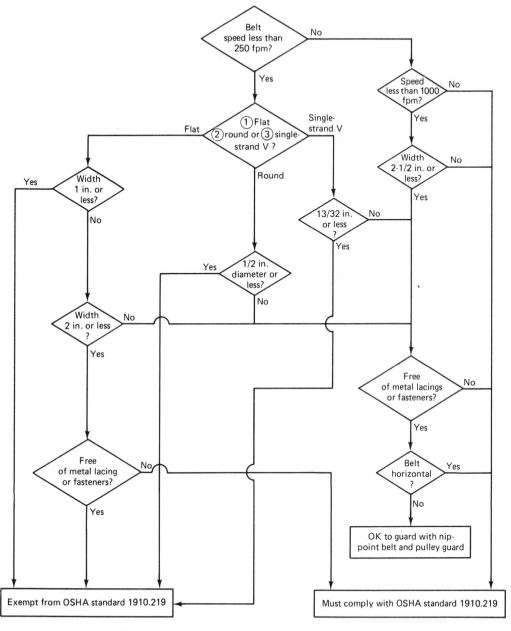

Figure 12.39. Decision diagram for OSHA belt guarding standard.

Related to belts and pulleys are shaft couplings, such as are typically found between a pump and the motor that drives it. The preferred method of eliminating hazards with these couplings is to design them such that any bolts, nuts, and setscrews are used *parallel* to the shafting and are countersunk, as shown in Figure 12.40. If these fasteners do not extend beyond the edge of the flange as shown in the figure, it is unlikely that they will cause injury. The greatest hazard with exposed setscrew heads is that they will catch parts of loose clothing and then draw the worker into the machine. The setscrews and other projections are typically invisible on the rapidly spinning shaft or flange, adding to the hazard. The installation of U-type guards, wherever needed throughout the plant, is an objective that needs to be set by the Safety and Health Manager. Once they are aware of the hazard, maintenance per-

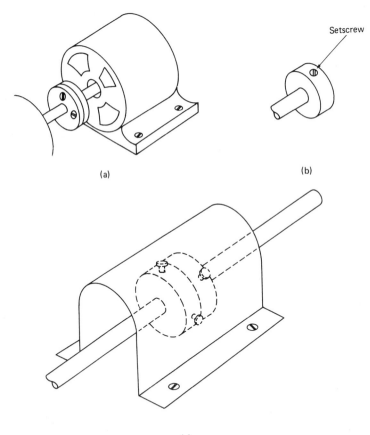

Setscrew

(a)   (b)

(c)

Figure 12.40. Safety with shaft couplings: (a) screws or bolts are mounted parallel to the shafting and are countersunk; (b) setscrew is mounted in the periphery of the flange, but is countersunk and does not extend beyond the flange; (c) exposed setscrews present no hazard because the shaft coupling is covered by a U-shaped guard.

sonnel can go about systematically taking measurements, having U-type guards fabricated in the sheet metal shop, and then installing these guards.

The possibility of safeguarding belts and pulleys by *location* should not be overlooked. Some belts and pulleys are located in a part of the machine that is protected from worker exposure. Some people believe that location is the answer to safeguarding the belt and pulley on motor-drive air compressors that operate intermittently. But because this equipment starts automatically, safeguarding by location may not be sufficient to eliminate the hazard.

One approach to guarding large air compressors is to place them in a room by themselves. The door should be kept locked. It is best that the room be kept small if heat dissipation methods permit so that it will not be used for storage or other purposes that will result in worker exposure. The safety of maintenance personnel who must enter the room to service the compressor should not be overlooked. Administrative procedures and training can be used to reduce hazards to these personnel.

It is important to call attention to the hazards of compressed air hoses used for cleaning. Compressed air hoses with nozzles are often used to spray chips away from machine areas. Safety standards specify that the air pressure used for such purposes must not exceed 30 psi. Most industrial compressed air systems operate at working pressures greater than 30 psi. Therefore, a reducer at the nozzle or a reducing nozzle is used to bring the air pressure within the specified maximum of 30 psi. Figure 12.41b shows one type of nozzle for this purpose.

Excessive air pressure from these nozzles can make flying chips hazardous. Even with proper air pressure, chip guarding and personal protective equipment are needed to protect the worker. If alternate means can be used to remove chips, it is usually in the interest of safety to discontinue the use of air nozzles. Unfortunately, metal chips are very sharp and hazardous to handle, making the cleaning process somewhat of a problem.

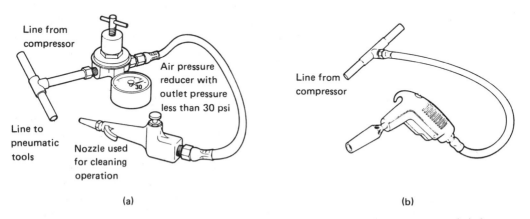

Figure 12.41. Two ways to comply with OSHA's requirement to reduce compressed air for cleaning to pressures less than 30 psi (a) pressure reducer in line; (b) special ventilating nozzle reduces pressure to less than 30 psi. (*Source:* From NIOSH.)

The hazard of flying chips is fairly obvious, but most people do not realize that compressed air hoses used for cleaning can even present hazards of *fatalities.* There have been a few recorded cases of fatalities where horseplay has failed to consider the hazards of compressed air. There is no way that the human body can contain without serious damage internal overpressures of even 30 psi. Unfortunately, workers have rarely been trained to respect the lethal pressures presented by the ordinary and seemingly harmless compressed air nozzle used for cleaning.

Jacks

A frequent weekend accident is the familiar fatality when a "shade tree mechanic" is killed under an automobile that has fallen from the jack supporting it. A jack is essential for lifting, but the supported load is usually much more stable if transferred to secure blocks, removing the load from the jack. What is unsafe for the shade tree mechanic is also unsafe for the industrial worker as far as jacks are concerned.

As with a crane or a hoist chain, the temptation is to use a jack until failure, but the result of this policy will eventually be a catastrophic failure, an unacceptable alternative for the end of life for a jack. Therefore, the only other alternative is to inspect the jack at intervals throughout its life to watch for signs that the jack either needs repair or is worn to the danger point.

SUMMARY

Machine guarding is a term almost synonymous with industrial safety and is a high-priority item for Safety and Health Managers. Although passed off by some health and safety professionals as old fashioned and nontechnical, machine guarding is actually a challenging task with new guarding technologies pressing the state of the art.

The most dangerous part of most machines is the point of operation, where the tool meets the workpiece. Unfortunately, it is also the most difficult part of the machine to guard in most cases. When guarding becomes impractical or infeasible, electromechanical devices have been devised for protecting the operator from the point of operation. Not to be overlooked are the indirect hazards at the point of operation such as flying chips or sparks.

Next in importance to the point of operation are belts, pulleys, and other power transmission apparatus. Usually more easily guarded than the point of operation, belts, pulleys, gears, shafts, and chains should receive plantwide attention. Guards can often be fabricated in-house with supervision of the Safety and Health Manager. Remember to consider the possibility of guarding by location or by distance.

One of the most important and dangerous of production machines is the mechanical power press. Safety requirements for guarding the point of operation of presses can be quite technical and complicated. Many of the guarding methods specified for presses can be used as principles for guarding machines in general.

The hazards of radial saws are exemplary of those of other types of saws. Kickback is a hazard with most saws, and various mechanical devices in addition to training workers in the mechanism of kickback are viable remedies.

OSHA's abrasive wheel machinery standard is quite complicated, but the principal problems are short and simple—guarding the wheel, nut, and flange; adjustment of the workrests on grinding machines; and adjustment of the tongue guards. The Safety and Health Manager should also give attention to such miscellaneous equipment as hand-held power tools, compressed air equipment, and jacks. Significant among these is the requirement to reduce air pressure to 30 psi or less when the air is used for cleaning.

EXERCISES AND STUDY QUESTIONS

12.1. Explain the term *point of operation.*

12.2. Name several types of mechanical hazards on machines in general. Which is the most important from a safety standpoint?

12.3. Identify several examples of in-running nip points.

12.4. Name two ways of safeguarding a machine that require no physical guard or device at all. What is the difference between these two methods of safeguarding?

12.5. What is a lockout? How does it differ from an interlock?

12.6. What are the disadvantages of nylon mesh guards for fans?

12.7. The stopping time of the ram on a certain part-revolution press has been measured to be 0.333 seconds. At what minimum safety distance should a presence-sensing device be placed?

12.8. A popular mechanical power press has a full-revolution clutch and 14 engagement points on the flywheel, which rotates at 90 rpm. At what minimum distance from the point of operation on this press should a two-hand trip device be placed?

12.9. Name some reasons why a machine might have bolt holes in its feet.

12.10. Name several types of safeguarding devices for the point of operation.

12.11. What is the difference between an interlocked barrier guard and a gate?

12.12. What is the difference between two-hand controls and two-hand trips?

12.13. What is the difference between Type A and B gates?

12.14. What is the advantage of Allen-head screws over wing nuts for machine guards?

12.15. A guard has a maximum opening size of ¾ inch. The guard openings are 6 inches from the danger zone. Does the opening size meet requirements?

12.16. What is an *awareness barrier?* What is a *jig guard?*

12.17. Explain the terms *full revolution* and *part revolution* as applied to press clutches. Which is safer?

12.18. What is "muting" of safeguarding devices, and when is it permitted?

12.19. What is the difference between pullbacks and restraints?

12.20. What is the biggest disadvantage of pullbacks?

12.21. Explain the difference between a *brake monitor* and a *brake stop-time measurement device.*

12.22. A part-revolution clutch press has a brake stop time of 0.37 second. At what minimum distance should two-hand controls be placed?

12.23. Where should a presence-sensing device be placed on the press of Question 12.22?

12.24. If the press of Question 12.22 had been a full-revolution press operating at 60 rpm and having four engagement points, at what distance should a two-hand trip device be placed?

12.25. Would a two-hand control offer any improvement in the press of Question 12.24? What about a presence-sensing device?

12.26. What are the three biggest problems with grinding machines? Why are they so important from a safety standpoint?

12.27. What is the "ring" test?

12.28. Compare hazards for table saws that are used as ripsaws versus those used as crosscut saws.

12.29. When do shaft couplings need no guards?

12.30. How can compressed air used for cleaning be dangerous? Under what circumstances is it permitted?

13

Welding

3%

Percent of OSHA
general industry citations
addressing this subject

After as broad a subject as machine guarding it may seem ridiculously specific to address a field as narrow as welding. However, it may surprise some to learn that welding processes present some of the greatest hazards to both safety and health. In terms of breadth of hazard, welding encompasses even more than the chapter on machine guarding, and for that matter, more than any other chapter of this book.

Although this chapter is entitled "Welding," this term is to be taken in a very broad sense to include gas welding, electric-arc welding, resistance welding, and even related processes such as soldering and brazing, which technically are not really welding processes. Welding processes are so diverse that before addressing the hazards of these processes, it is necessary to name them and to provide the Safety and Health Manager with the necessary background in welding process terminology.

PROCESS TERMINOLOGY

The key to understanding welding hazards is to know how the process itself works, and unless Safety and Health Managers have this knowledge, their credibility with their manufacturing and operations counterparts will be minimal. Everyone knows that welding requires that material melt or fuse to form a rigid joint. The first question to ask to determine the process is "What material melts?" If the melted material is of the parts to be joined themselves or of like filler material, the process is *welding*. If the material is some other material of lower melting temperature, the process is *brazing* or *soldering*. The breakpoint between brazing and soldering is 800 °F (427 °C), with brazing above 800 °F and soldering below.

Since welding requires materials to be melted, heat is required, usually applied intensely, to meet the demands of high melting points of welding materials. The method of applying this intense heat usually identifies the process. Excluding the unusual and exotic processes such as thermit and laser processes, the three basic categories of conventional welding are as follows:

- Gas welding
- Electric-arc welding
- Resistance welding

Gas welding is typified by the familiar oxyacetylene torch process, in which the very hot burning acetylene gas is made to burn even hotter by supplying pure oxygen to the flame. For welding lower-melting-point materials, alternate and safer gases such as natural gas, propane, or MAPP gas can be used. These alternate gases are often used for brazing and soldering.

The Safety and Health Manager should be careful not to confuse "gas welding" with some types of electric-arc welding that use an inert gas to facilitate the process. Indeed, some of these processes have names such as "gas metal arc welding" or "gas tungsten arc welding," but they are not gas welding. The telling feature of gas welding is that the gas must be used as a fuel for the process, not as an inert gas.

Even more diverse than the types of gas welding are the various types of electric-arc welding. Arc welding requires a small gap between electrodes, one of which is usually the workpiece itself. The intense heat is provided by the electric arc that forms between the electrodes. The process that typifies electric arc is "stick electrode" or shielded metal arc welding (SMAW),[1] shown in Figure 13.1. This highly portable and most popular operation is seen in welding structural steel for buildings, in repair of steel components, and in a wide variety of manufacturing processes. The *stick* is a

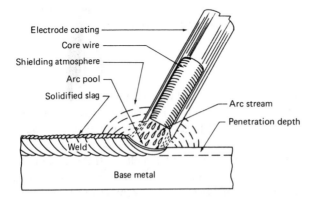

Electrode coating
Core wire
Shielding atmosphere
Arc pool
Solidified slag
Arc stream
Penetration depth
Weld
Base metal

Figure 13.1. "Stick electrode" or SMAW (shielded metal arc welding).

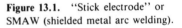

[1]American Welding Society recommended abbreviation

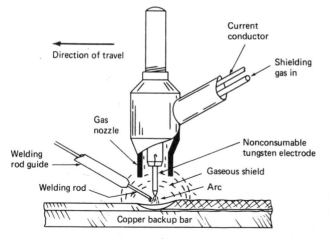

Figure 13.2. TIG (tungsten inert gas) welding or GTAW (gas tungsten arc welding).

piece of *welding rod* which is held by a gripper and is consumed by the process. The stick consists of a filler metal surrounded by a *flux,* a term to be explained later. Similar to SMAW welding is *flux-cored* arc welding (FCAW), in which the flux is on the inside of the rod, which is reminiscent of acid- or resin-core solder. Sometimes the welding process does not consume the electrode, a good example being the process commonly called TIG (tungsten inert gas) or GTAW (gas tungsten arc welding) shown in Figure 13.2. In a related process, GMAW (gas metal arc welding), the electric arc consumes a *flexible* electrode which is spooled on a reel and is continuously fed to the arc during welding.

The terms *flux* and *inert gas* have been used earlier and need some explanation. The extremely hot melting temperatures of steel and other metals make these metals very vulnerable to oxidation, which is harmful to the weld. Flux is typically a chemical compound which combines with impurities and with oxygen to prevent harmful oxidation of the hot metals. After combining with impurities while the flux is in the molten state, the resultant molten liquid is called "slag," which later solidifies and must be removed from the finished weld. With some processes, one of the inert gases such as argon or helium is used for the same purpose. The inert gas displaces the air ambient to the weld and thus keeps harmful oxygen away from the hot metals. Unfortunately, this inert gas also sometimes keeps the oxygen away from the *welder,* an obviously undesirable characteristic that will be explored subsequently in the section on hazards mechanisms.

One other electric-arc process should be mentioned at this point in the text: the "submerged arc" process, or SAW (submerged arc welding). In this process, the flux is granular and, as can be seen in Figure 13.3, the electric arc is hidden under the pile and puddle of granular and melted flux, respectively. This has great safety and health advantages, and submerged arc is growing in popularity. Automatic welding machines are often programmed to apply flux automatically and move the electrode over a long, straight path in the manufacture of large structural steel beams

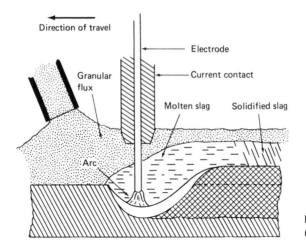

Direction of travel

Electrode

Granular flux

Current contact

Molten slag   Solidified slag

Arc

Figure 13.3. Submerged arc welding (SAW).

or "plate girders." However, SAW welding in overhead positions is a problem because the granular flux will fall by gravity instead of covering the weld.

Resistance welding (Figure 13.4) is one of the least hazardous of welding processes. It is widely used in mass-production manufacturing. However, resistance is generally restricted to relatively thin sheets of material. The concept of resistance welding is to pass electrical current *through* the material to be welded, enabling the heat so generated to melt the material. Physical pressure is also applied at the point of the weld. The attractive thing about resistance welding is that the melting generally occurs only where the mating surfaces meet. The outside and adjacent surfaces which are exposed to atmospheric contaminants harmful to the weld do not reach melting point, and damage to the material is minimal. This fact also precludes the need for a flux or inert gas to complicate both the production process and the safety and health aspects.

Resistance spot welding (RSW) is widely used to join sheet metal coverings, housings, guards, and shields on products as varied as space heaters and grain bins. Another important industry for spot welding is the automobile industry. Seam welding (RSEW) is generally preferred over spot welding for watertight seals because a pair of rollers apply continuous pressure and a series of electrical pulses make, in effect, a seam of overlapping spot welds.

Some of the more unusual or exotic processes should be mentioned because, although they may rarely be used or seen by the Safety and Health Manager, the nature of their hazards may be entirely different. The "thermit" method of welding (TW) employs a chemical reaction to produce the welding heat. Thermit is good for awkward applications or perhaps where the welding to be done is remote from convenient electrical power or gas sources. Laser beam welding (LBW) uses a concentrated laser light beam to generate the welding heat. Lasers pinpoint the weld so precisely that they are used for almost microscopic applications on tiny parts.

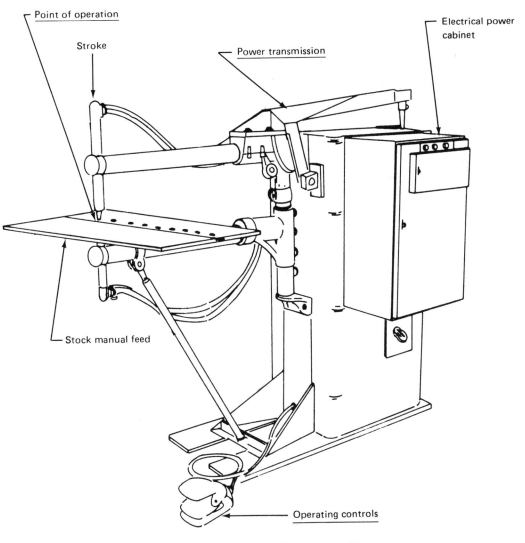

Point of operation

Stroke

Power transmission

Electrical power cabinet

Stock manual feed

Operating controls

Figure 13.4.   Resistance spot welder.

GAS WELDING HAZARDS

The myriad different processes for welding should give some clue as to why it was stated earlier in this chapter that the breadth of welding hazards is greater than that of any of the subjects of any of the other chapters of this book. We now examine these hazards, beginning with gas welding because it has given Safety and Health Managers the most problems.

Acetylene Hazards

Gas welding cylinders are so familiar that it is difficult to keep in mind their devastating destructive power. Acetylene gas, the fuel gas for most gas welding, is so unstable that its pressurization in manifolds to pressures greater than 15 psig (30 psia) is prohibited. Contrast this low pressure with the familiar oxygen, nitrogen, or other ordinary compressed gas cylinder, which contains pressure greater than 2,000 psi.

There are tricks to avoid the hazards of instability of acetylene gas; the most popular one is to dissolve the gas in a suitable solvent, usually acetone. Then the pressure can be raised to around 200 psi. As the acetylene gas is used from the acetylene cylinder, the pressure is lowered slightly, permitting a greater quantity of gas to bubble out of the solution, resulting in equilibrium at a pressure suitable for welding. In this fashion a relatively large quantity of acetylene can be stored in a reasonably portable cylinder.

Figure 13.5 shows the inside of an acetylene cylinder. Most people do not know that the cylinder contains a solid absorbent filler material for the acetone–acetylene solution. The contents are more liquid than gas, and this gives rise to a hazard mechanism. Acetylene cylinders should be kept valve end up both while in storage and while in use. There is no harm in tilting cylinders slightly, and in fact this is common practice in the use of hand trucks to handle cylinders connected for use. Good advice, however, is not to tip acetylene cylinders to an angle more than 45° from vertical.

If the cylinders are stored horizontally or valve end down, liquid acetone could enter the valve passages instead of the intended acetylene gas. Then later, when the

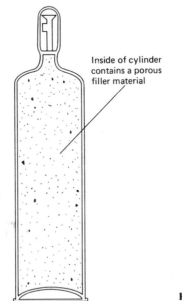

Inside of cylinder
contains a porous
filler material

Figure 13.5.   Acetylene cylinder.

valve is opened, the welder could get an unexpected flow of highly flammable liquid acetone instead of acetylene gas. Since the purpose of opening the valve is usually to ignite the torch with a sparking source, it is obvious that it would be easy to ignite the liquid acetone by accident. A burning quantity of spilled acetone is difficult to control and is quite dangerous. The reader is referred to Chapter 8 to consider the flammability characteristics of acetone.

Another way to get liquid acetone through the cylinder valve is to use the cylinder when it is nearly empty. Acetone coming through the valve or leaking elsewhere is easy to detect. The principal active ingredient in nail-polish remover is acetone, and the odor of nail-polish remover is familiar to everyone.

Acetylene cylinders have been known to leak around the valve stem, causing what welders call "stem fires." Another place to check for leaks is from the plug in the bottom of the cylinder. In one incident, fortunately not a serious accident, the welder was perplexed by occasionally hearing a small explosion, audible but causing no damage. The mystery was finally solved when it was discovered that the explosions were occurring in the small concave area beneath the bottom of the cylinder. The little explosions were harmless, but imagine the hazard that could have accumulated if there had been no ignition, and the defective cylinder had been allowed to slowly release acetylene, as in overnight storage.

It is important to be able to turn off the fuel flow quickly in an emergency, especially when the fuel is acetylene. Some acetylene cylinder valves are designed to accept a special wrench, but ordinary hammers and wrenches are in general inappropriate for opening cylinder valves. The special wrench should be kept available for immediate use.

A carbide "miner's lamp" burns on acetylene by dropping lumps of calcium carbide into water. The resulting chemical reaction is as follows:

$$CaC_2 + 2H_2O \longrightarrow Ca(OH)_2 + C_2H_2\uparrow$$

Thus acetylene gas slowly bubbles out of the water solution and fuels the lamp. The same chemical reaction can be used for generating acetylene gas for welding by means of an acetylene generator. This process avoids the hazards of large-scale storage of acetylene cylinders. But the process of storing acetylene in cylinders has been so perfected, so commercialized, and made relatively safe, that acetylene generators are rarely seen anymore. The cylinder supplier has a large acetylene generator somewhere, but this is of no concern to most Safety and Health Managers.

Before leaving the subject of acetylene hazards, we should consider alternative fuel gases. If a Safety and Health Manager really wants to win the approval of top management, he or she will offer a production innovation that will make the workplace safer, reduce the company's legal vulnerability, and cut production costs—*all at the same time!* It is difficult to do but not impossible, and selection of welding fuel gas presents a potential opportunity. It takes some "homework" and particular diligence to pursue the issue, but the rewards to the company and to everyone involved can be dramatic.

MAPP gas, natural gas, and propane were mentioned earlier as possible alternatives to acetylene. The first objection that the Safety and Health Manager will hear to the idea of using these gases is that they do not burn hot enough. It is true that acetylene excels when a very hot flame is needed, but many industrial applications do not need temperatures as high as welders want for *some* applications. Alternate gases are certainly hot enough for brazing and soldering, but they are also hot enough for some welding applications.

One reason that halfhearted attempts to switch to alternate gases do not work is that welders attempt to switch gases while using the same torch tips they used with acetylene. Special torch tips may be the secret to making the new idea a success. Welders may also need to be taught how to adjust their torches to achieve a "neutral flame" with the proper burning characteristics.

Natural gas is the cheapest alternative welding fuel of all in many cases because it is piped into the plant. On the negative side, welders will say that public-utility natural gas pressure levels of 4 oz/in^2 are too low to use for welding. But the author knows of one company that took this problem to the utility and negotiated a special arrangement in which the utility agreed to supply the welding manifold with natural gas at high pressures. The results were cost savings, a safer workplace, decreased exposure to acetylene regulations, which no longer applied because the acetylene was gone, a happy plant manager, and a *delighted* Safety and Health Manager.

Oxygen Cylinders

We have seen the hazards of highly unstable acetylene, and by comparison oxygen is much more stable; in fact it is almost completely inert if it is kept away from fuel sources. But ironically, oxygen cylinders are more dangerous than acetylene cylinders. The reason for this danger is the extremely high pressure contained by the oxygen cylinder. Figure 13.6 depicts the familiar oxygen cylinder; note its resemblance to the shape of a bomb or a rocket. It is difficult to comprehend the energy that can be released by the sudden rupture of the valve on an oxygen cylinder containing 2,000 psi pressure. There have been numerous accounts of cylinders becoming airborne and crashing into brick walls, demolishing the walls. If a cylinder valve ruptures while the cylinder is confined in a relatively small room, it may ricochet off walls until it kills anyone unfortunate enough to happen to be in the same room when the valve breaks. Consider the hazard of a heavy cylinder flying wildly around the room like a rapidly deflating balloon.

Workers often unwittingly drop oxygen cylinders onto the ground or bang them violently together. Oxygen cylinders are often seen standing alone and unsupported. Even though they are quite heavy, their small bases make them easy to tip over, with the resultant danger of knocking off the valve. The temptation to leave oxygen cylinders standing alone and unsecured needs to be dealt with in safety training sessions.

Another temptation with oxygen cylinders is that they seem to make perfect rollers for supporting and moving heavy items about. No matter whether the cylinders

Figure 13.6. Oxygen cylinder.

are full or "empty" (even a spent cylinder is not completely empty), use of cylinders as rollers or support can damage the cylinder and perhaps the valve. Furthermore, the large, heavy cylinders used as rollers present a problem of *control*. Once a heavy load begins to roll on a set of welding cylinders, it can even *run over* an unsuspecting victim.

A perennial problem is keeping track of the valve protection cap, which must be removed to use the cylinder but which also must be screwed back into place when the cylinder is in storage. This cap protects the important valve from damage, and if the cylinder is ever moved about without it, there is a risk of the cylinder falling and knocking off the valve, with disastrous results.

When an oxygen cylinder is strapped to a wheeled cart or hand truck along with its companion acetylene cylinder and the regulator assembly is in place for welding, the cylinder is generally considered to be in operational status, not in storage. Therefore, industry practice says that the valve protection caps do not have to be in place in these situations.

Looking carefully at the valve protection cap in Figure 13.6, a somewhat odd-looking vertical slot can be seen; actually, there are two, but one is hidden. These slots have a definite engineering purpose, but not the purpose most people think. If the valve comes off while the protection cap is screwed into place, the escaping gas will impact at high velocity on the closed top portion of the cap, tending to counter the force of the gas escaping at the valve. The slotted parts on the sides of the cap permit the gas to escape, but in directions exactly opposite to each other, balancing forces and leaving the cylinder relatively at rest.

Unfortunately, the existence of the slotted openings on the cap are an invitation to their misuse by the worker who is attempting to handle the cylinder. Cylinders are heavy and unwieldy, especially to the worker who has had to handle a lot of them in one day. Furthermore, in cold weather these cylinders have a tendency to become frozen to the ground, to a slab, or even to each other. The worker will want to find a means to break them apart from each other or from the slab to which they are frozen. The slotted opening on the cap seems to be an ideal place to insert a pry bar to obtain some leverage. But this is *not* the purpose of the slots, and misuse in this fashion is what can lead to a broken or damaged valve.

As if the extreme pressure hazards were not enough, oxygen presents additional hazards due to its chemical properties. As stated before, it is relatively stable in the absence of fuel sources, but the fire hazard of pure oxygen under pressure in the presence of a combustible substance is extraordinary. A substance as benign as ordinary grease can suddenly become explosively combustible in the presence of pure oxygen under pressure. A worker will often hold his or her hand over the valve opening when first opening the valve to test the cylinder. If grease is on the hands or if the worker is wearing greasy gloves, a hand can easily be lost in the explosive combustion that can follow.

We have examined the separate hazards of acetylene and oxygen, but when oxygen and acetylene cylinders are stored together, the hazards are multiplied. There is always the possibility that one or more cylinders will leak. The reader will perhaps recall the account earlier in this chapter describing small explosions from a leaking plug in the bottom of an acetylene cylinder. Acetylene is already highly flammable, and the presence of pure oxygen makes the situation about five times as serious. (Ordinary air is only about 20% oxygen; pure oxygen is 100%.) A noncombustible barrier at least 5 feet high must separate oxygen and acetylene cylinders, or they must be moved apart at least 20 feet.

Torches and Apparatus

Because of their vital role in safety, torches, manifolds, regulators, and related apparatus must be "approved," usually taken to mean by a recognized testing laboratory, such as Underwriter's or Factory Mutual.

The familiar torch, illustrated in Figure 13.7, is a more sophisticated piece of engineering than most people realize. The torch is often taken to be simply a handy double tube-and-valve assembly for delivering both oxygen and fuel gas to the weld flame, but it is more than that. Note in the figure that it has a *mixing chamber*. The torch is designed so that the mixing takes place at the right time and in the correct *total* volume. The welder controls the *proportion* of volumes of the mix by adjusting the torch valves for oxygen and/or acetylene, an oxidizing flame being oxygen-rich and a reducing flame being fuel-rich, but the total volume of the mixture is determined by the torch itself. The mixing chamber is mated to the various correct apertures for the approved torch tips, and this also is an important balance. If the balance is disturbed, flow rates may also be disturbed, and the flame may begin to travel

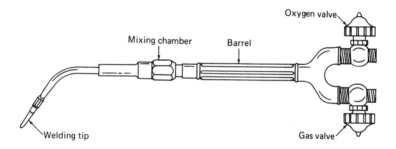

Figure 13.7. Oxyacetylene torch.

back up the mixture stream and begin to burn *inside* the torch! This is really not all that uncommon; all welders know this phenomenon and commonly call it "flashback." A popping or snapping sound is a warning that flashback is about to occur. Once the flashback begins, a distinctive humming sound can be heard. The heat being generated inside the torch will soon ruin it, and it also presents a safety hazard. The dangerous situation is alleviated by turning off both torch valves quickly.

Even with approved torches and tips, flashback can occur because of the deterioration of the equipment, especially the tips. The tips are close to the heat and naturally become brittle or burned and crack, or pieces may break off. If the tip is not replaced, flashback is likely.

Despite the importance and sophistication of the torch and tip, it is not uncommon to see the torch used as a hammer or chisel! The temptation arises due to the formation of a slag, a waste product of the flux mentioned earlier. This hard coating of slag generally covers the weld and sticks to it. To finish the job, the welder or helper must chip the brittle slag off the finished weld. The torch and tip are so handy for this purpose (take note of shape in Figure 13.7) that welders will often take the shortcut of using the torch as a chipping tool. This is a good way to ruin an expensive piece of apparatus and at the same time increase the likelihood of flashback due to a damaged torch or tip.

The torch assembly is expensive and may be owned personally by the welder. Even when the company owns the torch, the individual welder to whom it is assigned may rightfully be very possessive because of the importance of care of the apparatus. This can cause another safety problem. Welders may want to keep their torches in their locked toolboxes, but the danger here is that almost all toolboxes contain at least some grease or oily materials. Grease or oil on the torch is dangerous because of the oxygen hazards discussed earlier.

Another dangerous procedure is to lock the torch in a personal locker whose door is designed to allow the torch to remain connected to the hose. This might be all right if the welder is careful to close the valves at the regulator or tanks, in addition to the convenient torch valves, but if the regulator and tank valves are not closed, a tiny leak from the torch valve may allow an explosive mixture to gradually develop overnight inside the confined locker space. In one accident a welder opened his locker door one morning and was decapitated as the explosive mixture was apparently ignited by his cigarette.

The next most abused piece of welding apparatus is the hose for delivering the gas to the torch. To be practical, this hose must be flexible, and it is thus subject to the physical hazards of wear and tear and deterioration to which such materials are generally susceptible. Multiple hoses can easily get tangled, and because of this, welders commonly wrap the acetylene and oxygen hoses together with tape simply to keep them more orderly, but such taping practices may hide defects in the hose. It is a good idea to keep at least 8 of every 12 inches uncovered.

Manifolds are the rigid tubing networks that enable one or more cylinders to supply one or more torches. Manifold setups are sometimes found when a regular production operation requires long-term, regular use of welding gas. The principal purpose of welding manifolds is to increase gas welding volumes, but safety is usually enhanced by the more permanent arrangement.

Service Piping

Manifolds, described in the previous section, are not to be confused with "service piping," a more permanent arrangement. Some plants use so much welding gas that it becomes practical to pipe the gas to the work station, which can present some problems. The piping for acetylene must be either steel or wrought iron because copper might react with the acetylene to produce copper acetylide, a dangerous explosive. A danger with oxygen piping is again the possibility of contact with oil or grease. Fittings and pipe must be checked before assembly and thoroughly cleaned if necessary. A solution of hot water and caustic soda or trisodium phosphate is suggested for this purpose. A modern solvent such as chlorothane (1,1,1-trichloroethylene) is recommended by some engineers today. Oxygen piping systems should be purged after assembly by blowing out with oil-free nitrogen, or oil-free carbon dioxide. Chlorothane should be used to be sure that every trace of oil is removed.

Flashback can occur in service pipe systems, too, and flashback protection devices are specified, as are check valves in appropriate positions. One design for flashback protection devices involves a simple water-lock. But if the water freezes, the system will not work, so antifreeze protection is needed.

The things that can go wrong with a service piping system for welding are numerous and sometimes subtle. These problems involve inadequate emergency venting, improper joints, mistakes in installation in tunnels, and other items too numerous to mention here. The purpose of this book is to alert the Safety and Health Manager to the potential problems accompanying these systems so that he or she can be sure that personnel obtain and follow the appropriate standards for installing these systems.

ARC WELDING HAZARDS

Gas welding may have a more stormy safety record, but arc welding is a more popular process and is in many ways even more hazardous. This is one of the ironies of the subject. The major hazards of arc welding are health hazards, fires and explosions, eye (radiation) hazards, and confined space hazards. Although these hazards may be major for arc welding, they appear perhaps to a lesser degree in gas welding and other welding. Therefore, these subjects are addressed in sections of their own later in this chapter. Before these arc welding hazards are covered, the equipment used in arc welding needs to be understood.

Equipment Design

Industrial arc welding manufacturers do all they can, by means of federal standards they have helped to write and in other ways, to promote their equipment to the exclusion of lesser and cheaper models that might compete. But there is a logic to their efforts other than the profit motive. Small and relatively inexpensive models of arc welding machines are available which operate off ordinary household 110-volt current. But there are physical disadvantages with welding off ordinary household current. Welding requires large amounts of electrical power in the form of low-voltage, high-amperage circuits. Since household circuits are rated for amperages insufficient for effective welding, the small household machines make up in voltage what they lack in amperage. The high-voltage hazards of these small welding machines are admittedly difficult to understand because the industrial welding machines are supplied by much higher voltages—typically 240 to 480 volts! The key to this paradox is that the industrial machines step down the high voltage to less than 80 volts while raising the amperage to effective levels. Chapter 14 will return to the subject of voltage and amperage and will perhaps add clarity to the welding machine problem.

Grounding

Even with machines operating at proper voltages, the welder or other personnel can receive an electrical shock from contact with the machine if something goes wrong. The protection for this is to be sure that the frame of the welding machine is properly grounded. Thus if there is a dangerous short to the frame of the machine, the overcurrent protection mechanism on the circuit will be tripped, protecting personnel. Welding machine grounding needs to be strong, both physically and electrically, to meet the demands of the current that may be applied to it. This is an especially important consideration for portable machines.

Operation

In the case of welding equipment, safety and health training will pay dividends in longer service lives on the equipment, a point sometimes overlooked. Welding cable carries so much electrical current that it can overheat and damage the insulation. Coiling the cable, although convenient, contributes to this hazard. Coiled cable should be spread out before welding. Splices in the cable are not permitted within 10 feet of the electrode holder. The splices themselves must be properly insulated. Some judgment is required to determine when welding cables should be replaced. Certainly, damage to the extent that some conductors have bare spots is cause for replacement.

Care must be taken by the welder to prevent the wrong items from becoming a part of the welding circuit, either during welding or while the electrode holders are not in use. Voltage is not so much the danger as is heat produced by the potentially high amperages. Compressed gas tanks or cylinders must not be part of the electrical circuit, regardless of the flammability of their contents. The heat buildup caused by a high-amperage current through the conducting metal cylinder can cause pressure buildup in the cylinder which can exceed its design limits.

Some types of arc welding are safer than others, and they are gaining in popularity. This is especially true with respect to the hazards of fume generation and radiation, which will be discussed later.

RESISTANCE WELDING HAZARDS

The cleanest, most healthful, and probably safest form of welding is resistance welding. There are still hazards from electrical shock, but more important are the mechanical hazards surrounding the point of operation.

Shock Hazards

As in the spark coil of an automobile, many resistance welding machines build up electrical energy in a bank of capacitors for sudden release when the weld is made. The voltage may reach hundreds or even thousands of volts at peak. These voltages are not the empty variety seen in discharges of static electricity collected by walking on a thick carpet. The voltages may be at the same level, but the welding machine voltages carry with them the capacity to deliver a burning current. The capacitors that store this electrical energy should have interlocked doors and access panels. Not only must the interlock stop the power to the machine, it must also short-circuit all capacitors. Without this short-circuiting, the capacitors could deliver a lethal shock even with the power *disconnected*. Parenthetically, the same is true for the inside back of an ordinary television set.

Guarding

Spot and seam welding machines apply pressure to the materials when the weld is made. For spot welding machines this pressure makes the machine analogous to a power press, and the operator can be injured from the mechanical hazards alone.

Seam welding machines are not like power presses, but they, too, have point-of-operation hazards. The nature of the hazard for seam welders is that the opposing rotation of the rollers produces a pair of in-running nip points both above and below the material being welded. The reader may want to refer back to Figure 13.4 to study the operation of the spot welder. The hazards of seam welders do not seem to be as pressing as those of spot welders because seam welding is more often a part of an automatic or mechanized production operation than spot welding, and thus operator exposure is not as great.

FIRES AND EXPLOSIONS

Welding is one of the principal causes of industrial fires. Perhaps even more than for any other welding hazards, the Safety and Health Manager can have an impact upon this particular hazard because preventing welding fires is more of a procedural matter than anything else. This means that training becomes a very important element in the hazard prevention strategy. Fortunately for the Safety and Health Manager, there is a wealth of audiovisual aids, literature materials, and case studies on the subject of welding fires.

To use just one case history as an example, one of the most devastating and tragic industrial accidents in the nation's history occurred in Arkansas in the 1960s. A welder's spark started a fire in a missile silo, and 53 workers trapped inside the silo were killed. This accident dramatically illustrates how welding adds to other hazards, such as confined workspaces, flammable and combustible materials, and lack of ventilation. Welding on old oil drums or pipes that have contained asphalt or other petroleum products has resulted in a large number of explosions and senseless fatalities.

People do not seem to realize the ignition potentials of welding operations. Welding is not safe to watch directly because of the eye hazards, so unfortunately, except for welders themselves or their helpers, few people realize what kind of a fireworks display is really taking place. Some industrial movies are good for illustration of these fireworks. Sparks are flying everywhere—not just the benign variety as seen flying from a typical bench grinder, but visible chunks and spatters of red-hot molten metal that can burn a hole completely through heavy fabric, plastic containers, and cracks in floors. Welding is often a short repair operation, and the temptation is to take a few chances because of the short duration of the hazard.

Welding Permits

To counter the temptations to ignore the short-run hazards of welding repair operations, the Safety and Health Manager might do well to institute a procedure for mandatory permits to weld in general plant areas and in warehouses. There are so many special precautions to take that a signed checklist is a good idea. The responsible party is usually the supervisor of the area in which the welding is to take place, but in some instances the welder can make the necessary checks and sign the form. The Safety and Health Manager's responsibility is to set up the permit system and ensure that it is executed properly by actually checking permits occasionally when welding is seen to be taking place in areas of potential hazard. Any responsible party will give the matter some conscientious attention before signing a welding permit form, especially if personnel have received safety training that exposes them to the devastating hazards of welding fires such as the one that killed 53 in Arkansas. Such training will certainly make the supervisor or welder think twice before signing the form.

Judgment should be exercised to attempt to set up a permit system that will be considered reasonable by welders and plant personnel alike. The key to this reasonableness will be to set up blanket permits or exemptions from the permit system in those areas of the plant in which fire hazards are minimal. Welding done in the welder's own welding shop, for instance, should be his or her responsibility, and for the Safety and Health Manager to attempt to impose a permit system in the welding shop would obviously constitute unwise interference. For some plants the entire plant might be reasonably safe from welding fires, and no permit system may be needed at all. Hazard identification surveys and advance planning for exactly when and where the permit system is really needed will go a long way toward the establishment of a reasonable system.

EYE PROTECTION

Eye protection comes under the topic of personal protective equipment (Chapter 9), but eye protection for welding operations is so important that this chapter on welding would not be complete without a section on this subject.

Note the careful reference in the preceding paragraph to welding opera*tions,* not welding opera*tors.* The welders themselves need protection, of course, but it is easy to forget about welding helpers and others in the area. Almost every welder has already received an eyeburn some time during his or her career and knows to be careful to wear eye protection as much to prevent pain and discomfort as to prevent long-range eye injury. But less experienced personnel may need more supervision and administrative controls to ensure protection.

Remember when referring to welding lens shade numbers that the higher the shade number, the darker the shade (i.e., the more rays are shielded). The various arc welding methods produce much more intense radiation and thus require higher shade numbers than does gas welding. The radiation from arc welding, except submerged arc methods, is so intense that a helmet is necessary to protect the entire face area from painful burns. Gas welding is typically done while wearing goggles.

PROTECTIVE CLOTHING

Proper clothing is serious business to the professional welder. Virtually every experienced arc welder has, some time in his or her career, sustained a "sunburn" from the ultraviolet rays produced by the welding arc. Such a welder does not need to be told to use protective clothing to cover all skin areas that would otherwise be exposed to the arc rays. But ultraviolet burn is only one of several hazards against which protective clothing is intended. Hot, burning sparks or small pieces of molten metal can fall into openings around shoes or between pieces of clothing. Even when the clothing is seamless or when openings are well shielded, a hot piece of molten metal can be trapped in a fold and burn its way through to the welder.

Leather is the traditional favorite material for welding gloves, aprons, and leggings because of its superior thermal protective qualities. Wool is also very durable. But Nomex and other synthetic materials are also becoming popular for welders' protective clothing. Cotton fabric is attacked by the radiation and soon disintegrates even if it escapes ignition by the sparks.

Hazards to the welder from falling sparks and weld metal are greatly increased when the welding must be done overhead. Here the welder is virtually taking a shower in welding sparks. All clothing openings must be carefully shielded, and even under the helmet the welder's head must be protected from such misfortunes as a welding spark in the ear.

GASES AND FUMES

There are two extremes in degree of concern over welder breathing hazards. One extreme, call it position A, is taken generally by welders themselves, who often have no concern at all over chronic exposure to welding "smoke." Some welders even enjoy the smell of welding fumes in the air. The other extreme, position B, is the occasionally overzealous industrial hygienist who can find a hazard somewhere in almost all welding fume situations. Both extremes are only partially correct and can lead to dangerous errors in safety and health strategies.

The principal error in position A is that persons taking this extreme position are usually overlooking the long-term effects of chronic exposure. These persons tend to believe that if the welding smoke does not make them nauseated, dizzy, or give rise to some other acute symptom, the fumes are safe. Recalling the principles covered in Chapters 1 and 7, the chronic exposures can actually be the most dangerous because of their adverse effects on worker health.

Position B exaggerates the effects of tiny exposures to dangerous contaminants. It is terrifying to realize that some welding releases phosgene gas, the same gas that has been used in chemical warfare. But the exposures are generally very low and can be controlled by appropriate procedures. In the final analysis, no epidemiological studies have shown welding to be an unusually dangerous occupation. From a health standpoint, welders do not have significantly shorter life spans than workers in general. Having considered this perspective, let us categorize the hazards of welding atmospheres and examine rationally what should be done about them.

Contaminant Categories

Figure 13.8 diagrams the major kinds of welding atmosphere contaminants: particulates and gases. The particulates are dust particles or even tinier smoke particles. Metal fumes in the welding atmosphere are tiny particles of metal that have been vaporized by the arc and then resolidified into particles as they cool. The gases may be either already present, as in inert-shielding gases, or they may be chemical reaction products of the process.

The term *pneumoconioses* in Figure 13.8 was explained in Chapter 7; it is merely a general term which literally means "reactions to dust in the lungs." Everyone's lungs must deal with dust to some extent, and some welders' pneumoconioses are no more hazardous than would be caused by sweeping the floor. Some welding dusts

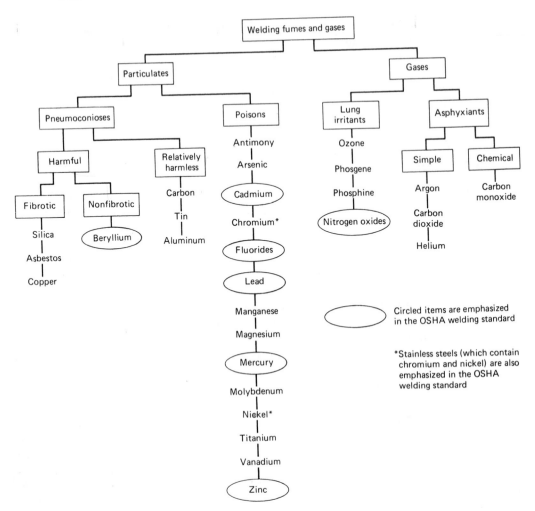

Figure 13.8.   Classification of welding fumes and gases by hazard.

are more hazardous, however, because they cause "fibrosis," or the building up of useless fibrous tissue in the lungs. The most harmful dusts are those in which the microscopic particles have the shape of fibers instead of more rounded particles. Asbestos and silica are good examples.

The "pulmonary irritants" are simpler in that they attack the lungs directly, whether the irritants be particulate or gas. The more insidious hazards, though, are from those particulates or gases that do not irritate the lungs directly, but through the lungs gain access to the rest of the body, where they act as systemic poisons.

It would be nice to have quantified tables of expected contaminant levels for various atmospheric constituents for various types of welding. There have been some experimental attempts at this (see ref. 30 and 88), but there are so many variables to be contolled that it is almost impossible to obtain reliable predictions of weld fume content. The best strategy is to be aware of potential hazardous contaminants and to know what conditions are most likely to produce these contaminants. Atmospheric sampling can then be used in suspect situations to establish whether contaminant levels are indeed excessive.

Hazard Potentials

The biggest contributor to atmospheric contaminants around welding is the coating or condition of the surfaces to be joined. It is true that welding on clean iron or ordinary construction steel produces fairly large concentrations of iron oxide fume, but fortunately siderosis, the pneumoconiosis resulting from iron oxide, is not really a very dangerous disease when it occurs alone. If the surface of the metal has a coating of material containing asbestos, however, this coating must be removed to prevent asbestos contamination of the air.

Even the act of cleaning the metal surfaces to be welded can result in secondary hazards. If chlorinated hydrocarbons, such as trichloroethylene, are used to clean the metal, these solvents must also be removed thoroughly before the welding takes place. The energy of the welding arc can cause decomposition of the solvent into dangerous phosgene gas.

Galvanized is a term referring to a zinc coating on the metal to prevent rust. Welding on galvanized steel needs extra caution and good ventilation because the welding arc can produce a fume of zinc or of zinc oxide. Zinc is not as dangerous as its relative, lead, but it can cause a brief but uncomfortable "metal fume fever." Daily exposure results in a sort of immunity, but this immunity is lost in a few days, even in just a weekend away from exposure. The next Monday morning the nausea and chills are back again, causing this disease to be known as "Monday morning sickness," although admittedly there are other things about Mondays that often make workers "sick."

Plating metals are usually much more dangerous to weld than the iron or steel upon which the plating is used. Cadmium is a plating metal whose welding fumes are considered very dangerous. This is one fume that has been known to be fatal in a single acute exposure. Even worse, acute exposures to cadmium usually do not display warning symptoms. Chronic exposures have been associated with emphysema and kidney impairment.

Stainless steel is one of the most dangerous materials to weld because of its high chromium content. Chromium trioxide is formed by the oxidation triggered by the welding heat, and chromium trioxide reacts with water to produce chromic acid. Ample sources of water can be found on human skin and the mucous membranes, resulting in chromic acid ulceration of these surfaces. Other chromic acid hazards were discussed in Chapter 9, where the phenomenon of "chrome holes" was discussed.

Welding in confined spaces complicates the atmospheric contamination problem. In confined spaces the hazards of gases increase dramatically. Nitrogen and argon are inerting agents for the protection of the weld, but they are also simple asphyxiants to the welder. Another simple asphyxiant in welding atmospheres is carbon dioxide. Contrasted with the simple asphyxiants is the chemical asphyxiant carbon monoxide, also present to some extent in welding atmospheres, especially for gas welding.

Nitrogen is not as inert as argon or helium, mentioned earlier as inerting agents, although it is true that nitrogen is a relatively stable element. But nitrogen can be oxidized, especially in the extremes of welding temperatures, creating oxides that can be harmful. Nitrogen dioxide (NO_2) is considered more dangerous than nitric oxide (NO); in fact, nitric oxide was formerly used as a dental anesthetic. But both oxides have been shown to have harmful effects.

Lead and mercury are well-known systemic poisons, and airborne fumes are the prime avenues for entry of these poisons into the body. Most welding does not involve these two metals. Soldering is used widely with lead alloys, but the low temperatures of soldering render the lead fumes relatively harmless.

Beryllium is a very useful alloy metal used in steel, copper, and aluminum. Unfortunately, the presence of the beryllium alloy in the material makes the metal very dangerous to weld. Since beryllium fume (particulate) hazards are both acute and chronic, most welders are wary of beryllium dangers.

Fluorine and fluorine compounds, usually fluorides, enter the welding atmosphere via welding flux or coverings. The popular shielded metal arc welding (SMAW) process is subject to hazards of fluorine compounds. The principal hazard is chronic, not acute, exposure, and long-term exposures cause abnormalities in the victim's bones. Other cleaning compounds and fluxes may also be hazardous, and personnel should check ingredients and heed manufacturer's instructions.

Before leaving the subject of welding gases and fumes, an important point is to be emphasized. None of the toxic materials or hazardous conditions described in this section is so dangerous as to prohibit welding. Welding atmospheres can be made safe by local or general exhaust ventilation or by personal protective equipment. The key is to recognize the potentially hazardous conditions, test atmospheres for excessive contaminant levels, and take corrective action if required.

SUMMARY

Welding represents a microcosm for the study of the entire field of occupational safety and health. It involves mechanical hazards, fire hazards, air contamination hazards, personal protective equipment considerations, and almost every other subject addressed in this book. Welding processes are many and varied, and most Safety and Health Managers know little about the technical aspects and terminology. A little study of the basics of welding, however, can open up opportunities for revision or substitution of processes that can enhance health and safety and at the same time improve efficiency and cut production costs. No other subject area seems to offer so much opportunity for Safety and Health Managers.

EXERCISES AND STUDY QUESTIONS

13.1. What are the three basic categories of conventional welding? Which of the three is the cleanest and most healthful?

13.2. What distinguishes soldering and brazing from welding?

13.3. What distinguishes soldering from brazing?

13.4. What is the commonly used name for the most popular process of arc welding? What is the official AWS designation for this process?

13.5. Identify the following welding processes:

GTAW

GMAW

SAW

RSEW

RSW

13.6. Why should acetylene cylinders be stored valve end up?

13.7. Why are oxygen cylinders charged at so much higher pressures than acetylene cylinders?

13.8. In what ways are greasy gloves of particular hazard to the oxyacetylene welder?

13.9. Describe a way in which Safety and Health Managers might cut production costs and at the same time avoid hazards.

13.10. Why are valve protection caps important to safety? Explain the purpose of the slots in the caps. How are the slots often misused?

13.11. Explain the phenomenon of flashback in welding operations.

13.12. How can taping welding hoses together to keep them orderly result in a hazard?

13.13. How can small arc-welding machines that operate off ordinary household current be more dangerous than industrial arc welding machines.

13.14. Why should welding cables be uncoiled before using?

13.15. What is the principal hazard of allowing metal tanks to become a part of a welding circuit?

13.16. What arc welding process is gaining in popularity because it is more healthful than other arc welding processes? What big disadvantage does it have?

13.17. What is the principal mechanical hazard of spot welders?

13.18. Give at least two reasons why people are psychologically inclined to risk the chance of welding fires.

13.19. Against what principal hazard are welding permits aimed?

13.20. Which of the following welding operations requires the most eye protection: SMAW, SAW, or RSEW?

13.21. What is the best natural material for welders' protective aprons, gloves, and leggings?

13.22. Name the pneumoconiosis resulting from exposure to iron oxide fume. Is it a severe hazard?

13.23. Why is it difficult to provide accurate tables of welding fume content?

13.24. What distinguishes welding fumes from toxic gases produced by the welding process?

13.25. Describe some conditions that, when present, make welding fumes and gases more dangerous to the welder.

13.26. Consider the conditions narrated in the following case study (obviously contrived) and describe apparent hazard mechanisms along with potential consequences.

> **Case Study**   A welder has constructed a manifolding arrangement for an assortment of oxygen and acetylene cylinders stored together lying on the floor. The manifold pressurizes the gaseous acetylene and depressurizes the oxygen to 50 psig for both. The room smells strongly of nail-polish remover. The welder is wearing greasy gloves and is chipping welding slag from the weld using his torch tip. The welding torch is connected to the manifold by two flexible hoses that are carefully wrapped together with duct tape, completely covering the hoses.

13.27. Nitrogen is a gas very widely used in welding operations, yet it is used principally in welding processes other than "gas welding." Explain.

13.28. In parts per million, calculate the approximate concentration of nitrogen in normal breathing air. How can nitrogen in breathing air be a hazard?

14

Electrical Hazards

13%

Percent of OSHA
general industry citations
addressing this subject

Year after year, the National Center for Health Statistics reports approximately 1,000 accidental electrocutions annually in the United States, with about 1 in 4 being industry and farm related. Everyone knows that electrical shock can be fatal, but the mechanism of the hazard is a mystery to most people. The mystery is due in large part to the fact that electricity is invisible. The use of electricity throughout our homes has led to a degree of complacency which is a factor in most electrocutions.

ELECTROCUTION HAZARDS

The first step toward safety from electrocution is to overcome the myth that "ordinary 110-volt circuits are safe." The truth is that ordinary 110-volt circuits can easily kill, and actually do kill, many more people than do 220-volt or 440-volt circuits, which nearly everyone respects. But the myth about 110 volts persists because almost everyone has sustained without serious injury an electrical shock around the home or on the job. An accident like this leads victims to the false conclusion that, although a 110-volt shock can be startling, it probably will not be fatal. Although they know that others have been killed by such shocks, they somehow feel resistant or too strong to be seriously injured.

It is true that some persons are more resistant to electrocution hazards than others, but a far more important factor is the set of conditions surrounding the accident. Wet or damp locations are known to be hazardous, but even body perspiration can provide the dampness which can make electrical contact fatal. Another important condition is the point of contact. If current flow enters the body through the fingers and passes out through a contact at the elbow, no vital organs receive direct exposure. But if the flow is from a hand through the body to the feet, vital organs such as the heart and lung diaphragm are affected, with possibly fatal results. Contact by the body torso to complete a circuit can also produce vital exposure to electrical current. Another factor can be the presence of wounds in the skin, which can result in a much higher current flow if contact is made where the skin is broken.

Physiological Effects

The central nervous system of our bodies is the conduit of signals between our brains and our muscles, including muscles of vital organs such as the heart and lung diaphragm. These signals are tiny electrical voltages which tell our muscles when to contract and when to relax. An external electric shock can send through the body currents that are many times greater than the tiny natural currents within our nervous systems. These larger currents can cramp or freeze muscles into a violent contraction—one that will not allow the victim to let go of the object contacted, or one that will stop breathing or stop the heart.

The heart is obviously our most important muscle. Its function is a rhythmic contraction and relaxation, which is timed by natural electrical pulses. The heart is thus very vulnerable to any pulsating electrical current. Common electric utility power supplies alternating current which cycles at a frequency of 60 hertz. It is ironic that 60 hertz is one of the most dangerous frequencies to which the heart can be exposed. This frequency tends to cause the heart to convulse weakly and irregularly at a rate too rapid to accomplish anything, a phenomenon known as *fibrillation*. Once fibrillation starts, death is almost a certainty, except that fibrillation has sometimes been stopped by controlled electric shocks to the heart muscle. The controlled electric shocks reestablish the heart's natural rhythms. Unfortunately, a defibrillation device is rarely available soon enough to save the life of an electrocution victim.

Stopped breathing from electric shock is due to cramped muscles responsible for respiration, such as the diaphragm and those controlling rib cage expansion. The first aid remedy is artificial respiration, the same as for near-drowning or other respiratory crises.

Just how much electrical current is fatal? There is no set answer to this question, but Figure 14.1 summarizes the opinions of several experts. The horizontal scale is logarithmic and is in units of milliamps or thousandths of an ampere. To put the chart in perspective, an ordinary table lamp with a 60-watt bulb draws about 500 milliamperes of current, far more than is needed to be fatal. An ordinary house circuit of 20 or 30 amperes will not trip the circuit breaker until there is a current flow of 20,000 to 30,000[1] milliamperes, respectively, about 100 to 1,000 times as much as the lethal dose.

With such lethal potential available from an ordinary 110-volt house circuit, it would appear that almost no one could survive an electrical shock from such a circuit. But the body, especially the skin, has resistance which limits the flow of electric current when exposed to 110-volt potential. To understand this resistance, some fundamentals of electricity may need to be reviewed.

[1] 1 ampere = 1,000 milliamperes.

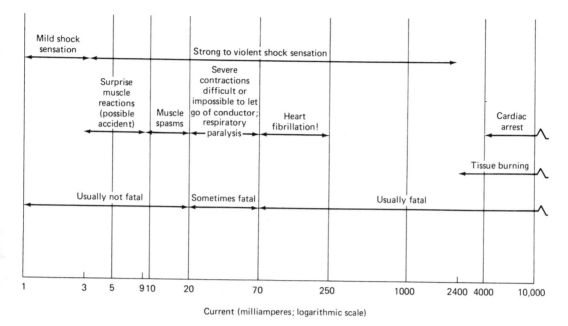

Figure 14.1. Effect of alternating electric current on the human body.

Ohm's Law

The basic law of electric circuits is Ohm's law, stated as:

$$I = \frac{V}{R} \tag{14.1}$$

where I = current in amperes
R = resistance in ohms
V = voltage in volts

The law can be rewritten as

$$V = IR \quad \text{or} \quad R = \frac{V}{I}$$

Wattage is a measure of power and can be computed from known quantities of current and voltage or resistance as follows:

$$W = V \times I \quad \text{and} \quad W = I^2R \tag{14.2}$$

where W is the power in watts.

Of principal concern are alternating current (AC) circuits, which are the predominant type in both domestic and industrial use. Standard AC circuits cycle 60 times

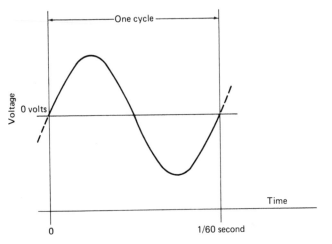

Figure 14.2.   Alternating current voltage.

per second (in the United States and Canada), as shown in Figure 14.2. Alternating currents are more convenient to generate and distribute than are direct currents. But computations of current, resistance, and voltage using Ohm's law are somewhat awkward for AC circuits because the voltage varies from zero to positive, back to zero, to negative, and back to zero again every cycle. For convenience, an "effective" current for an AC circuit is computed as a value somewhat less than the current peaks. A direct current operating through a given load is found to generate as much heat as an alternating current which has peak currents 41.4% higher than the direct current. Thus the ratio of effective current to peak current is computed as follows:

$$\frac{\text{effective current}}{\text{peak current}} = \frac{100\%}{100\% + 41.4\%} = 0.707 = 70.7\%$$

Effective voltages are computed by the same ratios as effective currents since they are related by Ohm's law. An ordinary 110-volt circuit then has an effective voltage of 110 volts, even though peaks of voltage over 150 volts occur every cycle.

The current drawn by an ordinary 60-watt lamp bulb can be computed by arranging equation (14.2) as

$$I = \frac{W}{V} = \frac{60 \text{ watts}}{110 \text{ volts}} = 0.55 \text{ ampere}$$

Since the wire and other parts of the circuit would consume some power, a fair approximation to the current flow in the 60-watt table lamp is ½ ampere, or 500 milliamperes, as stated earlier.

Returning now to the question of why more people are not killed by ordinary 110-volt circuits, we can use Ohm's law to determine how much the skin can limit the flow of electric current through our bodies. Human skin, if it is dry enough, is a good insulator and may have a resistance of 100,000 ohms or more. Using Ohm's law, a 110-volt exposure then would result in only a tiny current:

$$I = \frac{V}{R} = \frac{110 \text{ volts}}{100,000 \text{ ohms}} = 0.0011 \text{ ampere}$$
$$= \text{approximately 1 milliampere}$$

Referring to Figure 14.1, it can be seen that such a tiny current will probably not even be noticed. But add any perspiration or other moisture to the skin, and the resistance drops sharply. Due to perspiration alone the skin resistance can be reduced 200 times, to a level of about 500 ohms with good contact with the electrical conductor. Once inside the body the electrical resistance is very low and the current flows almost unimpeded. If the total resistance in the circuit is only 500 ohms, the current is calculated as

$$I = \frac{V}{R} = \frac{110 \text{ volts}}{500 \text{ ohms}} = 0.22 \text{ ampere}$$
$$= 220 \text{ milliamperes}$$

From Figure 14.1 it can be seen that an alternating current at this level passing through the body, including the heart, will most likely be fatal. Therefore, readers, if you have ever received an electrical shock, and most of us have, you can be glad that you were not perspiring enough, or that you did not have a good enough contact, or that the path of the current bypassed the trunk of your body, or that you were poorly grounded, or that some other resistance impeded the current. Otherwise you would have been killed by the ordinary 110-volt circuit, no matter how "tough" you are or how resistant you think you are to electrical shock.

Case Study 14.1 A worker is using a hand-held circular saw to cut extruded aluminum strips in the manufacture of storm windows. He is holding the workpiece firmly in his left hand and holding the saw in his right hand. Aluminum is an excellent conductor, and the workpiece is making solid electrical contact with ground. In an accident that happens frequently, the worker accidentally saws the electrical cord in half. What are the probable consequences in the following three sets of circumstances:

- *Case A:* The tool is grounded through the third prong of the electrical plug.
- *Case B:* The tool is double-insulated.
- *Case C:* The tool has a three-prong plug that is connected via an adapter to a two-hole wall socket; the tool is ungrounded.

Solution:

Case A. Current will flow through the metal case of the saw handle into two paths, one through the grounding circuit, and the other through the worker's right hand, through his body, crossing through his torso and through his left hand into the well-grounded workpiece. Although the resistance through each of these paths might be relatively low, the resistance through the third prong grounding conductor should be the lower of the two, on the order of 2 to 3 ohms. A resistance as low as 3 ohms would immediately trip a 15- or 20-amp circuit breaker, the type one would expect to find on such a circuit, as can be confirmed in the following calculation using Ohm's law:

$$I = V / R = 110 \text{ volts} / 3 \text{ ohms} = 36+ \text{ amps}$$

The current flow just calculated would be in addition to whatever current might flow through the man's body and other paths to ground, including perhaps some flow through the tool itself before the accident completely severs the cord. The total current would therefore easily trip any reasonable circuit breaker in the circuit and interrupt the flow of current, protecting the worker.

Case B. A double-insulated tool would have a nonconducting housing, resulting in no flow through the handle and the worker's body. The breaker would likely still be tripped as the metallic blade would make contact with both the hot wire and the well-grounded neutral. In addition, if the blade was cutting the aluminum workpiece at the time of the accident, another excellent path to ground would be through the metallic blade and into the grounded workpiece, causing an overcurrent to trip the breaker.

Case C. With no double-insulation to protect the worker and no grounding conductor to trip the breaker, conditions might be present to cause this common accident to result in an electrocution. The well-grounded left hand of the worker would permit a substantial flow of current through his upper torso, the danger zone for heart and lung exposure. A reasonable value for the resistance in a well-grounded path through the worker's left hand and the aluminum workpiece would be 600 ohms. The current in such a grounding circuit would be calculated as follows:

$$I = V / R = 110 \text{ volts} / 600 \text{ ohms} = 0.183 \text{ amp}$$

$$= 183 \text{ milliamps}$$

The current just calculated, only a small fraction of an ampere, would have no effect upon a normal 15- to 20-ampere circuit breaker. However small a 183-milliamp current may be for breaking the circuit, it is a very large and dangerous current to flow through the worker's upper body. Such a current is shown by Figure 14.1 to represent a strong to violent, usually fatal, shock capable of producing heart fibrillation. The metallic blade of the saw might provide a good grounding path via the severed neutral or the well-grounded workpiece, accommodating an overcurrent that would trip the breaker and save the worker's life. But such a grounding through the metallic blade would be dependent upon chance; without such grounding the accident would likely be fatal.

Case Study 14.2   A worker uses a "trouble light" suspended from the hood of an automobile while he repairs the engine. He leans across the fender of the car as he works so that his chest makes firm contact with the metallic fender, although that contact is resisted somewhat by a thin T-shirt he is wearing and, slightly, by the paint on the fender of the automobile. The light, which has been through many years of severe usage, has developed a worn connection at the point at which the flexible cord is connected to the lamp socket. As the worker adjusts the light's position, the worn connection results in accidental contact between the worker's index finger and the hot wire. Current passes through the man's finger and arm, continuing through multiple paths through his torso, most of it flowing to ground through his chest and the fender of the automobile and some through his feet and shoes. The contact between the hot wire and the man's finger is only partial, and the electrical resistance of the skin in the man's finger at the point of contact is about 800 ohms. If this resistance represents about half the effective total resistance in the short circuit, how much current would flow through the man's torso? Would the circuit breaker, rated at 15 amps, be tripped? Would the shock likely be fatal?

Solution: In this situation the current would flow through many parallel paths through the man's body, but for purposes of considering the total current flow due to the short circuit, one can consider the path to be equivalent to one effective path with a resistance of twice 800 ohms, or 1,600 ohms. Using Ohm's law:

$$I = V / R = 110 \text{ volts} / 1,600 \text{ ohms} = 0.069 \text{ amp}$$

$$= 69 \text{ milliamps}$$

Such a current flow is much too small to trip the 15-amp breaker, even if combined with the current flow through a 60-watt lighted lamp, which was calculated earlier in this chapter to be 0.55 ampere. If the entire 69-milliamp short circuit passes through the central part of the man's body, he is in critical danger of electrocution. Figure 14.1 reveals that 69 milliamps is in the region of "sometimes fatal" and "respiratory paralysis." The victim may survive if an alert bystander is trained in cardiopulmonary resuscitation and applies artificial respiration, and if the victim is fortunate enough to avoid heart fibrillation.

Grounding

In the previous discussion, the term *grounded* was used. Just what does this electrical term mean? A requirement for electrical current to flow is that its path make a complete loop from the source of electrical power through the circuit and back

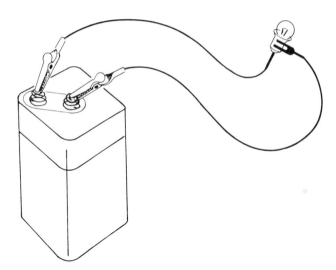

Figure 14.3. An electrical circuit makes a complete loop.

again to the power source. We understand this loop as we connect a lantern bulb to the posts of a lantern battery, as shown in Figure 14.3. Disconnection of the circuit at any point in the complete loop stops the flow of current. This means that there must always be two conductors: one to carry the current to the device (usually called "load") which uses it and another to carry the current from the load back to the electrical source. However, a trick makes the long trip back to the electrical source very simple in most applications of electrical power.

The earth for the most part is a fairly good conductor of electricity. Besides this, it is so massive that it is difficult for a human-made source of electricity to affect it much one way or another. Thus no matter what we do on the surface of the earth, the earth maintains a relatively even potential or charge. This means that if we drive two stakes firmly into the earth, even at great distances from each other, we may consider the resistance between them to be nil. The current flow may not be directly from one stake to the other because there are millions of electrical contacts to the earth at all times. Some of these contacts are positive and some are negative, but the total result is zero or earth potential. Thus any electrical conductor driven into the earth immediately assumes the zero reference potential of the earth. This is a very convenient characteristic of the earth because it enables us to use it as one great common conductor back to the source of power. Figure 14.4 illustrates the use of the ground as a return conductor.

A careful examination of Figure 14.4 reveals that the power company provides a separate neutral conductor for the completion of the circuit back to the source. There are conditions that make dependence on the common potential of the earth somewhat unreliable. A very dry season, for instance, may make the surface of the earth lose its conductivity. This is especially a problem if the area is dry around the grounding conductor stake which has been driven into the ground. The neutral conductor then ensures the completion of the circuit regardless of conditions.

The use of ground in electrical circuits is so advantageous as to be considered

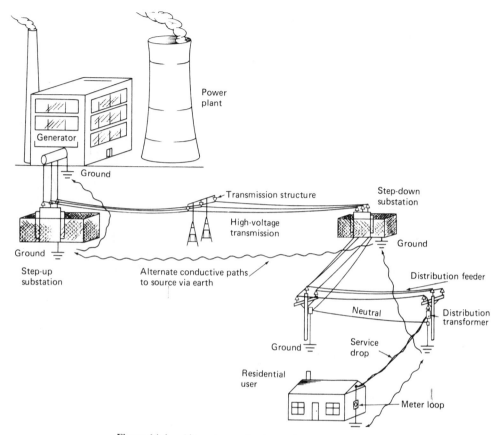

Figure 14.4. Alternate conductive paths to source via earth.

indispensable. However, the very convenience and proximity of the ground everywhere present a hazard. If a person contacts an energized conductor and at the same time is in contact with the ground or some other object that has a conductive path to ground, that person completes the electrical circuit loop by passing electric current through his or her body. A major portion of the *National Electrical Code®* [2] is devoted to prevention of this hazard.

The principal way in which persons are protected from becoming a part of the path to ground is by insulation of conductors. In addition, exposed conductive surfaces are given a good connection to the ground, usually by means of the ground wire, so that opportunity for a person's body to be the path to ground in minimal. Paradoxically, in some rare instances the *National Electrical Code®* takes exactly the opposite approach. For some systems it makes more sense to *isolate* the entire structure from

---

[2]The *National Electrical Code®* , commonly abbreviated NEC, is published regularly by the National Fire Protection Association (NFPA), 470 Atlantic Avenue, Boston MA 02201.

ground. If the structure is isolated, workers are protected by not being in contact with conductors that could connect them with ground.

Wiring

A typical 110-volt circuit has three wires: "hot," "neutral," and "ground." Sometimes the neutral is called the "grounded" conductor, in which case the ground is called the "grounding" conductor. The purpose of the hot wire (usually a black insulated wire) is to provide contact between the power source and the device (load) that uses it. The neutral (usually a white insulated wire) completes the circuit by connecting the load with ground. Both the hot and the neutral normally carry the same amount of current, but the hot is at an effective voltage of 110 volts with respect to ground, whereas the neutral is at a voltage of nearly zero with respect to the ground.

The third wire is the ground wire and is usually either green or is simply a bare wire. The purpose of the ground wire is safety. If something goes wrong so that the hot wire makes contact with the equipment case or some other conductive part of the equipment, the current in its path to ground can bypass the load and take a short-cut, commonly called a "short." Since the load is bypassed, the short is a very low resistance path to ground and by Ohm's law draws a very high current. This high current in a properly protected circuit will almost immediately either "blow" a fuse or "trip" a circuit breaker, depending on the type of overcurrent protection provided in the circuit, and stop all flow of current in the circuit.

It is possible, of course, to have a short without a ground wire. The equipment may be naturally grounded by its location or installation, or the hot wire can somehow contact the neutral. Sometimes the short to ground is only partial because there is considerable resistance in the short path to ground. Such a short may go undetected because the short flow of current is of insufficient amperage to cause the total circuit current to trip the overcurrent protection in the circuit. In this case current will continue to flow, and equipment loads will continue to operate in the presence of these shorts, or "ground faults," as this type of short is sometimes called. Such ground faults can be especially dangerous on construction sites. This hazard is the basis for ground-fault circuit-interruptor (GFCI) devices on construction sites. The GFCI protection is in addition to overcurrent protection such as circuit breakers or fuses.

Figure 14.5 explains how a GFCI works. Whenever the current flow in the neutral is less than the current flow in the hot wire, a ground fault is indicated, and the current flow is stopped by a switch that breaks the entire circuit. One difficulty with GFCIs is that some leakages to ground are almost impossible to prevent, especially when conditions are wet or extension cords are extremely long. This causes the GFCI to trip even when no hazard exists, a condition known in the construction industry as "nuisance tripping." An alternative to GFCIs is for the employer to test, inspect, and keep records of the condition of equipment grounding conductors.

One misconception about shorts is the idea that a good fuse or circuit breaker is sufficient to stop the flow of a dangerous short through a person's body. A re-examination of Figure 14.1 shows that a person will almost certainly be killed by

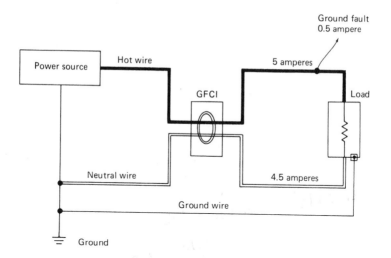

Figure 14.5. Ground-fault circuit-interruptor (GFCI). The 0.5-ampere fault to ground causes a current flow imbalance between the hot and neutral. This imbalance triggers the GFCI to break the circuit.

exposure to a current that would not blow even the smallest popular household fuses (i.e., 15- or 20-ampere fuses). A fuse or breaker rated at 15 amperes will handle up to 15,000 milliamperes before blowing, several times as great as the fatal current shown in Figure 14.1. The beauty of the third wire or grounding wire is that it provides a very low-resistance, high-current short to ground, which will go ahead and trip the fuse or breaker immediately, before *other* short paths to ground (such as through a person's body) can do their damage.

Double Insulation

Unfortunately, less than half of the electric hand tools in actual use are properly grounded. Studies of equipment returned to the factory for repair have shown that a large number of units have been altered so that the grounding system is no longer intact. A common alteration is to cut off the third prong of the plug so that it can be plugged into an old two-wire receptacle. To counter this practice, the use of "double-insulated" tools is permitted in lieu of equipment grounding. A second covering of insulation gives an extra measure of protection to the operator of double-insulated tools in case of a short to the equipment case.

Most double-insulated tools have a plastic, nonconductive housing, but this is not a fully reliable indication that the tool is double-insulated. The second covering of insulation must be applied according to precise specifications before the tool can qualify to receive the designation "double-insulated." Qualifying tools have the manufacturer's mark "double-insulated" or a square within a square ⊡ to indicate double insulation.

Miswiring Dangers

In the original wiring job electricans sometimes make mistakes or use slipshod practices that increase hazards. One of these practices is to "jump" (connect) the ground wire to the neutral wire. Actually, this is a trick that will work, and usually no one will be the wiser, but the practice does cause hazards. Figure 14.6 shows how a circuit is correctly wired, revealing that both the neutral and grounding wire are connected directly to ground. So in Figure 14.7, where the ground is jumped to the neutral, a third wire is not used in the wiring system.

The main hazard in jumping the ground to neutral is that it can create low voltages on exposed parts of equipment. The equipment case or housing is connected to the ground wire. Since normally no current flows through the ground wire, it serves as an excellent method of keeping the voltage on the equipment case close to zero with respect to ground. But the neutral does carry considerable current. Using Ohm's law, it can be determined that this large current on the neutral may cause the neutral terminal to have a low voltage with respect to ground, especially if the neutral wire must travel a long distance back to the ground at the meter. If the circuit is carrying a current of 20 amperes and the resistance of the neutral wire is ½ ohm, the voltage on the equipment case is calculated to be

$$V = IR = 20 \times \tfrac{1}{2} = 10 \text{ volts}$$

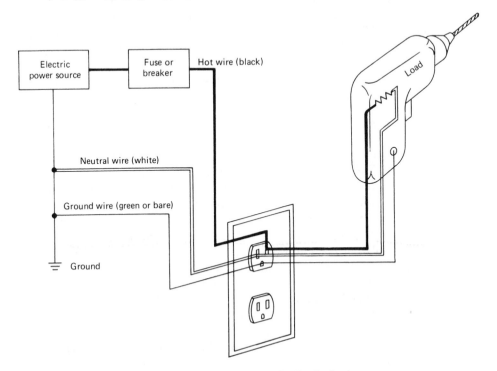

Figure 14.6. Correctly wired 110-volt circuit.

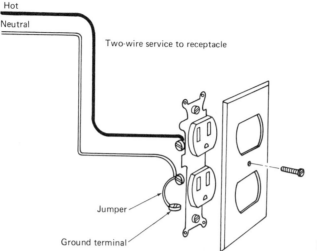

Hot

Neutral

Two-wire service to receptacle

Jumper

Ground terminal

Figure 14.7. Ground jumped to neutral.

This is a low voltage but is theoretically capable of producing a fatal current through a person's body if conditions are just right (rather, just wrong). The real hazard, though, is that a loose or corroded connection somewhere in the neutral circuit would increase its resistance, perhaps to 4 or 5 ohms, causing the voltage to increase several times.

Another common wiring mistake is "reversed polarity," which simply means that the hot and neutral wires are reversed. This is another subtle problem because most equipment will operate perfectly well with reversed polarity. One hazard of reversed polarity is that the designated leads (black lead, hot; white lead, neutral) become reversed, and the confusion could bring on an accident to an unsuspecting technician. Another hazard is that a short to ground between the switch and the load could cause the equipment to run indefinitely, independent of whether the switch is on or off (see Figure 14.8). Finally, bulb sockets can become hazardous when the polarity is reversed. In Figure 14.9a, a correctly wired socket shows the screw threads to be neutral. But in a reversed polarity socket, as shown in Figure 14.9b, the exposed screw threads become hot, and the button, which is naturally more protected at the bottom of the socket, becomes neutral.

Perhaps the most common wiring mistake of all is to fail to connect the ground terminal to a ground wire, a condition known as "open ground" or "ground not continuous." This is another error that can easily go unnoticed because equipment attached to circuits miswired in this way will usually operate normally. But if an accidental short to the equipment case occurs, the worker is in danger of electrocution.

The three cases of miswiring discussed in this chapter are not the only mistakes that can be made in wiring electrical circuits; they are not even the most hazardous. But because they permit electrical circuits to "work normally," they go unnoticed by uninformed users of equipment attached to such circuits. Because the errors are not immediately crippling to the function, they are frequently committed. Some simple checks with inexpensive testers can easily demonstrate the problems. Later in this chapter these testers will be examined.

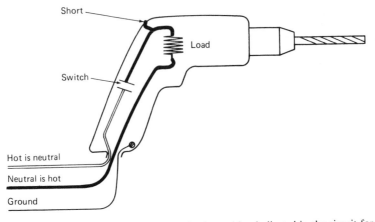

Figure 14.8. Reversed polarity. A short in the position indicated in the circuit for this drill will cause the drill to operate continuously, independent of the switch.

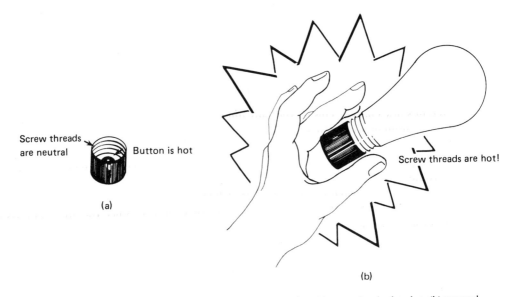

Figure 14.9. Hazards of reversed polarity in lamp socket: (a) correctly wired socket; (b) reversed polarity socket.

FIRE HAZARDS

Most people think of electrocution when they think of electrical safety, but electrical codes have as much to do with fire hazards as they do with electrocution. Many systems, such as fuses or circuit breakers, protect against both fire and electrocution, but their primary function is fire prevention.

Wire Fires

One of the most common causes of fires of electrical origin are wires that become overheated because they conduct too much current. Wire diameters (gauges) must be properly sized to handle the current load expected, and overcurrent protection (fuses or breakers) must ensure that these loads are not exceeded. Substitution of fuses with copper pennies is a common method of defeating the overcurrent protection so that the circuit will handle larger loads. If no fuse is present to burn in two, the wire itself may act as the next weakest link. If the wire becomes hot enough to burn in two, any contact with combustible material along the wire run is likely to produce a fire.

Arcs and Sparks

Whenever two conductors make a physical contact to complete a circuit, a tiny (or not so tiny) electric arc jumps the air gap just prior to contact. This arc may be so small as to be undetectable, but it is hot enough to ignite explosive vapors or dusts within their dangerous concentration ranges.

When the electric arc is an instantaneous discharge of a statically charged object, it is sometimes called a "spark." Such sparks are capable of igniting an explosive mixture, the spark plug of an automobile engine being ample testimony. Sparks are prevented by electrically connecting, or "bonding," two objects that may be of different static charge. This is especially important when pouring flammable liquids from one container to another.

The arc that occurs when an ordinary electrical circuit is completed is virtually impossible to prevent. This means that switches, lights, receptacles, motors, and almost any electrical device, even telephones, are a source of ignition to hazardous concentrations of explosive vapors or dusts. Chapter 8 discussed the ranges of explosive vapors and defined the LEL and UEL explosive limits. Since the arc is impossible to prevent, some means must be used to separate the arc from the hazardous concentrations in the air. This is done by using wire, conduit, or equipment that is either vapor tight or strong enough to contain and prevent the propagation of an explosion inside the conduit or equipment. This is an expensive undertaking, and it is tempting to take shortcuts. The *National Electrical Code®* has a strict code for electrical wiring and equipment designed for hazardous locations. The Safety and Health Manager should be able to identify the operations or hazardous locations within the plant that require special wiring and electrical equipment. Accordingly, this identification scheme is discussed next.

Hazardous Locations

One of the most difficult tasks in the field of industrial safety is the definition of various industrial locations that require special wiring and equipment to prevent explosions. Industrial processes are so diverse as to defy a general definition. In addition, the ignition mechanisms are different for different materials. For instance, the

hazard of heat buildup on electric equipment housings and bearings coated with ignitable dusts is altogether different from the hazard of spark ignition of explosive
vapors derived from flammable liquids. Difficult as the problem is, it must be dealt
with because some industrial locations are just too dangerous to allow exposure to
electric ignition sources.

The *National Electrical Code®* meticulously defines various conditions to classify
hazardous locations roughly into six categories. Within these classifications are various
groups that identify the substance group that is causing the hazard.

The major classification is according to the physical type of dangerous material
present in the air and is designated "Class." The next classification is called "Division" and relates to the extent of the hazard by considering the relative frequency
with which the process releases hazardous materials into the air. The criteria for "Division" are subjective, not quantitative, except around paint spray areas. This subjectiv-
ity introduces problematic gray areas.

Figure 14.10 is a decision chart that attempts to simplify the complicated process
of classification of hazardous locations. The chart is approximate only because strict
definition would require enumeration of pages of exceptions and conditions, many
of which would overlap. The thing to remember is that the "Class" is the material
and the "Division" is the extent of the hazard. Thus one can state that Division 1
locations are more hazardous than Division 2 locations but cannot state absolutely
that Class I locations are more hazardous than Class II or III locations.

Since classification is so complex, industry often relies on examples common
in similar industries to decide whether a location is Division 1 or 2 or not of sufficient hazard to classify at all. Examples of some common hazardous locations are
as follows:

Description	Classification
Paint spray areas (flammable paint)	Class I, Division 1
Areas adjacent to but outside of paint spray booth	Class I, Division 2
Areas of open tanks or vats of volatile flammable solvents	Class I, Division 1
Storage areas for flammable liquids	Class I, Division 2
Inside refrigerators containing open or easily ruptured containers of volatile flammable liquids	Class I, Division 1
Gas generator rooms	Class I, Division 1
Grain mills or processors	Class II, Division 1
Grain storage areas	Class II, Division 2
Coal pulverizing areas	Class II, Division 1
Powdered magnesium mill	Class II, Division 1
Parts of cotton gin locations	Class III, Division 1
Excelsior storage areas	Class III, Division 2
Closed piping for flammable liquids (piping has no valves, checks, meters, or similar equipment)	Not classified

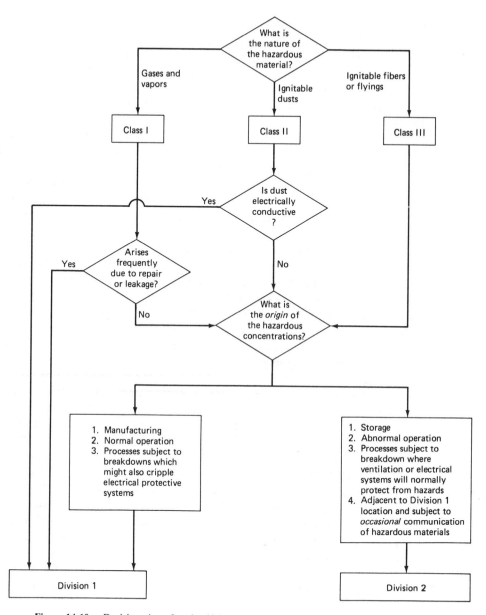

Figure 14.10. Decision chart for classifying "hazardous locations" that are dangerous from the standpoint of ignition of materials in the air.

Equipment qualifying for hazardous locations usually qualifies for both Division 1 and 2. When in doubt, most industries want to install equipment approved for Division 1 in order to be prepared for the worst. Most violations of code are not for selection of Division 2 equipment when Division 1 should have been selected;

most violations are from the use of thin-walled conduit and conventional electrical equipment in Division 1 or 2 locations.

"Approval" of electrical equipment for use in hazardous locations means that the manufacturer's design has been tested and approved by a recognized testing laboratory such as Underwriters' or Factory Mutual. A label must be on the equipment designating its classification if that equipment is to be used in hazardous locations.

Figure 14.11 shows several examples of "explosion-proof" equipment approved for Class I, Division 1 locations. Explosion-proof conduit looks more like pipe than conventional thin-walled conduit, which looks more like tubing. Explosion-proof junction boxes are castings, as opposed to conventional formed sheet metal boxes. The complicated structures for telephones and even light switches make obvious the fact that explosion-proof equipment costs several times as much as conventional equipment.

In Division 1 locations it is recognized that there is no way to ensure that the vapors will be kept out of the conduit and equipment. During maintenance, installation, or other open periods, vapors will enter the system. Therefore, the electrical equipment designer assumes this inescapable reality and designs the Division 1 equipment to withstand an internal explosion and cool the explosion gases as they escape before they can ignite the entire area in a devastating explosion.

By contrast, Division 2 electrical equipment enjoys some isolation from dangerous explosive vapors *most* of the time. Therefore, if the Division 2 equipment can be properly sealed with gaskets to make it vapor-tight, it will be safe. Class I, Division 2 equipment is characterized as "vapor-tight," whereas Class I, Division 1 equipment is *not* vapor-tight but is "explosion-proof." Of course, if equipment is classified as Class I, Division 1 (explosion-proof), it is also acceptable for use in Class I, Division 2 areas even though the equipment is not vapor-tight.

The foregoing comparison of Division 1 and Division 2 equipment and locations acknowledged the generalization that any equipment approved for Division 1 locations is also acceptable for Division 2 locations *of the same classification*. However, it should be noted here that equipment approved for Class I locations is not necessarily approved for Class II or III. The hazard mechanisms of ignitable Class II dusts or Class III fibers are somewhat different from the hazards of Class I vapors. Dusts and fibers can settle on warm equipment, insulating it from necessary heat dissipation during operation. Such insulation can cause a substantial heat buildup on the equipment which can result in a smoldering dust ignition and subsequent explosion.

A common error when selecting electrical receptacles for hazardous locations is to mistake weatherproof electrical outlets for approved equipment. Ordinary weatherproof outlets, as shown in Figure 14.12, are not approved for any type of hazardous locations in either Division 1 or 2. The spring-loaded cover protects the receptacle from weather when the receptacle is not in use, but when a plug is inserted into the receptacle, the receptacle is exposed as much as any conventional one.

The Safety and Health Manager may be baffled when reading equipment labels

Figure 14.11. Explosion-proof electrical equipment approved for Class I, Division 1 hazardous locations. Note heavy-duty, machined components. (a) Electrical outlet plug and receptacle; (b) wall switches. (Courtesy Appleton Electric Co.)

to find that equipment is classified and labeled by Class and *Group*, rather than Class and *Division*. Like the "Class" designation, the "Group" designation identifies the type of material present in the atmosphere, but the "Group" classification is more detailed. Four of the Groups belong to Class I and three to Class II and are summarized as follows:

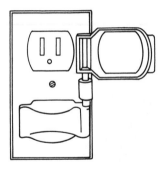

Figure 14.12. Ordinary weatherproof outlets not approved for hazardous locations.

Class	Group	Description	Examples	Industries
I	A		Acetylene	Welding fuel generators
I	B	Highly flammable gases and some liquids	Hydrogen	Chemicals and plastics
I	C	Highly flammable chemicals	Ethyl ether, Hydrogen sulfide	Hospitals, chemical plants
I	D	Flammable fuels, chemicals	Gasoline	Refineries, chemical plants, paint spray areas
II	E	Metal dusts	Magnesium	Chemical plants
II	F	Carbon, coke, coal dust	Carbon black	Mines, steel mills, power plants
II	G	Grain dusts	Flour, starch	Grain mills and elevators

A typical classification label will state "approved for Class I, Groups A and B," omitting the Division designation. When the Division is omitted, the classification is invariably acceptable for both Division 1 and 2 locations. If the equipment is merely vapor-tight and approved only for Division 2, the Division designation should be displayed on the label. Class I equipment is usually either approved for Groups C and D or is approved for all four Class I Groups: A, B, C, and D.

TEST EQUIPMENT

Pursuant to general electrical safety, there are a few inexpensive items of test equipment that the Safety and Health Manager should have access to for occasional in-house inspections. Brief descriptions of these items follow.

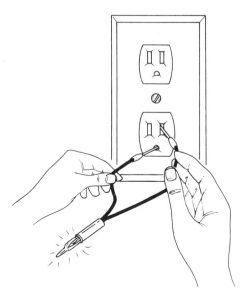

Figure 14.13. Circuit tester.

Circuit Tester

A circuit tester (Figure 14.13) simply has two wire leads connected by a small lamp bulb, usually neon. Whenever one of the leads touches a hot wire and the other lead touches a grounded conductor, completing the circuit, the bulb glows. The tester works only for a given voltage range, but most are able to handle both 110- and 220-volt circuits. The circuit tester is somewhat of a safety device in itself, enabling maintenance workers to be sure that power is turned off before touching hot wires. The circuit tester can also be used to determine whether various exposed parts of machines or conductors are "live" or "energized."

Receptacle Wiring Tester

One of the simplest and most widely used test devices is the receptacle wiring tester, which can be carried around in one's pocket (see Figure 14.14). No battery or cord is required. The user simply plugs the device into any standard 110-volt outlet and interprets the arrangement of indicator lights to determine whether the receptacle is wired improperly. Note carefully the use of the word *improperly* instead of the word *properly* in the preceding sentence. Only certain errors will be detected, and a "correct" indication of the lights means only that certain errors were not found. The receptacle could still be wired incorrectly.

Figure 14.15 reveals that the receptacle wiring tester is nothing more than three simple circuit testers combined into one tester. The lower portion of the figure shows types of wiring errors that can be interpreted from the indicator lights. One of the most common wiring errors—"ground jumped to neutral"—is shown by the tester to be wired correctly.

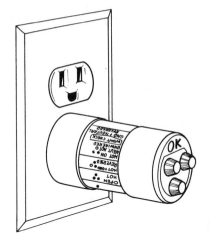

Figure 14.14. Receptacle wiring tester.

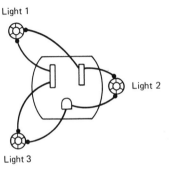

Light 1

Light 2

Light 3

Figure 14.15. Schematic of receptacle wiring tester.

Indicator lights

Condition description	1	2	3

Correct wiring

Reversed polarity

Open ground

Ground jumped to neutral
or
Neutral and ground reversed

Both of these errors go
undetected by the tester
because the tester will
indicate:

(i.e., correct wiring!)

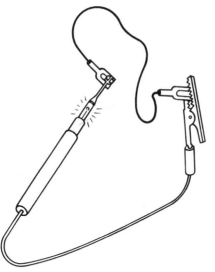

Figure 14.16.   Continuity tester.

Continuity Tester

It is sometimes useful to check a dead circuit simply to see if all the connections are complete or to see whether a break in a conductor has occurred. Figure 14.16 illustrates a simple continuity tester, which is similar to a circuit tester except that a small battery is included to provide power for the light. The continuity tester also has much less electrical resistance than the circuit tester because it is used on dead circuits, not live ones. If the continuity tester is used on a live 110-volt circuit, it will immediately blow a fuse or throw a breaker in the circuit.

An important application for a continuity tester is to check the path to ground. One terminal of the tester can be connected to the exposed equipment case of a machine in question and the other terminal to a known-grounded object. If the lamp lights, the machine is known to be grounded. If the lamp does not light, there is a break somewhere in the path to ground.

FREQUENT VIOLATIONS

Having been versed in the principles of electrical hazards, the Safety and Health Manager should know what to look for when making an inspection. But for review, the remainder of this chapter will describe frequently cited violations of the *National Electrical Code®*.

Grounding of Portable Tools and Appliances

It should come as no surprise that the most frequently cited electrical code is for grounding. This is the code cited for ungrounded electric drills, sanders, saws, and other portable hand-held equipment. Nonportable equipment, such as refrigerators, freezers, and air conditioners, must also be grounded. Guarded motors and the metal frames of certain electrically heated appliances are not excepted. Special attention should be given to the use of portable electric tools used in damp or wet locations and, of course, to plugs that have the ground pin broken off.

Exposed Live Parts

Almost as frequent as ungrounded portable equipment is the existence of exposed live parts. If the actual equipment conductors cannot be insulated and terminals covered, locked rooms should be used to prevent exposure to workers. One of the most frequent observations of exposed live parts is sloppy electrical installations in which covers are left off junction boxes or receptacle cover plates are missing. A switchbox, fuse, or breaker box in which the door is open also constitutes "exposed live parts."

Improper Use of Flexible Cords

Makeshift or temporary electrical installations are prohibited as substitutions of flexible cords for fixed wiring. Obvious instances are flexible cords run through holes in the walls, ceilings, floors, doorways, or windows.

Marking of Disconnects

This is an easy thing to correct. The disconnect box or switch panel for motors and appliances must be identified so that equipment can be quickly and confidently disconnected. Also, the origins of branch circuits, such as in the breaker box, must be labeled to indicate their purposes. If the location or arrangement of the disconnect makes it obvious which machines or circuits it controls, marking may not be necessary. Very few of the disconnect marking violations are designated as "serious."

Connection of Plugs to Cords

Here is another extremely simple item. When repairing a plug to an extension cord, appliance cord, or any other flexible cord, be sure to tie a knot in the cord or do something that will prevent a pull on the cord from being transmitted directly to joints or terminal screws. This is just a basic principle of electrical maintenance.

SUMMARY

This chapter began with some sobering thoughts of what electricity, even in small amounts, can do to the human body. The greatest hazard is with 110-volt circuits, not 220-volt or higher circuits, because of the popularity of 110-volt circuits and the complacency of the people who use them. Besides electrocution hazards, electricity also presents fire hazards, not to speak of burns and other electrical exposure hazards.

Simple testers can demonstrate quickly some of the frequent wiring mistakes that might otherwise go undetected in normal operations. But some of these testers are too simple and overlook common errors such as "ground jumped to neutral."

Providing for electrical equipment in industrial processes that produce flammable vapors, dusts, or fibers is a difficult and expensive task. The definitions and codes for hazardous locations with explosive atmospheres are also complicated and tricky. The Safety and Health Manager is advised to look for required markings on electrical equipment installed in hazardous locations to be sure that the equipment is approved for the Class and Division of the location in which it has been installed.

Violations of electrical code are often for conditions that are easy to correct. The largest single item to remember is grounding—grounding of both portable and fixed equipment and the circuits that serve them. Of less frequency but of great seriousness are the OSHA citations for unapproved conduit and electrical equipment used in explosive atmospheres of flammable vapors or dusts.

[handwritten margin notes: G / Live / Flex / Discon / C]

EXERCISES AND STUDY QUESTIONS

14.1. What are the two principal hazards of electricity?

14.2. Approximately how many people die each year of electrocution in the United States?

14.3. Compare 110-, 220-, and 440-volt hazards of electricity.

14.4. At approximately what current levels through the heart and lungs does the sustained flow of electricity become fatal?

14.5 What part does grounding play in an electrical circuit? Why do safety regulations require equipment to be grounded?

14.6. Compare the functions of hot, neutral, and ground wires.

14.7. What is a GFCI, and where is it used?

14.8. Explain the term *double insulation*.

14.9. What are the hazards of reversed polarity?

14.10. Compare electrical "arcs" and "sparks."

14.11. Explain the difference between Class I, II, and III hazardous locations.

14.12. What do the terms *Division* and *Group* mean in the classification of hazardous locations?

14.13. Explain the difference between a continuity tester and a circuit tester.

14.14. Name some of the most frequent violations of electrical code.

14.15. A certain string of Christmas tree lights has eight bulbs, each rated at 5 watts. If there is no ground fault, how much current flows through the hot wire at the receptacle plug? How much flows through the neutral?

14.16. In Question 14.15, if all the current flowing through the hot wire would pass through a person's heart muscle, would the current likely be fatal?

14.17. A maintenance worker wires a 110-volt electric lamp without disconnecting the circuit. With no load energized in the circuit, the worker accidentally contacts the bare hot wire. What will probably happen? Under what conditions would the worker be likely to be killed?

14.18. In Question 14.17, suppose that the wire contacted were the neutral wire. What would probably happen? Are there conditions under which the worker would be killed?

14.19. In Question 14.17, suppose that the wire contacted were the equipment ground wire. What would probably happen?

14.20. Portable "trouble lights" used in automobile repair have been involved in many fatalities. Explain probable hazard mechanisms.

14.21. The following fatality actually occurred. A factory worker was electrocuted when he grasped the small chuck key (Figure 14.17) tied with a wire to the electrical cord of his portable electric drill. The twisted wire tie had worn through the cord insulation during constant use. Comment on how this fatality might have been prevented.

14.22. In the fatality case history of Question 14.21, suppose that the ground pin had been broken off the electric plug for the drill. How would this code violation have affected this fatality?

14.23. A worker was electrocuted when he drilled through a wall and the drill bit contacted an electric wire inside the wall and cut through the insulation. The ground pin of the electric plug for the drill was broken off. Describe several ways in which this fatality could have been prevented.

14.24. The broken-off ground pin represented a code violation in the fatality case described in Question 14.23. How did this contribute to the hazard? What would probably have happened had the ground pin been intact?

14.25. An architectural design and engineering firm seeks your advice regarding federal safety standards for wiring for a new process being set up to manufacture dyestuffs. The process uses chlorobenzene (flashpoint 85 °F) and releases ignitable concentrations in the vicinity of the process equipment. Specify the appropriate Class, Division, and Group classification for wiring and electrical equipment located in this area.

14.26. For the dye manufacturing process described in Exercise 14.25, an alternative supplier of process equipment proposes a completely closed system in which the only releases would occur during repair or in event of a leak. Such occurrences arise frequently out of a need to regularly clean the in-feed mechanism. Would the closed system have an impact upon safety and upon the prescribed wiring system?

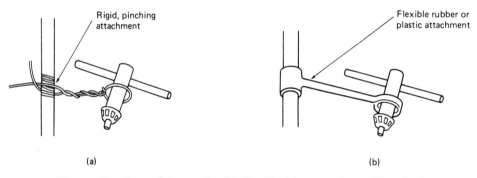

Figure 14.17.   Cause of electrocution fatality: Chuck key secured to cord by twisted wire tie. (a) Unsafe, improvised attachment of chuck key. Wire or tape will eventually damage cord insulation. (b) Safer method of attaching chuck key.

14.27. For each of the following situations, explain the nature of the hazard and describe probable outcomes:

    (a) A worker brushes his pantleg across a 120-volt terminal strip with screw terminals exposed. Every other screw is hot, with neutral screws intervening.

    (b) A worker's boot contacts a 440-volt bus bar. A ground path occurs through the worker's foot and the nails in the sole of the boot to the concrete floor. The resistance of the ground path is 10,000 ohms.

    (c) A worker wires a 220-volt circuit "hot" using a wood handle screwdriver. The screwdriver slips, and the shaft of the screwdriver makes a direct short across the hot terminal and the adjacent neutral.

    (d) A worker changes a 120-volt wall receptacle in a garage while standing on a concrete floor. The circuit is energized, but the worker is careful not to touch both the hot wire and the neutral wire at the same time.

14.28. A right-handed worker is repairing a 120-volt lamp socket using a screwdriver with an insulated handle. He is wearing a short-sleeve shirt, and his bare right arm is braced against a water pipe. A bare hot wire contacts the metal housing of the socket assembly the worker is holding in his hand while he is using the screwdriver in his other hand. A ground path occurs with total resistance of 600 ohms. Calculate the current flow and describe its probable path. Will the breaker likely be tripped? Is there a risk of electrocution? If so, what factors contribute to the risk?

15

Construction

28%

Percent of all OSHA
citations addressing
construction

The Safety and Health Manager for a construction company will find this chapter
of primary importance, but even the general industry Safety and Health Manager
should find the chapter useful. Almost every industry has occasional remodeling or
expansion projects that result in construction. Even if the company contracts out
the construction project, what goes on is still important to the company because its
own employees can become exposed to hazards created by the contract construction
personnel. Such hazards can be both physical and legal.

GENERAL FACILITIES

Lighting

Curiously, there are standards for adequate lighting on construction sites, specifying
minimum illumination intensities for various areas, whereas general industry has no
such table of intensities. The reason perhaps has to do with trip hazards and pitfalls
(in the literal sense of the word) common to construction sites, hazards that are in-
tensified by poor illumination. The minimum illumination standard of 5 foot-candles
is really quite low for general construction area lighting, and the standard drops even
lower, to 3 foot-candles for concrete placement, excavation and waste areas, ac-
cessways, active storage areas, loading platforms, and refueling and field maintenance
areas. Lighting requirements are higher for most construction *shops* and indoor areas.

Materials Handling and Storage

The real structural test of most buildings is during their own construction. The heaviest load a floor will probably ever support is the stacked materials used during its own construction. Planning is needed to prevent overloading and possible collapse. All nails should be withdrawn from used lumber before it is stacked.

Rigging equipment for material handling such as chain, rope, and wire rope is, unfortunately, often used until failure. If failure does occur on a construction site, the Safety and Health Manager must be prepared to explain why, because rigging equipment is required to be inspected prior to use on *each shift.*

Disposal of scrap material requires vigilance throughout the construction phase. It is difficult to concentrate on scrap and waste removal when working on a tight project completion schedule, but haphazard scrap accumulation is not only unsafe, it also slows progress. Enclosed chutes are needed for dropping materials from distances of greater than 20 feet.

PERSONAL PROTECTIVE EQUIPMENT

Construction requirements for personal protective equipment are similar to those for general industry, the principal difference being one of emphasis.

Hardhats

At the top of the list is head protection; indeed, the hardhat is a symbol of the construction industry. So obvious is the absence of a hardhat in a construction work crew that the hardhat rule can be a source of embarrassment to worker and manager alike. Only general conditions are laid out to describe *when* hardhats are needed. This lays the decision in the Safety and Health Manager's lap, and as was pointed out in Chapter 9, discretion is advised.

Hearing Protection

It may surprise some readers that hearing protection is a concern for *construction,* but construction work often involves damaging levels of noise. Consider the noise levels and durations of exposures of the compressed-air "jackhammer," for example.

Eye and Face Protection

The greatest concern for construction workers' eyes is for mechanical injury, as from using structural steel riveters, grinders, powder-actuated tools, woodworking tools, concrete nozzles, and other spark- and chip-producing equipment. Surprisingly, construction workers can even be exposed to lasers used as tools for checking steel girder alignment and deflection in bridges and buildings.

Fall Protection

On the bases of both fatalities and injuries, falls are probably the greatest hazard in construction work. Where general industry has a permanent wall, construction may have only a guardrail. Where general industry has a permanent guardrail, construction may have a temporary guardrail or perhaps will have no protective structure at all. Where general industry has a permanent stairway, construction may have a temporary ladder. General industry fixed ladders may have cages or ladder safety devices; construction ladders often have no safety devices.

Personal protective equipment is the answer to many construction industry fall hazards, simply because protection by other means may be awkward or even impossible. Safety belts and lanyards tied to lifelines are essential to the safety of construction workers subject to fall hazards.

One mistake made in selecting fall protection equipment is to improvise with ordinary leather belts and ropes. An ordinary leather belt and hardware will not satisfy the specified 4,000-pound tensile test for safety belt hardware. No one weighs 4,000 pounds, but what matters in a fall is the shock load, which can be several times as great as the ordinary dead weight. A falling 200-pound person can therefore result in ½ ton to 1 ton of force upon the belt and lifeline. Using a safety factor of approximately 4, it is easy to see why the standard specifies a 4,000-pound tensile load limit.

The lanyard is that part of the fall protection system which attaches to the safety belt on one end and the lifeline or structure on the other. The lanyard must have a nominal breaking strength of 5,400 pounds. The applicable standard specifies "½-inch nylon or equivalent." Beware of substituting materials of equivalent breaking strength to ½-inch nylon. Tensile or breaking strength is not the only consideration in selecting a lanyard. A certain resiliency or elasticity exists with artificial fiber ropes which lessens the shock load when arresting a fall.

An important point with safety lanyards is that they must not be too long. The standard specifies "a maximum length to provide for a fall of no greater than 6 feet." The rationale is that there is no point in breaking a fall with the lanyard if the worker has already fallen so far that the shock of the rope will be lethal. However, the standard is often misunderstood on this point. Note carefully that the wording just quoted does not limit the lanyard length to 6 feet. Figure 15.1 clarifies this point.

One difficulty is attempting to attach a lanyard to a vertical lifeline, especially when the lanyard must be adjusted upward or downward on the lifeline, such as in use with a scaffold. It is necessary for the attachment to slide easily when this is intended, but the attachment must lock and hold if the worker falls. There are mechanical devices, but an easy-to-tie knot for this purpose is the triple rolling hitch shown in Figure 15.2.

Fall protection is usually considered from the standpoint of height, but working over or near water presents a different hazard. Even the best of swimmers will have difficulty with a fall into water if he or she is fully dressed and perhaps burdened by tools, equipment, or materials such as rivets or bolts. If the water is rather cold, the danger of hypothermia increases the drowning hazard.

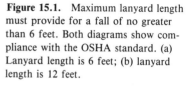

(a) Lanyard length is 6 feet

(b) Lanyard length is 12 feet

Figure 15.1. Maximum lanyard length must provide for a fall of no greater than 6 feet. Both diagrams show compliance with the OSHA standard. (a) Lanyard length is 6 feet; (b) lanyard length is 12 feet.

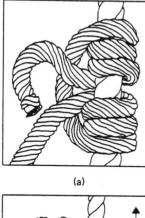

(a)

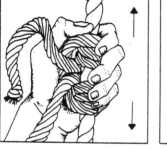

(b)

(c)

Figure 15.2. (a) Triple rolling hitch tied with free end of lanyard; (b) easily raised and lowered slipping on cable or lifeline; (c) when the lanyard is pulled tight as in a fall the triple rolling hitch will not slip on the lifeline.

Drowning hazards are taken very seriously in federal standards, which require

1. Life jackets or buoyant work vests *and*
2. Ring buoys every 200 feet *and*
3. A lifesaving skiff

whenever employees work over or near water and a danger of drowning exists.

FIRE PROTECTION

From a property-loss standpoint, fires are more dangerous after a building is completed, but to protect construction workers fire hazards must also be controlled *during* construction. Construction sites have somewhat more latitude in distributing fire extinguishers than do general industries. Even an ordinary garden hose may be used in place of fire extinguishers on construction sites. But there are so many restrictions on such use of garden hose that most Safety and Health Managers will regret having considered this alternative.

The biggest problem with fire prevention during construction is the handling of flammable liquids. For ordinary flammable liquids such as gasoline, quantities handled must be no more than 1 gallon unless approved metal safety cans are used. Approved metal safety cans must be used even for quantities up to 1 gallon unless the flammable liquid is used out of its *original* container. Keep the containers off stairways and away from exits and aisles.

TOOLS

It has been well publicized that mushroomed heads on chisels, wedges, and other impact tools are unsafe. The hazard is that a sliver of metal can break off and cause a severe eye injury, even total loss of sight. Another problem with hand tools is defective handles, in particular loose hammer heads.

Construction sites may employ pneumatic tools, such as jackhammers, staplers, or nailers. Pneumatic tools need to be secured to the hose by some positive means to prevent accidental disconnection. Hoses larger than ½ inch inside diameter need a pressure reducer device to prevent whip action in case of hose failure. Figure 15.3 is a diagram of a pressure reducer device for this purpose. It is a simple in-line device, usually placed between the hose and the compressor. Unfortunately, the device reduces the overall capacity of the system. Operating at full capacity, as from several tools

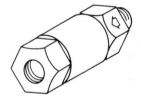

Figure 15.3.   In-line device for preventing whip action in event of pneumatic hose failure.

operating at once, the pressure downstream from the device becomes so low that the device (a spring-loaded valve) begins to close as if the downstream line had ruptured. This shuts off or nearly shuts off the supply of air to the tools, making the system useless at that capacity. The result is that the worker removes the device from the line, and often as not it is soon lost. This is a sore point with many construction Safety and Health Managers.

Some construction applications are being found for hydraulically operated tools, especially in the public-utility construction field. Hydraulic tools operate under the same principle as pneumatic tools but use liquids instead of air as the medium. Fluid pressures in these tools can reach 3,000 psi gauge, approximately 20 times the maximum pressure achievable with pneumatic tools. Such tremendous pressures give hydraulic tools much greater power than pneumatic tools, but a hazard can exist if operating pressure limits of the equipment are exceeded. Under the criterion of occupational noise, however, hydraulic tools have a safety and health advantage over pneumatic tools. Additional hazards of hydraulic fluids are electrical conductivity and fire. These hazards tend to conflict in that the more fire resistant fluids are electrical conductors. When working in construction and alteration of electric utility transmission and distribution systems, the hazard of electrical conductivity is more serious than the hazard of fire. The hydraulic fluids used for the insulated sections of derrick trucks, aerial lifts, and hydraulic tools which are used on or around energized lines and equipment for power transmission and distribution are required to be of the insulating type. The fluids for hydraulic tools used in other applications are required to be fire resistant.

Powder-actuated tools carry an explosive charge to provide the driving force. The applications of these tools are increasing because they are both fast and effective. Driving fasteners into concrete, masonry, or steel demands large, accurately placed impact forces. Powder-actuated tools are able to provide these forces in a convenient way, speeding up production on construction projects. But together with this speed, force, and convenience come safety hazards.

A powder-actuated tool looks and operates very much like a handgun, as can be seen in Figure 15.4. Even the powder cartridges look like bullets for a gun. In powder-actuated tools, however, the projectile is separate from the cartridge, as shown in Figure 15.4.

In some ways, powder-actuated tools are even more dangerous than handguns. Powder-actuated tools are capable of handling a variety of cartridges with a wide range of power ratings. These cartridges must be selected with care by a knowledgeable person. Insufficient power will fail to do the job, but too much power may drive the fastener completely through the material and kill a co-worker in another room (this has actually happened). To protect against such hazards and for convenience, cartridges are color coded for easy identification. Metal-cased cartridges are available in a range of 12 different power ratings, as shown in Table 15.1. It may be necessary to back the material with a substance that will prevent the fastener from passing completely through.

Knowledge and judgment are also required in avoiding very hard or brittle

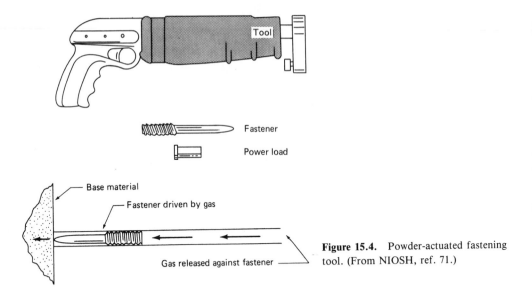

Figure 15.4. Powder-actuated fastening tool. (From NIOSH, ref. 71.)

materials, such as cast iron, glazed tile, surface-hardened steel, glass block, live rock, face brick, or hollow tile. If the material is spalled or cracked by an unsatisfactory previous fastening, the new fastening must be driven elsewhere. Fasteners driven too close to the edge of the material can cause explosive chipping of the material at the edge. Obviously, eye protection is always needed when using powder-actuated tools.

TABLE 15.1 Color Identification for Cased Power Loads for Powder-Actuated Tools

Power Level	Case Color	Load Color	
1	Brass	Gray	Lightest
2	Brass	Brown	
3	Brass	Green	
4	Brass	Yellow	
5	Brass	Red	
6	Brass	Purple	
7	Nickel	Gray	
8	Nickel	Brown	
9	Nickel	Green	
10	Nickel	Yellow	
11	Nickel	Red	
12	Nickel	Purple	Heaviest

ELECTRICAL

Construction workers are often in close contact with ground and frequently work in wet locations under adverse conditions. Electrocution ranks with falls near the top of the list of causes of fatalities among construction workers.

A principal requirement on construction sites is that all 15- and 20-ampere outlets have either ground fault circuit interruptor (GFCI) protection or a program of equipment ground conductor assurance, including inspection, testing, and recordkeeping. The Safety and Health Manager in a construction company is faced with a decision between the two alternatives, and the following paragraphs are intended to assist in this decision.

The purpose and operating principle of the GFCI were discussed in Chapter 14. The hazards of electrical shock by faults to ground are greater on construction sites than in the general industrial workplace. Because of the increased hazards, the *National Electrical Code®* specified GFCIs for construction but not for general industry, although they would be of value in any electric circuit serving hand-held appliances or tools. Many individuals have installed them in their homes.

It seems that almost every safety device has its drawbacks, and the GFCI is no exception. The GFCI closely monitors any difference in current flow between the ground and neutral conductors, as low as fractions of a milliampere. But there are ways in which these currents can be unbalanced when no hazard exists. Tiny amounts of current leakage to ground occur for quite innocent reasons. Damp or weakened insulation might produce tiny currents in various locations. Even an extension cord that is too long can create a condition of capacitance between the conductor and the ground, resulting in a tiny leak. Although none of these conditions of itself is a significant hazard, the cumulative effect is one that may be great enough to trip the GFCI, shutting down the entire circuit. This is so-called "nuisance tripping" and has made the GFCI a controversial issue in the construction industry.

Chapter 14 mentioned a permissible alternative to the GFCI: the careful maintenance of the grounding conductors of electrical equipment. Such maintenance includes regular inspections and records of these inspections. The idea behind the assured equipment grounding conductor program is that if an electrical tool shorts to the case or handle, the third-wire grounding system will shunt the current to quickly throw the circuit breaker. A good grounding conductor can therefore provide protection similar to the GFCI.

The assured equipment grounding conductor program is appealing to many construction companies because they can avoid buying the GFCI equipment. They can also avoid the nuisance tripping of the GFCIs discussed earlier. But although the costs are less tangible, they nevertheless exist with the grounding assurance alternative. It takes instruments and time to test the grounding conductors, and the business of recordkeeping always involves intangible costs. An economic impact analysis of the two alternatives has been made which estimated a cost of compliance of $87.5 million for purchase, installation, and first-year maintenance of GFCIs. The study estimated for the alternative assured equipment grounding conductor program a similar cost of $36 to $43.8 million. Inflation changes absolute annual cost estimates, but the relative difference between the costs of the two alternatives suggests that the assured equipment grounding conductor program is cheaper.

Temporary lighting is an electrical problem on construction sites more than in general industry. Often seen are ordinary incandescent bulbs suspended from electrical cords. Cords and lights that are suspended in this way must actually be *designed* for this purpose; all electric cords for temporary lighting must be heavy duty, and insulation must be maintained in safe condition. To prevent accidental contact with bulbs, the bulbs must be guarded unless the construction of the reflector is such that the bulbs are deeply recessed.

A construction site is a profusion of temporary conditions, and electrical cords or extension cords strung about the area are a common sight. Unfortunately, the area is also visited by heavy-duty vehicles such as excavation equipment, heavily loaded trucks, and very heavy concrete delivery trucks. The situation is too hazardous to permit electrical cords to pass through work areas unless covered or elevated to protect from hazardous damage. No splices are permitted in flexible cord unless properly molded or vulcanized.

LADDERS AND SCAFFOLDS

The "care and use" provisions for ladders are most important for construction ladders, just as they are for ladders used in general industry covered in Chapter 6. Construction ladders have some differences, however, that have caused some problems.

Job-Made Ladders

Construction companies often make their own ladders, and such ladders are not illegal if made properly. The first requirement is to determine how many persons will be needing the ladder. If simultaneous two-way traffic is anticipated, a conventional ladder will not work, and a double-cleat ladder as shown in Figure 15.5 should be used. In fact, if the ladder is the only means of access or exit from a working area for 25 or more employees, the double-cleat ladder is *mandatory*. However, care must be taken not to use double-cleat ladders over 24 feet long (see Table 6.1). If greater length is needed, the solution is to provide landings with guardrails and toeboards

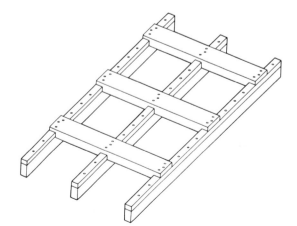

Figure 15.5. Double-cleat ladder for simultaneous two-way traffic.

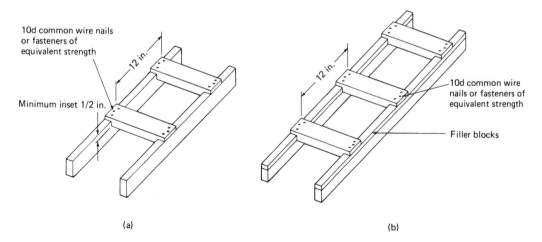

Figure 15.6. Two acceptable ways of setting cleats in construction job-made ladders: (a) cleats inset into side rails; (b) cleats braced by filler blocks.

on exposed sides. To be effective, each landing must offset two sections of ladders so that the length of a fall is minimized by the position of the landing.

The biggest mistake made in building job-made or homemade ladders is to fail to inset the cleats into the side rails (see Figure 15.6). It is much more trouble to inset the cleats or use filler blocks than to simply nail the cleats in place, but the security and stability of the cleats are increased many times by this additional effort to make the ladder safe.

Most commercial ladders are flared top to bottom. It is permissible to flare job-made ladders top to bottom, but it is also permissible to make the side rails parallel. The flare must be very small if used at all and must not be more than ¼ inch for each 2 feet in length. This determines limits for the width of the bottom cleats because top cleats must be at least 15 inches but no more than 20 inches wide. Thus the bottom cleat must be at least 15 inches wide but no more than given by the following formula:

$$W_B = W_T + \frac{1}{2}\left(\frac{L}{2}\right)$$

where W_B = bottom cleat width in inches

W_T = top cleat width in inches

L = ladder length in feet

Scaffolds

The subject of scaffolds can become somewhat technical, and these technical details can be very important. The Safety and Health Manager will find it useful, and in some cases imperative, to obtain the services of a Registered Professional Engineer. This is one area in which the credential as well as the knowledge can be quite useful.

One of the technical aspects of scaffolds is safety factor. Design safety factor for scaffolds and their components is a factor of 4. This factor increases to a factor of 6 for the suspension ropes supporting the suspended type of scaffolding. The application of counterbalances, tie-downs, footings, and the allowance for wind loading all can be quite technical, and an engineering evaluation is advisable.

The Safety and Health Manager may be frustrated by the many bewildering names of scaffolds listed in applicable construction standards. But the majority of the scaffolds with unfamiliar names, such as "window-jack scaffolds," "outrigger scaffolds," and "chicken ladders," are rarely seen. The most popular scaffolds are the following types:

- Welded frame (or "bedstead" scaffolds)
- Manually propelled mobile scaffolds (on casters)
- Two-point suspension (or "swinging" scaffolds)
- Tube and coupler scaffolds

Some scaffolds, such as tube and coupler scaffolds and welded frame scaffolds, are supported by structure on the ground and must have sound footings. If the ground slopes, scaffold jacks may be necessary to assure that the footings are level. Certain engineered "cribbing" may be acceptable, but such unstable objects as barrels, boxes, loose brick, or concrete blocks are criticized. Concrete blocks are a real problem because most people feel that they are very strong and rigid. But scaffolds direct very highly concentrated loads on their relatively tiny feet, and such loads can break through a molded concrete block. Once a scaffold support breaks through its footing, a major shift can occur aloft, and the consequences can be extremely serious. Such a mishap is most likely to occur at the worst time (such as when personnel are on the scaffold).

For scaffolds suspended from above, such as from two-point suspension (swinging) scaffolds, the security of the attachment on the roof is of obvious importance. Since rooftops vary in structure and design, an engineer is very useful in ensuring a safe anchor point. Cornice hooks are designed to hook over the edge, not into it. In addition, tiebacks are needed as a secondary means of support. Sometimes a sound structure for the tieback is not available on the rooftop, and the only solution is to cross the entire roof and go down to the ground on the other side of the building. Tying a scaffold to an ordinary vent pipe on the roof is asking for trouble.

Safety belts and lifelines for personnel on suspended swinging scaffolds must be tied to the building, not to the scaffold. Thus if the scaffold falls, the personnel can still be saved. For guidance on the attachment of the lanyard to the lifeline, refer back to the discussion of fall protection covered earlier in this chapter.

The floor of the scaffold is also important. Loose planks can be especially hazardous if the overhang beyond the support is insufficient for security. But too much overhang can also be dangerous because a worker might step beyond the support and cause the plank to tip like a seesaw. Figure 15.7 illustrates minimums and maximums for plank overhang. Figure 15.8 illustrates the minimum for overlap.

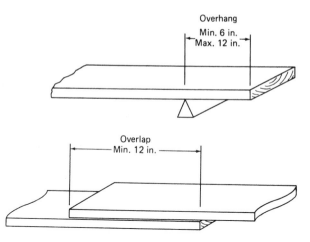

Figure 15.7. Scaffold plank overhang specifications.

Figure 15.8. Scaffold plank overlap specifications.

FLOORS AND STAIRWAYS

As in general industry, the standard for guarding of open-sided floors and platforms is one of the most frequently cited standards in construction. Curiously, though, instead of setting the vertical fall distance at 4 feet as in general industry, the specification for construction is *6 feet.* Thus an open-sided floor or platform 6 feet or more above the adjacent floor level is required to be guarded by a standard railing. Runways 4 feet high or more must be guarded.

During construction of buildings with stairways, construction employees use the new stairways during finishing operations for the building. If the building's stairways are properly designed (see Chapter 6), the Safety and Health Manager for the construction company has little to worry about.

A pitfall needs mentioning on the subject of use of new building stairways during construction. Many stairways and landings today are of steel construction with hollow pan-type treads that are filled on-site with concrete or other materials. Contractors often save until last the job of pouring the treads. Meanwhile, during building construction, workers are walking on the unfilled treads and are subject to the trip hazard created by the steel nosing along the leading edge of the riser. Of course, exposure to the hazard is unavoidable during the actual construction of the stairway itself. But after the stairways have been installed, lumber or other temporary material can be used to fill the hollow space and eliminate the trip hazard during completion of the building.

CRANES AND HOISTS

Cranes, hoists, and other material- or personnel-handling equipment are essential tools of the construction industry. Chapter 11 investigated the hazards of these machines in detail. This chapter discusses the subject from the viewpoint of the construction industry, the most important user of these machines.

One of the hazards addressed in Chapter 11 was two-blocking. Although two-blocking can be a hazard with any crane or hoist, most of the fatalities resulting from this hazard have occurred in the construction industry. A crane can be two-blocked in many ways. Hoisting the load, exending the boom, or even *lowering* the boom on a crane with a stationary winch mounted to the rear of the boom hinge can lead to two-blocking. It is difficult for the crane operator to avoid all of the ways in which a crane can be two-blocked, and consequently fatalities have occurred even when experienced crane operators are at the controls. In 1973 the ANSI standard[1] for mobile hydraulic cranes incorporated a requirement for a "two-blocking damage prevention feature" on telescoping boom cranes with less than 60 feet of extended boom. Safety and Health Managers should beware of the temptation to purchase old equipment that might not be built to safe standards. And for equipment that has been purchased, the feasibility for retrofit should be considered.

On construction sites the simultaneous execution of many parts of the overall project means that personnel will often be working around or near a crane in operation. The crane operator will avoid moving the bucket or other load over personnel and will attempt to keep the cab from striking personnel, but it is impossible for the crane operator to watch all moving parts at all times. Particularly hazardous is the rear of the cab, which on many models swings outside the crawler or other substructure when the cab and boom rotate, as shown in Figure 15.9. This motion generally occurs on every cycle of a crane's operation and is a constant threat to personnel on the ground. The hazard is a very serious one, and accidents carry a high likelihood of fatality. The employee may be either struck or crushed between the cab and some other object, such as a building wall, stack of materials, or another vehicle.

Another serious hazard with construction cranes is the possibility of contact with exposed live overhead utility lines. Crane boom contact with high-voltage transmission lines results in fatalities every year. Figure 15.10 compares safety standards for general industry and construction regarding crane clearance from high-voltage electric lines.

A common worker practice is "riding the headache ball." Figure 15.11 identifies the headache ball as the ball-shaped weight used to keep a necessary tension on the wire rope when the hook is not loaded. It is possible for workers to stand on this ball and take a ride, using the crane as an elevator, a practice that usually horrifies uninitiated passers-by. On crawler cranes used in construction, however, OSHA does not expressly prohibit the practice of riding the headache ball. What is needed to make the ride safe for the passenger is to attach a safety belt lanyard to the wire rope above the ball while riding. Also the worker should be seated on the ball and the attachment of the lanyard should be above the attachment of the ball. Riding the headache ball should be a method of last resort, to be used in only those unusual situations in which a more normal and safe method of transporting workers is impossible.

[1]ANSI B30.15-1973.

Swing

Figure 15.9.   Rear of crane cab is a hazardous area due to swing radius. (*Source:* Courtesy of the Construction Safety Association of Ontario.)

Hammerhead tower cranes are large structures that take advantage of counterweights on the end of the jib opposite the work (see Figure 15.12). This crane is often used for large building construction. Sometimes construction workers will have duties *on* the horizontal jib of a hammerhead tower crane. This presents a fatal fall hazard, and guardrails or safety belts, lanyards, and lifelines are needed to protect the worker.

Helicopters are sometimes used in construction for such operations as the placement of a steeple. Although the use of helicopters in construction is very rare, the nature of the hazards of the operation is somewhat strange and therefore deserves some mention. Ordinary tag lines, used with all cranes to control the load from below, can be a hazard to helicopters because the lines can be drawn up into the rotors, resulting in a tragedy. Tag lines must be of a length that will not permit their being drawn up into the rotors.

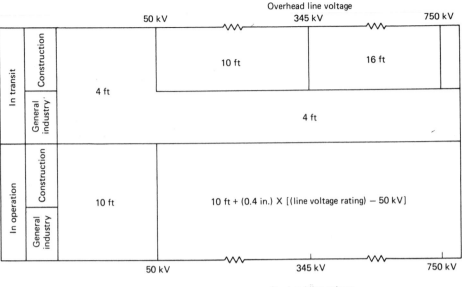

Figure 15.10.   Clearance for cranes from electric transmission lines.

Cargo hooks are another problem with helicopters used as cranes. With ordinary cranes the sole concern is that the hook will hold and not release at the wrong time. With helicopters there is this concern and the additional concern that the hook might *not* release at the right time. Cargo hooks for helicopter cranes need to have an emergency mechanical control to release the load in case the electrical release fails.

Another unusual effect that can represent a hazard is the generation of a static electrical charge on the load. This charge is developed by the friction of the air on the rotor and other moving parts. To deal with this hazard, a grounding device can be used to dissipate the charge before ground personnel touch the load, or protective rubber gloves can be worn. Once the load actually touches down, the static charge is generally dissipated through the load itself directly into the ground.

An indirect hazard that can develop from the use of helicopters is fire on the ground. The rotating blades generate such a wind on the ground that it is considered unsafe to have open fires in the path of a low-flying helicopter.

Material and Personnel Hoists

Temporary external elevators are often used on construction sites to move workers and materials and must be designed, maintained, and used properly to avoid a serious hazard. The seriousness of this hazard is quite personal to me because my brother narrowly escaped a fatal accident on such a hoist when two of five members of his engineering inspection section were killed when a personnel hoist failed in an oil refinery. Two others were totally and permanently disabled, and the fifth, my brother,

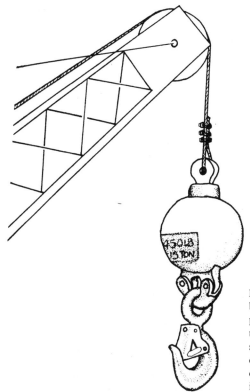

Figure 15.11. Headache ball. These balls range in weight from less than 100 pounds to a ton or more. The weight of the ball overcomes the friction of the sheaves for the running rope of the crane. The headache ball should not be confused with the much heavier wrecking ball.

Figure 15.12. Hammerhead tower crane.

was uninjured because he had been asked to remain in the office that fateful day to complete an engineering drawing.

The design requirements for personnel hoists and material hoists are different, and one of the principal safety factors is maintaining the distinction between the two hoists during use. Material hoists must be conspicuously marked "No riders allowed." On the other hand, it is permissible to move material on a personnel hoist provided that rated capacities are not exceeded.

Latched gates are needed to guard the full width of the landing entrance for both material and personnel hoists. In the case of personnel hoists, an electrical interlock must not allow movement of the hoist when the door or gate is open. Furthermore, on personnel hoists, the hoistway doors or gates must have mechanical locks that are accessible only to persons in the car.

Aerial Lifts

An alternative to scaffolds, ladders, and hoists is needed for high and awkward locations on a construction site. The use of vehicle-mounted boom platforms or aerial "buckets" is becoming increasingly popular. The boom is usually *articulating* (capable of bending in the middle) or is *hydraulically extensible* (telescoping) or both.

The biggest problem with aerial lifts is not their construction but the way they are used. Anyone in an aerial bucket needs to recognize the difference between his or her perch and terra firma. The floor of the bucket is the only place to stand—not on a ladder or plank carried aloft. Sitting on the bucket's edge is also dangerous. Even if the worker *does* stand properly in the bucket, a body belt with lanyard tied to the boom or bucket is needed to protect against a hazard such as a surprise encounter with a tree limb.

HEAVY VEHICLES AND EQUIPMENT

Next to falls and electrocutions, more construction fatalities involve vehicles, tractors, and earthmoving equipment than any other hazard source. The fatality hazard is both to drivers of the equipment and to their co-workers. Vehicle *rollovers* are the principal cause of driver fatalities, whereas vehicle *runovers* are the problem for co-workers. Not to be excluded, however, is a significant number of fatalities from repairing tires for these vehicles.

ROPS

The acronym ROPS (rhymes with "hops") represents the term "rollover protective structures" and is a major change in construction vehicle design brought about by federal safety standards. The purpose of the ROPS, illustrated in Figure 15.13, is to protect the operator from serious injury or death in the event the vehicle rolls over. The following kinds of construction equipment require ROPS:

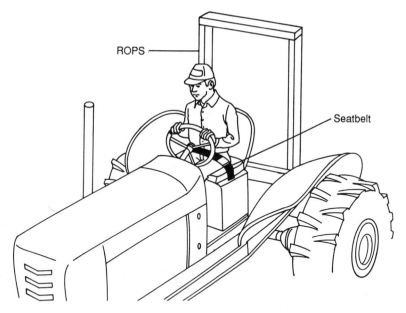

ROPS

Seatbelt

Figure 15.13.   Rollover protective structures for construction vehicles prevent fatalities
when used with seatbelts.

- Rubber-tired, self-propelled scrapers
- Rubber-tired, front-end loaders
- Rubber-tired dozers ·
- Wheel-type agricultural and industrial tractors
- Crawler tractors
- Crawler-type loaders
- Motor graders

Exempted are sideboom pipelaying tractors.

To be effective, the ROPS system must be able to withstand tremendous shock loads, which increase as the weight of the vehicle increases. The standards are quite specific regarding the structural tests to which ROPS systems must be subjected in order to qualify. For all wheel-type agricultural and industrial tractors used in construction, either a laboratory test or a field test is required to determine whether performance requirements are met. The laboratory test may be either static or dynamic. In the static test, the stationary tractor chassis is gradually loaded while strain is measured by deflection instruments. The required input energy is a function of tractor weight, which in turn is required to be a function of tractor horsepower. In other words, the tractor's gross weight cannot be lightened below rated horsepower limits in order to meet ROPS tests.

The dynamic test, an alternative to the static test, uses a 2-ton[2] pendulum that provides an impact load on the rear and side of the ROPS in successive tests. The height from which the pendulum is dropped is dependent on the calculated tractor weight based on horsepower, as in the static test just described. Deflection limits must not be exceeded.

If the field test is used, the tractor is actually rolled over, both rearward and sideways, as both types of accidents can easily occur. If the actual weight of the tractor is less than specified for its horsepower, ballast must be added for the tests.

For all ROPS tests it is a good idea to remove protective glass and weather-shields, which would probably be destroyed during the test. If there is any question whether such shields may absorb some of the energy, thereby assisting ROPS to pass the test, the shields *must* be removed.

Having read thus far, the reader can see that retrofitting an old tractor to meet current requirements for ROPS is not an easy task. The Safety and Health Manager is cautioned against taking the tractor to a local welder and requesting a ROPS system to be fabricated. Unless the tractor is actually going to be subjected to the standard ROPS test, the frame should be of a design identical to a frame actually tested for the model tractor in question. Very old or rare model tractors are obviously a problem. If a qualified ROPS system is removed for any reason, it must be remounted with bolts or welding of equal or better quality than those required for the original. The ROPS must be permanently labeled with manufacturer's or fabricator's name and address and the machine make, model, or series number that the structure is designed to fit. This labeling requirement is a sobering thought for the welder or fabricator who is asked to retrofit an old tractor with a ROPS system. All told, it is easy to see why most companies dump the old equipment on the used-equipment market, and many are shipped to foreign countries that do not require ROPS.

After the Safety and Health Manager has ensured that all appropriate construction equipment has been equipped with ROPS systems, the next task is to ensure that the equipment is used properly. Essential to the effectiveness of the ROPS system is that the operator wear a seat belt. If the operator is thrown out of the vehicle, the ROPS will afford no protection at all and may actually contribute to a fatality. Passengers are another hazard unless the vehicle is equipped with seat belts for passengers. Hitchhiking on heavy equipment at construction sites is a dangerous practice.

Runover Protection

Most of the balance of fatalities with heavy construction equipment is due to personnel being run over by the equipment. Confrontation of this major fatality category has two main thrusts: operator visibility and pedestrian awareness.

Operator visibility as good as the visibility in a private automobile is simply not feasible for a huge piece of earthmoving machinery. It is no wonder that runovers occur frequently on construction sites. The operator needs all the help affordable,

---

[2]More specifically, 4,410 lb. (2,000 kg).

but ironically, some of the poorest windshield conditions occur on construction equipment. In the morning the operator, foreman, and everyone concerned is anxious to get equipment rolling, but if the morning is cold, defrosting and defogging are essential. Often the defrosting or defogging equipment is ineffective. Dirty or cracked windshields are also a common sight in the harsh environment of the construction job.

The second link in the hazard prevention chain for runovers is the horn used by the operator to warn personnel when visibility *is* good enough to notice the endangered worker on the ground. Personnel are generally distributed all over a construction site, and the operator needs a good operable horn to warn them when they are dangerously close.

Besides the ordinary horn, many construction vehicles also need "backup alarms." Earthmoving equipment and construction vehicles that have an obstructed view to the rear need these backup alarms if they are used in reverse gear. The term *obstructed view* may be somewhat vague, but most safety and health professionals are taking the position that earthmoving machines of all types need these backup alarms. Too many people have been killed by machines backing over them to take this requirement lightly. This is said although it is acknowledged that the steady beep-beep-beep of the backup alarms can be very monotonous on a construction site and perhaps even lead to a certain complacency on the part of the personnel endangered. An alternative to the beeping backup alarms is the use of an observer standing behind the machine to alert others every time the machine backs up. This is expensive, though, and has the disadvantage of being an "administrative" or "work practice" control, instead of the preferred "engineering control" represented by the backup alarm.

Dump Trucks

One more hazard with construction vehicles and equipment needs emphasis. Dump trucks can cause a terrible accident if the raised dump body falls while the driver or some other worker has crawled into the exposed area for maintenance or inspection work. Sometimes all that holds the dump body aloft is the pressure in a hydraulic line, which can be suddenly lost due to any of a variety of failure modes. For this reason, the safety of the maintenance or inspection worker inside the exposed area demands that the truck be equipped with some positive means of support, permanently attached and capable of being locked into position.

TRENCHING AND EXCAVATIONS

A major cause of construction fatalities is the sudden collapse of the wall of a trench or excavation. It is difficult to imagine the drama of digging for a co-worker who has been literally buried alive in such a cave-in. Before becoming involved in the field of occupational safety and health, I coincidentally became an eye-witness to such a drama in Tempe, Arizona. The trench was located directly below a public stairway landing on which I happened to be standing, and thus my vantage point was directly over the scene of the cave-in. The impression was unforgettable, and its memory would motivate anyone to try to prevent such accidents in the future, a point supporting the principles of hazard avoidance set forth in Chapter 3.

Let it be understood without doubt that OSHA *has* had an impact on the reduction of fatalities due to cave-ins. Recognizing the seriousness of this hazard, OSHA undertook a comprehensive program of "special emphasis" on trenching and excavation cave-ins in the mid-1970s. The result was an impressive reduction in fatalities from this hazard source.

All trenches are excavations, but not all excavations are trenches. Trenches are narrow, deep excavations; the depth is greater than the width, but the width is no greater than 15 feet, according to the standard definition. A trench is more confined and generally more dangerous than other excavations, especially because both walls can collapse, trapping the worker. However, the walls of a trench are easier to shore than the walls of an excavation. Both are dangerous if over 5 feet deep and will easily snuff out the life of anyone who stands in the path of a collapsing wall. The hazard is not simply one of suffocation. A cave-in generally represents tons of falling earth, which can crush the body and lungs of the worker even if the face and breathing passages are left clear.

The "angle of repose" is defined as the greatest angle above the horizontal plane at which a material will lie without sliding. The angle naturally varies with the material, and approximate angles are shown in Figure 15.14. The science of soil slides is not exact, and the uncertainty thwarts attempts to control the hazard. It is difficult to say whether a particular soil type is "typical" or "compacted angular gravels," or something between the two. The specifications for trench shoring are more detailed, as can be seen in Table 15.2.

Adding to the uncertainty of the cave-in hazard are certain hazard-increasing factors, such as:

- Rainstorms, which soften the earth and promote slides
- Vibrations from heavy equipment or street traffic nearby

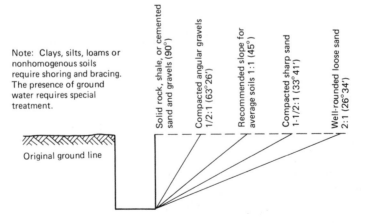

Figure 15.14. Approximate angle of repose for sloping the sides of excavations.

TABLE 15.2 Trench Shoring—Minimum Requirements

Depth of Trench (ft)	Kind or Condition of Earth	Uprights Minimum Dimension (in.)	Uprights Maximum Spacing (ft)	Stringers Minimum Dimension (in.)	Stringers Maximum Spacing (ft)	Size and Spacing of Members — Cross Braces[a] (in.) — Width of Trench (ft) Up to 3	3–6	6–9	9–12	12–15	Maximum Spacing (ft) Vertical	Horizontal
5–10	Hard, compact	3 × 4 or 2 × 6	6			2 × 6	4 × 4	4 × 6	6 × 6	6 × 8	4	6
	Likely to crack	3 × 4 or 2 × 6	3	4 × 6	4	2 × 6	4 × 4	4 × 6	6 × 6	6 × 8	4	6
	Soft, sandy, or filled	3 × 4 or 2 × 6	Close sheeting	4 × 6	4	4 × 4	4 × 6	6 × 6	6 × 8	8 × 8	4	6
	Hydrostatic pressure	3 × 4 or 2 × 6	Close sheeting	6 × 8	4	4 × 4	4 × 6	6 × 6	6 × 8	8 × 8	4	6
10–15	Hard	3 × 4 or 2 × 6	4	4 × 6	4	4 × 4	4 × 6	6 × 6	6 × 8	8 × 8	4	6
	Likely to crack	3 × 4 or 2 × 6	2	4 × 6	4	4 × 4	4 × 6	6 × 6	6 × 8	8 × 8	4	6
	Soft, sandy, or filled	3 × 4 or 2 × 6	Close sheeting	4 × 6	4	4 × 4	6 × 6	6 × 6	8 × 8	8 × 8	4	6
	Hydrostatic pressure	3 × 6	Close sheeting	8 × 10	4	4 × 6	6 × 6	6 × 8	8 × 8	8 × 10	4	6
15–20	All kinds or conditions	3 × 6	Close sheeting	4 × 12	4	4 × 12	6 × 8	8 × 8	8 × 10	10 × 10	4	6
Over 20	All kinds or conditions	3 × 6	Close sheeting	6 × 8	4	4 × 12	8 × 8	8 × 10	10 × 10	10 × 12	4	6

[a]Trench jacks may be used in lieu of, or in combination with, cross braces. Shoring is not required in solid rock, hard shale, or hard slag. Where desirable, steel sheet piling and bracing of equal strength may be substituted for wood.

Source: Code of Federal Regulations 29 CFR 1926.652.

- Previous disturbances of the soil, as from previous construction or other excavations
- Alternate freezing and thawing of the soil
- Large static loads, as from nearby building foundations or stacked material

Although judgment is required in deciding whether to employ shoring, the hazard is so serious that it is wise to adopt a conservative policy, well clear of the marginal area where a cave-in might or might not occur. One thing is certain: *After* the cave-in *does* occur, and there is a fatality, the OSHA officer will come to the scene, and everyone (including the OSHA officer) will conclude that the shoring or cave-in protection was insufficient.

A trench shoring system is shown in Figure 15.15. The trench jacks may be either screw-type or hydraulically operated. They need to be secured to prevent falling or sliding if they loosen as the sidewalls adjust slightly. Care must be taken to level the jacks and also to be sure that they are below the plane of the surface of the surrounding earth. A trench jack placed too high can be subjected to bending stresses, as shown in Figure 15.16. This can damage or even ruin the trench jack. The shoring system should be removed slowly and carefully. It is sometimes necessary to remove braces or jacks by means of ropes from above after everyone has cleared the trench.

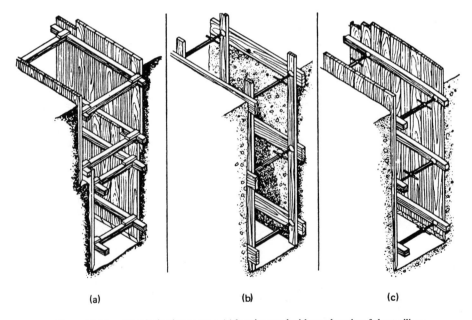

      (a)   (b)   (c)

Figure 15.15. Trench shoring system: (a) bracing used with two lengths of sheet piling; (b) bracing with screw jacks, hard soil; (c) screw jacks used with complete sheet piling. (*Source:* Courtesy of the National Safety Council, Chicago, Illinois; used with permission.)

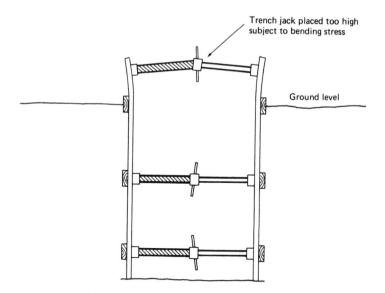

Trench jack placed too high
subject to bending stress

Ground level

Figure 15.16.   Improperly placed trench jack.

A cave-in is not the only hazard to working in trenches and excavations. All workers in the ditch are subject to the hazards of falling rocks, tools, timbers, or pipes. Head protection is needed for workers below, and good housekeeping is important along the edges of the excavation.

Even when personnel are not inside the excavation and machines do all the work, another hazard presents itself. Utility lines are often broken, resulting in fire, explosion, or inhalation hazards. Such accidents are not only dangerous but are always costly and require additional coordination with the utility company—coordination that would have been better handled in advance of the excavation. The Safety and Health Manager should install a procedure that ensures that someone stops and checks with utility companies before proceeding "blind" into an excavation project.

Signs of an imminent cave-in, rupture of a utility line, dangerous accumulations of toxic gases in deep excavations, or other possible emergency situations dictate the need for quick and easy exit. A ladder, steps, or other adequate means of exit should be located so as to require no more than 25 feet of lateral travel for escape.

CONCRETE WORK

Perhaps the most dramatic industrial accident in history was the partial collapse of a vertical wall of concrete. The year was 1978, and the location was Willow Island, West Virginia, where a huge cooling tower was under construction for a nuclear power plant. The poured concrete walls of the structure supported scaffolds for workers 170 feet above the ground. At the time of the accident the "green" concrete wall was insufficiently cured to accept the load. The wall failed, dropping the scaffold; 51 workers fell to their deaths.

The pressure to keep a construction project on schedule is antagonistic to the careful curing of concrete. But the consequences of rushing the job are serious even if no one is injured. The memory of the tragic Willow Island accident serves to remind concrete project managers of these consequences. Concrete curing time is a function of both time and temperature. About 70 °F is ideal, and temperatures either too hot or too cold can delay curing.

Even before the concrete is poured there are hazards to the placement of the reinforcing steel "rebars." For vertical structures, rebars need guys or other support to prevent collapse. Another hazard is the protruding points of exposed vertical rebars. It may sound farfetched, but workers have actually been *impaled* by falling on these bars. The most serious hazards are for workers on ladders when these ladders are placed over protruding rebars. But even at floor or ground level, a trip and fall on an exposed rebar can be fatal. I know of at least one individual case in which a worker stumbled at ground level and fell on an exposed rebar, which impaled his neck and ruptured his jugular vein. The life of the worker was saved by the quick action of a trained first-aid person. The answer to the exposed rebar problem is to bend down the ends, cover them with plywood, or wrap them with canvas until ready for pouring.

Concrete forms need carefully designed shoring systems to prevent collapse together with hazards reminiscent of the excavation cave-in hazards discussed earlier. The hydrostatic pressure of wet concrete can be very great just after it is poured. Then, at a time when the forms' stress is greatest, vibrating equipment is often applied to ensure even distribution, adding to the stresses on the forms. Overdesign is necessary to prevent the hazard of forms "kickout."

Concrete is often poured by buckets handled by cranes. Unfortunately, vibrator crews have to work with the freshly poured concrete in close proximity to the moving bucket. The pouring strategy should be to keep the vibrator crews away from the overhead path of the bucket. Riding the concrete bucket is prohibited.

STEEL ERECTION

Who has not marveled at the daring of the highrise steelworker "walking the beam" hundreds of feet up the steel superstructure of a new building under construction? Perhaps this work will always be dangerous, but the hazard has been mollified somewhat by requiring safety nets to be installed whenever the fall distance exceeds two stories or 25 feet. An alternative is to use scaffolds or temporary floors.

A safety railing around the perimeters of temporary floors is now required for tier buildings and other multifloored structures. During structural steel assembly, however, the use of ½-inch wire rope approximately 42 inches high is permitted for the safety railing. To be effective, the wire rope should be checked frequently to be sure that it remains taut.

To maintain structural integrity on the way up, the permanent floors should follow the structural steel as the work progresses. The general rule is no more than eight stories between the erection floor and the uppermost permanent floor. No more than four floors or 48 feet of unfinished bolting or welding is permitted above the foundation or uppermost permanently secured floor.

Structural steel erection sites are subject to constant hazards of falling objects. Rivets, bolts, and drift pins are required to be stored in secured containers, and if all else fails, there is the hardhat to protect the worker below. For some objects at the steel erection site, the hardhat is no protection, however. I am reminded of an instance in Chicago in which a falling steel beam flattened a parked car from left taillight to right headlight. The car was parked next to a sidewalk along which I walked to and from a meeting. The accident occurred during the meeting; luckily, no one was injured.

DEMOLITION

Some would say that the subject of demoliton does not belong in a chapter entitled "Construction," but demolition and construction are actually closely related. If the construction site is not clear, demolition of previous structures may be the first step in construction of a new building. Many of the tools and equipment, such as cranes and bulldozers, are the same.

People identify skill, knowledge, and quality as important to construction jobs, but most people do not think of demolition as requiring skill and knowledge. But often the engineering expertise required of a demolition job far exceeds the engineering for the original construction. Buildings to be demolished have often been previously damaged by fire or may have been condemned for some serious reason, such as structural damage. Required for every demolition operation is a written report of an engineering survey conducted in advance.

A demolition operation begins with manual operations, such as disassembly of salvage items, and then proceeds to material teardown and dumping to street level. Dangers exist in the debris dumping operation. The area below needs protection if it is outside the walls of the structure. Well-designed chutes capable of withstanding impact loads, together with substantial discharge gates, are needed to control the dropping material. One hazard is that personnel can fall down the chute while dumping debris. A substantial guardrail about 42 inches high is needed to protect against this hazard. A toeboard or bumper is also needed if wheelbarrows are used to prevent losing the wheelbarrow down the chute.

Once light teardown operations are completed, heavier demolition equipment such as cranes with wrecking balls are used. Most walls are unstable without lateral support, so they should not be allowed to stand alone at heights greater than one story. No unstable standing wall should be left at the end of a shift.

One sensational demolition technique that is gaining in popularity is controlled explosive demolition, illustrated in Figure 15.17. In this method, carefully engineered explosive charges are detonated to precipitate a catastrophic failure of the building structure, resulting in an immediate and total collapse. The operation has been carried out successfully upon many downtown buildings in US cities, typically triggered at the quiet time of dawn on Sunday mornings. Although the operation is dramatic and seems dangerous, it is really quite safe and avoids many of the hazards of a slow teardown process.

Figure 15.17. Building collapses under controlled explosive demolition. (*Source:* Courtesy of Jim Wolfe Photo, Tulsa, Oklahoma.)

EXPLOSIVE BLASTING

Demolition is only one application for blasting; the construction industry has others. The preparation of roadway cuts is the most important. The chief concern with construction blasting is the safe handling, storage, and transportation of the explosives themselves. The reader might want to review some of the concepts of explosives handling covered in Chapter 8 of this book.

Almost everyone has witnessed the familiar warning to "turn off two-way radio" when in a blasting area. The chance is remote, but an electrically fired blasting cap could be detonated by a small stray current induced by a radio transmitter. Lightning is even more of a hazard, and all blasting operations should cease when an electrical storm is near. Radar, nearby power lines, or even dust storms can also be sources of stray currents.

Good visibility reduces many hazards, and explosive blasting hazards are in this category. Aboveground blasting should be conducted in the daytime only. Black powder blasting has been replaced by safer modern methods, and black powder blasting is now prohibited in construction.

The transportation of explosive materials is subject to Department of Transportation (DOT) regulations familiar to suppliers and most construction operators who use explosives. Signs saying **EXPLOSIVES** in large (4 inch) red letters are required on all four sides of the vehicle. Blasting caps should be transported in a separate vehicle from the vehicle transporting other explosives, and both should be transported separate from other cargoes.

Vehicles for explosives need a good fire extinguisher rated at least 10-ABC on board. It would be foolhardy to attempt to control a fire in the cargo compartment of a vehicle transporting explosives. However, most vehicle fires begin in the engine compartment or outside but adjacent to the vehicle. Such fires can sometimes be controlled with a good fire extinguisher wielded by a trained operator, thereby averting a major explosives catastrophe.

ELECTRIC UTILITIES

A specialized type of construction is the erection and modification of electric transmission and distribution lines and equipment. The efficient transmission of usable levels of electrical power necessitates very high voltages. The rules for handling high voltages are quite different from the rules for handling ordinary household and industrial and commercial voltages. For instance, with ordinary voltages, a danger is the contact with exposed live parts. With high voltages, it can be dangerous even to approach the *vicinity* of live parts, as is reflected in Table 15.3 based on the OSHA standard. For voltages in the kilovolt range, the atmosphere may not be an effective insulator, and arcing becomes a hazard. Therefore, safety distances must be maintained. Of course, the distances shown in Table 15.3 apply safety factors. The actual physical arcing distances are much smaller, but there is an element of uncertainty due to such factors as humidity and barometric pressure. Furthermore, the electric-utility lineman may not be able to estimate precisely his distance from the high-voltage line or equipment, making safety factors essential.

Personal protective equipment for high-voltage work takes on a new dimension, that of *degree* of protection. Ordinary insulators on tools, protective gloves, and other insulating equipment which are effective insulators for ordinary applications might completely break down in high-voltage exposures. The whole business of working with energized high-voltage lines is a strange world to the uninitiated—a world fraught with curious physical effects. The electric-utility industry offers a prime example of an industry in which training in awareness and understanding of hazards is the key to a safe workplace, echoing the principles set forth in Chapter 3.

TABLE 15.3   Minimum Clearance Distances for Live-Line Bare-Hand Work (Alternating Current)

Voltage Range (Phase-to-Phase) (kV)	Distance (Feet and Inches) for Maximum Voltage	
	Phase to Ground	Phase to Phase
2.1–15	2–0	2–0
15.1–35	2–4	2–4
35.1–46	2–6	2–6
46.1–72.5	3–0	3–0
72.6–121	3–4	4–6
138–145	3–6	5–0
161–169	3–8	5–6
230–242	5–0	8–4
345–362	7–0[a]	13–4[a]
500–552	11–0[a]	20–0[a]
700–765	15–0[a]	31–0[a]

[a]For 345–362, 500–552, and 700–765 kV, the minimum clearance distance may be reduced provided that the distances are not made less than the shortest distance between the energized part and a grounded surface.

Source: Code of Federal Regulations 29 CFR 1926.955.

SUMMARY

The construction industry deserves special consideration because it is so dangerous and also because OSHA has watched construction more closely than general industries. The Safety and Health Manager for construction jobs should remember that the principal task is to avoid *fatalities*. The top five categories of fatalities in the construction industry are:

- Falls
- Electrocutions
- Vehicle rollover
- Personnel runover by vehicle
- Excavation cave-ins

If the Safety and Health Manager keeps these fatality categories in mind, it will help to place overall efforts in proper perspective on the construction site.

Critics of OSHA should recognize one clear fact: OSHA has had a significant effect on reducing the number of fatalities from trenching and excavation cave-ins in the construction industry. Most trench workers are young males, and many of them are fathers with several dependents. The reduction in the number of deaths from this cause has a far-reaching value that is difficult to measure.

EXERCISES AND STUDY QUESTIONS

15.1. What is the minimum illumination level permitted for general construction areas? In what areas may lighting be reduced to 3 foot-candles?

15.2. How are lasers used in construction?

15.3. What tensile strength is specified for safety belt hardware? Give two reasons that the specification is so much higher than the weight of any human being.

15.4. What is a safety belt lanyard? What nominal breaking strength is specified for lanyards?

15.5. What is a triple rolling hitch?

15.6. Compare hydraulic versus pneumatic power tools from a safety and health standpoint.

15.7. How are construction cranes particularly hazardous to persons on the ground?

15.8. Why are helicopter hooks (for load attachment) more complicated than those for ordinary construction cranes?

15.9. What does the acronym ROPS represent?

15.10. What are the two principal strategies for preventing personnel from being run over by construction equipment?

15.11. What is the difference between a trench and an excavation?

15.12. How can trench jacks be damaged by improper placements?

15.13. What is a rebar? Why is it dangerous?

15.14. When must steel erection workers be protected against falls with safety nets?

15.15. What type of safety rail construction is permitted during steel erection?

15.16. Why is an engineering survey required prior to demolition of a building?

15.17. Is it a good idea to carry a fire extinguisher aboard a truck that transports explosives? Why or why not?

15.18. A crawler-type construction crane is operating near a 550-kilovolt power line. What is the minimum distance the boom should approach the line?

15.19. On the way to the worksite, the crane passed under this same power line at a different location. What minimum distance is specified for this situation?

15.20. Suppose in Question 15.18 that the crawler crane had been employed in general industry instead of construction. What minimum distance would be permitted in general industry?

15.21. What minimum distance would be permitted for the crane in transit?

15.22. A painter is standing on a work platform that is 27 feet above ground level. For fall protection, the worker's safety belt is attached to a 12-foot safety line, which serves as a lanyard and is securely fastened to the structure at a point 40 feet above ground level. Does this arrangement violate standards for fall protection? Explain.

15.23. A convenient and secure attachment point for fall protection lines is located 35 feet above ground level on the exterior wall of a building. A 20-foot safety line, which can be used as a belt lanyard, is available for connection to this attachment point. What is the lowest and highest level at which a work platform can be safely positioned for workers to be protected by a 20-foot belt lanyard fastened to this attachment point? State any assumptions necessary to your solution.

Bibliography

1. *Accident Facts*. Chicago: National Safety Council, 1981.
2. *Arc Welding and Gas Welding and Cutting*, Safety and Health in (NIOSH No. 78-138). Cincinnati, OH: U.S. Department of Health, Education and Welfare (NIOSH), 1978.
3. *Arkansas Workers' Compensation Laws and Rules of the Commission* (rev. ed.). Little Rock, AR: Workers' Compensation Commission, June 1986.
4. *Auto and Home Supply Stores*, Health and Safety Guide for (NIOSH No. 76-113). Cincinnati, OH: U.S. Department of Health, Education and Welfare (NIOSH), 1975.
5. Bland, Jay, ed., *The Welding Environment*. Miami, FL: American Welding Society, 1973.
6. "Brake Monitoring," *Manufacturing Engineering*, February 1976.
7. Chaffin, D. and K. Park, "Biomechanical Evaluation of Two Methods of Manual Load Lifting," *AIIE Transactions*, 6, No. 2, June 1974.
8. Chapnik, Elissa-Beth, and Clifford M. Gross, "Evaluation, Office Improvements Can Reduce VDT Operator Problems," *Occupational Health and Safety*, 56, No. 7, July 1987.
9. Christensen, Herbert E., et al., eds., *Registry of Toxic Effects of Chemical Substances* (1975 ed.). Rockville, MD: U.S. Department of Health, Education and Welfare (NIOSH), 1975.
10. Christensen, Herbert E., et al., eds., *Suspected Carcinogens*. Rockville, MD: U.S. Department of Health, Education and Welfare, 1975.
11. *A Common Goal*. Little Rock, AR: Arkansas Department of Labor, 1975.
12. *Concepts and Techniques of Machine Safeguarding* (OSHA No. 3067). Washington, DC: U.S. Department of Labor, Occupational Safety and Health Administration, 1980.
13. *Concrete Products Industry*, Health and Safety Guide for (NIOSH No. 75-163). Cincinnati, OH: U.S. Department of Health, Education and Welfare (NIOSH), 1975.
14. *Construction and Related Machinery Manufacturers*, Health and Safety Guide for (NIOSH No. 78-103). Cincinnati, OH: U.S. Department of Health, Education and Welfare (NIOSH), 1977.

15. *Construction Industry OSHA Safety and Health Standards* (29 CFR 1926/1910; OSHA No. 2207) (rev.). Washington, DC: U.S. Department of Labor, Occupational Safety and Health Administration, 1979.

16. *Cost of Government Regulation Study.* Chicago: Arthur Andersen & Company, 1979.

17. Covan, John M., *Electrical Hazards Control Manual.* Little Rock, AR: Arkansas Department of Labor, 1977.

18. *Cranes*, A Guide to Good Work Practices for Operators (NIOSH No. 78-192). Cincinnati, OH: U.S. Department of Health, Education and Welfare (NIOSH), 1978.

19. Cross, Rich, "EPA's Tank Rule: A Compliance Nightmare for Motor Carriers," *Commercial Carrier Journal*, November 1988.

20. Dickie, D.E., *Crane Handbook.* Toronto, Ontario: Construction Safety Association of Ontario, 1975.

21. *Drug Testing Monitor* (brochure), Washington, DC: Traffic World, 1989.

22. *Engineering Control of Welding Fumes* (NIOSH No. 75-115). Cincinnati, OH: U.S. Department of Health, Education and Welfare (NIOSH), 1975.

23. *Fabricated Structural Metal Products Industry*, Health and Safety Guide for (NIOSH No. 78-100). Cincinnati, OH: U.S. Department of Health, Education and Welfare (NIOSH), 1977.

24. *Fatal Facts*, Washington, DC: U.S. Department of Labor, No. 36, 1988.

25. *Fiberglass Layup and Sprayup* (NIOSH No. 76-158). Cincinnati, OH: U.S. Department of Health, Education and Welfare (NIOSH), 1976.

26. *Fire Protection Handbook* (15th ed.). Quincy, MA: National Fire Protection Association, 1981.

27. *Flash Point Index of Trade Name Liquids* (9th ed.). Boston: National Fire Protection Association, 1978.

28. *Foundries*, Health and Safety Guide for (NIOSH No. 76-124). Cincinnati, OH: U.S. Department of Health, Education and Welfare, 1976.

29. Fox, Stephen, "Manhole Cover Industry Hurt by Industry Imports," *Northwest Arkansas Times*, 1981. (An Associated Press interview with Mr. Jim Pinkerton, Pinkerton Foundry, Lodi, CA.)

30. *Fumes and Gases in the Welding Environment.* Miami, FL: American Welding Society, 1979.

31. *General Industry OSHA Safety and Health Standards* (29 CFR 1910; OSHA No. 2206) (rev.). Washington, DC: U.S. Department of Labor, Occupational Safety and Health Administration, 1989.

32. Garg, A., D. B. Chaffin, and G. D. Herrin, "Prediction of Metabolic Rates for Manual Materials Handling Jobs," *American Industrial Hygiene Association Journal*, 39 (1978), pp. 661–674.

33. *Grain Mills*, Health and Safety Guide for (NIOSH No. 75-144). Cincinnati, OH: U.S. Department of Health, Education and Welfare (NIOSH), 1975.

34. Grimaldi, John V., and Rollin H. Simonds, *Safety Management* (3rd ed.). Homewood, IL: Richard D. Irwin, 1975.

35. Hamilton, Robert W., "The Role of Nongovernmental Standards in the Development of Mandatory Federal Standards Affecting Safety or Health," *Texas Law Review*, 56, No. 8 (November 1977).

36. Hammer, Willie, *Occupational Safety Management and Engineering* (2nd ed.). Englewood Cliffs, NJ: Prentice-Hall, 1981.

37. *Handbook of Organic Industrial Solvents*, Technical Guide No. 6 (5th ed.). Chicago: Alliance of American Insurers, 1980.

38. *Health and Safety Guide for Public Warehousing*. Cincinnati, OH: U.S. Department of Health, Education and Welfare (NIOSH), 1978.

39. Heinrich, H. W., *Industrial Accident Prevention* (4th ed.). New York: McGraw-Hill, 1959.

40. *Hotels and Motels*, Health and Safety Guide for (NIOSH No. 76-112) Cincinnati, OH: U.S. Department of Health, Education and Welfare (NIOSH), 1975.

41. Hyatt, E. C., et al., "Effect of Facial Hair on Worker Performance," *American Industrial Hygiene Association Journal*, April 1973.

42. Imre, John, Unpublished doctoral dissertation, Michigan State University, East Lansing, MI, 1974.

43. *The Industrial Environment: Its Evaluation and Control*. Cincinnati, OH: U.S. Department of Health, Education and Welfare (NIOSH), 1973.

44. *Industridl Noise Control Manual* (NIOSH 79-117) (rev. ed.). Cincinnati, OH: U.S. Department of Health, Education and Welfare, 1978.

45. *Industrial Ventilation* (15th ed.). Lansing, MI: Committee on Industrial Ventilation, American Conference of Governmental Industrial Hygienists (ACGIH), 1978.

46. *Lithographic Printing Industry*, Employee Health and Safety (NIOSH No. 77-223). Cincinnati, OH: U.S. Department of Health, Education and Welfare (NIOSH), 1977.

47. Lovett, John N., *Industrial Noise Control Manual*. Little Rock, AR: Arkansas Department of Labor, 1976.

48. *Machine Guarding: Assessment of Need* (NIOSH No. 75-173). Cincinnati, OH: U.S. Department of Health, Education and Welfare (NIOSH), 1975.

49. *Manual of Respiratory Protection against Radioactive Materials* (NUREG-0041). U.S. Atomic Energy Commission, Directorate of Regulatory Standards, 1974.

50. *Manufacturers of Paints and Allied Products*, Health and Safety Guide for (NIOSH No. 75-179). Cincinnati, OH: U.S. Department of Health, Education and Welfare (NIOSH), 1975.

51. McClay, Robert E., "Toward a More Universal Model of Loss Incident Causation," *Professional Safety*, 34, No. 1, January 1989.

52. (a) McElroy, Frank E., ed., "Administration and Programs," *Accident Prevention Manual for Industrial Operations* (8th ed.). Chicago: National Safety Council, 1981.

53. (b) McElroy, Frank E., ed., "Engineering and Technology," *Accident Prevention Manual for Industrial Operations* (8th ed.). Chicago: National Safety Council, 1981.

54. *Metal Stamping Operations*, Health and Safety Guide for (NIOSH No. 75-174). Cincinnati, OH: U.S. Department of Health, Education and Welfare (NIOSH), 1975.

55. *Method of Recording and Measuring Work Injury Experience* (ANSI Z16.1-1967). New York: American National Standards Institute, 1967.

56. Mills, Wilbur, (interview), *Overview* (the journal of Overlook Hospital, Summit, NJ), 1, No. 3, Fall 1980.

57. *NFPA Inspection Manual* (3rd ed.). Boston: National Fire Protection Association, 1970.

58. *NIOSH Certified Equipment*; Cumulative Supplement (NIOSH No. 77-195). Morgantown, WV: U.S. Department of Health, Education and Welfare (NIOSH), 1977.

59. *National Electrical Code®* (NEC). Boston: National Fire Protection Association, 1971 and 1981.

60. *Occupational Diseases: A Guide to Their Recognition*. Cincinnati, OH: U.S. Department of Health, Education and Welfare, 1951.

61. *Occupational Exposure Sampling Manual* (NIOSH No. 77-173). Cincinnati, OH: U.S. Department of Health, Education and Welfare (NIOSH), 1977.
62. *Occupational Safety and Health Cases*, Vols. 1–9. Washington, DC: Bureau of National Affairs, Inc., 1974–1982.
63. *Occupational Safety and Health Directory* (NIOSH No. 80-124). Cincinnati, OH: U.S. Department of Health and Human Services (NIOSH), 1980.
64. *Occupational Safety and Health in Vocational Education* (NIOSH No. 79-125). Cincinnati, OH: U.S. Department of Health and Human Services (NIOSH), 1979.
65. *Occupational Safety and Health Reporter* (weekly periodical), Vols. 1–11. Washington, DC: Bureau of National Affairs, Inc., 1971–1982.
66. Olishifski, Julian B., and Frank E. McElroy, eds., *Fundamentals of Industrial Hygiene*. Chicago: National Safety Council, 1971.
67. *Overhead and Gantry Cranes* (ANSI B.30.2.0-1976). New York: American National Standards Institute, 1976.
68. Parker, Sandra C., C. Ray Asfahl, and Susan Johnsen, "An Industrial Chemical Hazards Database with Natural Language Interface: An Application of Artificial Intelligence," *Proceedings, Tenth Annual Conference for Computers and Industrial Engineering*, Dallas, TX, 1988.
69. Peterson, Donald R., and David B. Thomas, *Fundamentals of Epidemiology*. Lexington, MA: Lexington Books, 1978.
70. *Plumbing, Heating and Air Conditioning Contractors*, Health and Safety Guide for (NIOSH No. 76-127). Cincinnati, OH: U.S. Department of Health, Education and Welfare (NIOSH), 1975.
71. *Power Actuated Fastening Tools* (NIOSH No. 78-178A). Cincinnati, OH: U.S. Department of Health, Education and Welfare (NIOSH), 1978.
72. *A Prescription for Battery Workers* (NIOSH No. 76-153). Cincinnati, OH: U.S. Department of Health, Education and Welfare (NIOSH), 1976.
73. *Public Law 91-596* (The Williams-Steiger Occupational Safety and Health Act of 1970), U.S. Congress, December 19, 1970.
74. *Recordkeeping Requirements under the Occupational Safety and Health Act of 1970*. Washington, DC: U.S. Department of Labor, Occupational Safety and Health Administration, 1978.
75. *Respiratory Protection*, A Guide for the Employee (NIOSH No. 78-193B). Cincinnati, OH: U.S. Department of Health, Education and Welfare (NIOSH), 1978.
76. *Respiratory Protection*, An Employer's Manual (NIOSH No. 78-193A). Cincinnati, OH: U.S. Department of Health, Education and Welfare (NIOSH), 1978.
77. *Rigging Manual*. Toronto, Ontario: Construction Safety Association of Ontario, 1975.
78. Ryan, Joseph, P., "Power Press Safeguarding: A Human Factors Perspective," *Professional Safety*, 32, No. 8, August 1987, pp. 23–26.
79. *Safety Requirements for Woodworking Machinery* (ANSI 01.1-1975). New York: American National Standards Institute, 1975.
80. *Safety Standard for Monorail Systems and Underhung Cranes* (ANSI B30.11-1973). New York: American National Standards Institute, 1973.
81. Saulter, Gilbert J., Regional Administrator, OSHA Region VI, in a speech entitled *OSHA Update*, July 27, 1988, in Little Rock, AR.
82. Sax, N. Irving, *Dangerous Properties of Industrial Materials* (5th ed.). New York: Van Nostrand Reinhold, 1975.

83. Sensidyne Product Literature, "The First Truly Simple Precision Gas Detector System," Sensidyne, Inc., Largo, FL, 1984.
84. *Sign and Advertising Display Manufacturers*, Health and Safety Guide for (NIOSH No. 76-126). Cincinnati, OH: U.S. Department of Health, Education and Welfare (NIOSH), 1976.
85. Smith, Harry C., "BLEVE Can Blow You to Oblivion," *Ohio Monitor*, 52, No. 10, October 1979, p. 16.
86. *Threshold Limit Values* (TLV Booklet). Cincinnati, OH: American Conference of Governmental Industrial Hygienists.
87. *Title III Fact Sheet*. U.S. Environmental Protection Agency, Washington, DC: U.S. Government Printing Office Document 1987-718-872-1302/1280, 1987.
88. *The Welding Environment*. Miami, FL: American Welding Society, 1973.
89. Wilkinson, Bruce S. "Substance Abuse Programs," *Proceedings, American Society of Safety Engineers Annual Professional Development Conference and Exposition*, Baltimore, MD, June 14–17, 1987.
90. *Wooden Furniture Manufacturing*, Health and Safety Guide for (NIOSH No. 75-167). Cincinnati, OH: U.S. Department of Health, Education and Welfare (NIOSH), 1975.
91. *Worker Exposure to AIDS and Hepatitis B*, Washington, DC: U.S. Department of Labor, Occupational Safety and Health Administration, 1987.
92. *Work Injury and Illness Rates*. Chicago: National Safety Council, 1981.
93. *Work Practices Guide for Manual Lifting* (NIOSH No. 81-122). Cincinnati, OH: U.S. Department of Health and Human Services (NIOSH), 1981.
94. *Houston Chronicle*, January 10, 1984.
95. *Occupational Injuries and Illnesses in the United States by Industry*. Washington, DC: U.S. Department of Labor, Bureau of Labor Statistics, 1985.

Appendix A

OSHA Permissible Exposure Limits

APPENDIX A.1 GENERAL TABLE OF PELs

Limits for Air Contaminants

Substance	CAS No.†	Transitional Limits PEL* ppm[a]	mg/m³[b]	Skin Designation	Final Rule Limits** TWA ppm[a]	mg/m³[b]	STEL[c] ppm[a]	mg/m³[b]	Ceiling ppm[a]	mg/m³[b]	Skin Designation
Acetaldehyde	75–07–0	200	360	—	100	180	150	270	—	—	—
Acetic acid	64–19–7	10	25	—	10	25	—	—	—	—	—
Acetic anhydride	108–24–7	5	20	—	—	—	—	—	5	20	—
Acetone	67–64–1	1000	2400	—	750	1800	1000	2400	—	—	—
Acetonitrile	75–05–8	40	70	—	40	70	60	105	—	—	—
2-Acetylaminofluorine; see OSHA std 29 CFR 1910.1014	53–96–3										
Acetylene dichloride; see 1,2-Dichloroethylene											
Acetylene Letrabromide	79–27–6	1	14	—	1	14	—	—	—	—	—
Acetylsalicylic acid (Aspirin)	50–78–2	—	—	—	—	5	—	—	—	—	—
Acrolein	107–02–8	0.1	0.25	—	0.1	0.25	0.3	0.8	—	—	X
Acrylamide	79–06–1	—	0.3	X	—	0.03	—	—	—	—	X
Acrylic acid	79–10–7	—	—	—	10	30	—	—	—	—	—
Acrylonitrile; see OSHA std 29 CFR 1910.1045	107–13–1										
Aldrin	309–00–2	—	0.25	X	—	0.25	—	—	—	—	X
Allyl alcohol	107–18–6	2	5	X	2	5	4	10	—	—	X
Allyl chloride	107–05–1	1	3	—	1	3	2	6	—	—	—
Allyl glycidyl ether (AGE)	106–92–3	(C)10	(C)45	—	5	22	10	44	—	—	—
Allyl propyl disulfide	2179–59–1	2	12	—	2	12	3	18	—	—	—
alpha-Alumina	1344–28–1										
Total dust		—	15	—	—	10	—	—	—	—	—
Respirable fraction		—	5	—	—	5	—	—	—	—	—
Aluminum (as Al) Metal	7429–90–5										—

Substance	CAS No.	Transitional Limits ppm	Transitional Limits mg/m³	Skin	Final Rule Limits ppm (TWA)	mg/m³ (TWA)	STEL ppm	STEL mg/m³	Skin
Total dust		—	15	—	—	15	—	—	—
Respirable fraction		—	5	—	—	5	—	—	—
Pyro powders		—	—	—	—	5	—	—	—
Welding fumes***		—	—	—	—	5	—	—	—
Soluble salts		—	—	—	—	2	—	—	—
Alkyls		—	—	—	—	2	—	—	—
4-Aminodiphenyl; see OSHA std 29 CFR 1910.1011	92-67-1	—	—	—	—	—	—	—	—
2-Aminoethanol; see Ethanolamine		—	—	—	—	—	—	—	—
2-Aminopyridine	504-29-0	0.5	2	—	0.5	2	—	—	—
Amitrole	61-82-5	—	—	—	—	0.2	—	—	—
Ammonia	7664-41-7	50	35	—	—	—	35	27	—
Ammonium chloride fume	12125-02-9	—	—	—	—	10	—	20	—
Ammonium sulfamate	7773-06-0								
Total dust		—	15	—	—	10	—	—	—
Respirable fraction		—	5	—	—	5	—	—	—
n-Amyl acetate	628-63-7	100	525	—	100	525	—	—	—
sec-Amyl acetate	626-38-0	125	650	—	125	650	—	—	—
Aniline and homologs	62-53-3	5	19	X	2	8	—	—	X
Anisidine (o-, p-isomers)	29191-52-4	—	0.5	X	—	0.5	—	—	X
Antimony and compounds (as Sb)	7440-36-0	—	0.5	—	—	0.5	—	—	—
ANTU (alpha Naphthylthiourea)	86-88-4	—	0.3	—	—	0.3	—	—	—
Arsenic, organic compounds (as As)	7440-38-2	—	0.5	—	—	0.5	—	—	—
Arsenic, inorganic compounds (as As); see OSHA std 29 CFR 1910.1018	7440-38-2	—	—	—	—	—	—	—	—
Arsine	7784-42-1	0.05	0.2	—	0.05	0.2	—	—	—
Asbestos; see OSHA std 29 CFR 1910.1001 and 29 CFR 1910.1101	Varies	—	—	—	—	—	—	—	—
Atrazine	1912-24-9	—	—	—	—	5	—	—	—
Azinphos-methyl	86-50-0	—	0.2	X	—	0.2	—	—	X
Barium, soluble compounds (as Ba)	7440-39-3	—	0.5	—	—	0.5	—	—	—

APPENDIX A.1 GENERAL TABLE OF PELs

Limits for Air Contaminants

Substance	CAS No.[†]	Transitional Limits PEL[*]			Final Rule Limits[**] TWA		STEL[c]		Ceiling		Skin Designation
		ppm[a]	mg/m³[b]	Skin Designation	ppm[a]	mg/m³[b]	ppm[a]	mg/m³[b]	ppm[a]	mg/m³[b]	
Barium sulfate	7727–43–7										
Total dust		—	15	—	—	10	—	—	—	—	—
Respirable fraction		—	5	—	—	5	—	—	—	—	—
Benomyl	17804-35-2										
Total dust		—	15	—	—	10	—	—	—	—	—
Respirable fraction		—	5	—	—	5	—	—	—	—	—
Benzene; see OSHA std 29 CFR 1910.1028	71–43–2	See Table A.2 for the limits applicable in the operations or sectors excluded in OSHA std 29 CFR 1910.1028[d]									
Benzidine; see OSHA std 29 CFR 1910.1010	92–87–5										
p-Benzoquinone; see Quinone											
Benzo(a)pyrene; see Coal tar pitch volatiles											
Benzoyl peroxide	94–36–0	—	5	—	—	5	—	—	—	—	—
Benzyl chloride	100–44–7	1	5	—	1	5	—	—	—	—	—
Beryllium and beryllium compounds (as Be)	7440–41–7	See Table A.2			See Table A.2	—					—
Biphenyl; see Diphenyl											
Bismuth telluride, Undoped	1304–82–1										
Total dust		—	15	—	—	15	—	—	—	—	—
Respirable fraction		—	5	—	—	5	—	—	—	—	—
Bismuth telluride, Se-doped		—	—	—	—	5	—	—	—	—	—
Borates, tetra, sodium salts											
Anhydrous	1330–43–4	—	—	—	—	10	—	—	—	—	—
Decahydrate	1303–96–4	—	—	—	—	10	—	—	—	—	—
Pentahydrate	12179–04–3	—	—	—	—	10	—	—	—	—	—

Substance	CAS No.	ppm	mg/m³	Skin	ppm	mg/m³	ppm	mg/m³	ppm	mg/m³	Skin
Boron oxide	1303-86-2										
Total dust		—	15	—	—	10	—	—	—	—	—
Boron tribromide	10294-33-4	—	—	—	—	—	—	—	1	10	—
Boron trifluoride	7637-07-2	(C)1	(C)3	—	—	—	—	—	1	3	—
Bromacil	314-40-9	—	—	—	—	10	—	—	—	—	—
Bromine	7726-95-6	0.1	0.7	—	0.1	0.7	0.3	2	—	—	—
Bromine pentafluoride	7789-30-2	—	—	—	0.1	0.7	—	—	—	—	—
Bromoform	75-25-2	0.5	5	X	0.5	5	—	—	—	—	X
Butadiene (1,3-Butadiene)	106-99-0	1000	2200	—	1000	2200	—	—	—	—	—
Butane	106-97-8	—	—	—	800	1900	—	—	—	—	—
Butanethiol; see Butyl mercaptan											
2-Butanone (Methyl ethyl ketone)	78-93-3	200	590	—	200	590	300	885	—	—	—
2-Butoxyethanol	111-76-2	50	240	X	25	120	—	—	—	—	X
n-Butyl-acetate	123-86-4	150	710	—	150	710	200	950	—	—	—
sec-Butyl acetate	105-46-4	200	950	—	200	950	—	—	—	—	—
tert-Butyl acetate	540-88-5	200	950	—	200	950	—	—	—	—	—
Butyl acrylate	141-32-2	—	—	—	10	55	—	—	—	—	X
n-Butyl alcohol	71-36-3	100	300	—	—	—	—	—	50	150	X
sec-Butyl alcohol	78-92-2	150	450	—	100	305	—	—	—	—	—
tert-Butyl alcohol	75-65-0	100	300	—	100	300	150	450	—	—	—
Butylamine	109-73-9	(C)5	(C)15	X	—	—	—	—	5	15	X
tert-Butyl chromate (as CrO₃)	1189-85-1	—	(C)0.1	X	—	—	—	—	—	0.1	X
n-Butyl glycidyl ether (BGE)	2426-08-6	50	270	X	25	135	—	—	—	—	—
n-Butyl lactate	138-22-7	—	—	—	5	25	—	—	—	—	—
Butyl mercaptan	109-79-5	10	35	—	0.5	1.5	—	—	—	—	—
o-sec-Butylphenol	89-72-5	—	—	—	5	30	—	—	—	—	X
p-tert-Butyltoluene	98-51-1	10	60	—	10	60	20	120	—	—	—
Cadmium fume (as Cd)	7440-43-9	See Table A.2				0.1				0.3	

APPENDIX A.1   GENERAL TABLE OF PELs

Limits for Air Contaminants

Substance	CAS No.†	Transitional Limits PEL* ppm[a]	mg/m³[b]	Skin Designation	Final Rule Limits** TWA ppm[a]	mg/m³[b]	STEL[c] ppm[a]	mg/m³[b]	Ceiling ppm[a]	mg/m³[b]	Skin Designation
Cadmium dust (as Cd)	7440-43-9	See Table A.2		—	—	0.2	—	—	—	—	0.6
Calcium carbonate	1317-65-3										
Total dust		—	15	—	—	15	—	—	—	—	—
Respirable fraction		—	5	—	—	5	—	—	—	—	—
Calcium cyanamide	156-62-7	—	—	—	—	0.5	—	—	—	—	—
Calcium hydroxide	1305-62-0	—	—	—	—	5	—	—	—	—	—
Calcium oxide	1305-78-8	—	5	—	—	5	—	—	—	—	—
Calcium silicate	1344-95-2										
Total dust		—	15	—	—	15	—	—	—	—	—
Respirable fraction		—	5	—	—	5	—	—	—	—	—
Calcium sulfate	7778-18-9										
Total dust		—	15	—	—	15	—	—	—	—	—
Respirable fraction		—	5	—	—	5	—	—	—	—	—
Camphor, synthetic	76-22-2	—	2	—	—	2	—	—	—	—	—
Caprolactam	105-60-2										
Dust		—	—	—	—	1	—	3	—	—	—
Vapor		—	—	—	5	20	10	40	—	—	—
Captafol (Difolatan[R])	2425-06-1	—	—	—	—	0.1	—	—	—	—	—
Captan	133-06-2	—	—	—	—	5	—	—	—	—	—
Carbaryl (Sevin[R])	63-25-2	—	5	—	—	5	—	—	—	—	—
Carbofuran (Furadan[R])	1563-66-2	—	—	—	—	0.1	—	—	—	—	—
Carbon black	1333-86-4	—	3.5	—	—	3.5	—	—	—	—	—
Carbon dioxide	124-38-9	5000[e]	9000	—	10,000	18,000	30,000	54,000	—	—	—
Carbon disulfide	75-15-0	See Table A.2		—	4	12	12	36	—	—	X
Carbon monoxide	630-08-0	50	55	—	35	40	—	—	200	229	—
Carbon tetrabromide	558-13-4	—	—	—	0.1	1.4	0.3	4	—	—	—
Carbon tetrachloride	56-23-5	See Table A.2		—	2	12.6	—	—	—	—	—
Carbonyl fluoride	353-50-4	—	—	—	2	5	5	15	—	—	—

Substance	CAS No.										
Catechol (Pyrocatechol)	120-80-9	—	—	—	—	—	5	—	—	20	X
Cellulose	9004-34-6	—	—	—	—	—	—	—	—	—	—
Total dust		—	15	15	—	—	—	—	—	15	—
Respirable fraction		—	5	5	—	—	—	—	—	5	—
Cesium hydroxide	21351-79-1	X	—	—	—	—	—	—	—	2	X
Chlordane	57-74-9	X	0.5	0.5	—	—	—	—	—	0.5	X
Chlorinated camphene	8001-35-2	—	0.5	0.5	—	1	—	1	—	0.5	—
Chlorinated diphenyl oxide	55720-99-5	—	0.5	0.5	—	—	—	—	—	0.5	—
Chlorine	7782-50-5	—	(C)3	(C)1	—	1	—	3	—	1.5	—
Chlorine dioxide	10049-04-4	—	0.3	0.1	0.3	0.9	—	—	—	0.3	—
Chlorine trifluoride	7790-91-2	—	(C)0.4	(C)0.1	—	—	0.1	0.4	—	—	—
Chloroacetaldehyde	107-20-0	—	(C)3	(C)1	—	—	1	3	—	—	—
a-Chloroacetophenone (Phenacyl chloride)	532-27-4	—	0.3	0.05	—	—	0.05	—	—	0.3	—
Chloroacetyl chloride	79-04-9	—	—	—	—	—	—	—	—	0.2	—
Chlorobenzene	108-90-7	—	350	75	—	—	—	—	—	350	—
o-Chlorobenzylidene malononitrile	2698-41-1	X	0.4	0.05	—	—	0.05	0.4	—	—	X
Chlorobromomethane	74-97-5	—	1050	200	—	—	—	—	—	1050	—
2-Chloro-1, 3-butadiene; see b-Chloroprene											
Chlorodifluoromethane	75-45-6	—	—	—	—	—	—	—	—	3500	—
Chlorodiphenyl (42% Chlorine) (PCB)	53469-21-9	X	1	—	—	—	—	—	—	1	X
Chlorodiphenyl (54% Chlorine) (PCB)	11097-69-1	X	0.5	—	—	—	—	—	—	0.5	X
1-Chloro, 2, 3-epoxypropane; see Epichlorohydrin											
2-Chloroethanol; see Ethylene chlorohydrin											
Chloroethylene; see Vinyl chloride											
Chloroform (Trichloromethane)	67-66-3	—	(C)50	(C)240	—	—	2	9.78	—	—	—
bis(Chloromethyl) ether; see OSHA std 29 CFR 1910.1008	542-88-1	—	—	—	—	—	—	—	—	—	—

APPENDIX A.1 GENERAL TABLE OF PELs

Limits for Air Contaminants

Substance	CAS No.†	Transitional Limits PEL* ppm^a	mg/m³^b	Skin Designation	Final Rule Limits** TWA ppm^a	mg/m³^b	STEL^c ppm^a	mg/m³^b	Ceiling ppm^a	mg/m³^b	Skin Designation
Chloromethyl methyl ether; see OSHA std 29 CFR 1910.1006	107–30–2										
1-Chloro-1-nitropropane	600–25–9	20	100	—	2	10	—	—	—	—	—
Chloropentafluoroethane	76–15–3	—	—	—	1000	6320	—	—	—	—	—
Chloropicrin	76–06–2	0.1	0.7	—	0.1	0.7	—	—	—	—	X
beta-Chloroprene	126–99–8	25	90	X	10	35	—	—	—	—	—
o-Chlorostyrene	2039–87–4	—	—	—	50	285	75	428	—	—	—
o-Chlorotoluene	95–49–8	—	—	—	50	250	—	—	—	—	—
2-Chloro-6-trichloro-methyl pyridine	1929–82–4										
Total dust		—	15	—	—	15	—	—	—	—	—
Respirable fraction		—	5	—	—	5	—	—	—	—	—
Chlorpyrifos	2921–88–2	—	—	—	—	0.2	—	—	—	—	X
Chromic acid and chromates (as CrO₃)	Varies with compound	See Table A.2			—	—	—	—	0.1	—	—
Chromium (II) compounds (as Cr)	Varies with compound	—	0.5	—	—	0.5	—	—	—	—	—
Chromium (III) compounds (as Cr)	Varies with compound	—	0.5	—	—	0.5	—	—	—	—	—
Chromium metal (as Cr)	7440–47–3	—	1	—	—	1	—	—	—	—	—
Chrysene; see Coal tar pitch volatiles											
Clopidol	2971–90–6										
Total dust		—	15	—	—	15	—	—	—	—	—
Respirable fraction		—	5	—	—	5	—	—	—	—	—

Substance	CAS No.	ppm	mg/m³	Skin	ppm	mg/m³	Skin
Coal dust (less than 5% SiO₂), Respirable quartz fraction	—	—	See Table A.3	—	—	2	—
Coal dust (greater than or equal to 5% SiO₂), Respirable quartz fraction	—	—	See Table A.3	—	—	0.1	—
Coal tar pitch volatiles (benzene soluble fraction), anthracene, BaP, phenanthrene, acridine, chrysene, pyrene	65966–93–2	—	0.2	—	—	0.2	—
Cobalt metal, dust, and fume (as Co)	7440–48–4	—	0.1	—	—	0.5	—
Cobalt carbonyl (as Co)	10210–68–1	—	—	—	—	0.1	—
Cobalt hydrocarbonyl (as Co)	16842–03–8	—	—	—	—	0.1	—
Coke oven emissions; see OSHA std 29 CFR 1910.1029							
Copper	7440–50–8						
Fume (as Cu)		—	0.1	—	—	0.1	—
Dusts and mists (as Cu)		—	1	—	—	1	—
Cotton dust (raw),	—	—	1	—	—	1	—
Crag herbicide (Sesone)	136–78–7						
Total dust		—	15	—	—	10	—
Respirable fraction		—	5	—	—	5	—
Cresol, all isomers	1319–77–3;	5	22	X	5	22	X
Crotonaldehyde	123–73–9; 4170–30–3	2	6	—	2	6	—
Crufomate	299–86–5	—	—	—	—	5	—
Cumene	98–82–8	50	245	X	50	245	X
Cyanamide	420–04–2	—	—	—	—	2	—

This 8-hour TWA applies to respirable dust as measured by a vertical elutriator cotton dust sampler or equivalent instrument. The time-weighted average applies to the cotton waste processing operations of waste recycling (sorting, blending, cleaning, and willowing and garnetting). See also OSHA std 29 CFR 1910.1043 for cotton dust limits applicable to other sectors.

APPENDIX A.1   GENERAL TABLE OF PELs

Limits for Air Contaminants

Substance	CAS No.†	Transitional Limits PEL* ppmᵃ	mg/m³ᵇ	Skin Designation	Final Rule Limits** TWA ppmᵃ	TWA mg/m³ᵇ	STELᶜ ppmᵃ	STELᶜ mg/m³ᵇ	Ceiling ppmᵃ	Ceiling mg/m³ᵇ	Skin Designation
Cyanides (as CN)	Varies	—	5	—	—	5	—	—	—	—	—
Cyanogen	460–19–5	—	—	—	10	20	—	—	—	—	—
Cyanogen chloride	506–77–4	—	—	—	—	—	—	—	0.3	0.6	—
Cyclohexane	110–82–7	300	1050	—	300	1050	—	—	—	—	—
Cyclohexanol	108–93–0	50	200	—	50	200	—	—	—	—	X
Cyclohexanone	108–94–1	50	200	—	25	100	—	—	—	—	X
Cyclohexene	110–83–8	300	1015	—	300	1015	—	—	—	—	—
Cyclohexylamine	108–91–8	—	—	—	10	40	—	—	—	—	X
Cyclonite	121–82–4	—	—	—	—	1.5	—	—	—	—	X
Cyclopentadiene	542–92–7	75	200	—	75	200	—	—	—	—	—
Cyclopentane	287–92–3	—	—	—	600	1720	—	—	—	—	—
Cyhexatin	13121–70–5	—	—	—	—	5	—	—	—	—	—
2, 4-D (Dichlorylphenoxyacetic acid)	94–75–7	—	10	—	—	10	—	—	—	—	—
Decaborane	17702–41–9	0.05	0.3	X	0.05	0.3	0.15	0.9	—	—	X
Demeton (Systox®)	8065–48–3	—	0.1	X	—	0.1	—	—	—	—	X
Dichlorodiphenyltrichloroethane (DDT)	50–29–3	—	1	X	—	1	—	—	—	—	X
Dichlorvos (DDVP)	62–73–7	—	1	X	—	1	—	—	—	—	X
Diacetone alcohol (4-Hydroxy-4-methyl-2-pentanone)	123–42–2	50	240	—	50	240	—	—	—	—	—
1, 2-Diaminoethane; see Ethylonediamine											
Diazinon	333–41–5	—	0.1	—	—	0.1	—	—	—	—	X
Diazomethane	334–88–3	0.2	0.4	—	0.2	0.4	—	—	—	—	—
Diborane	19287–45–7	0.1	0.1	—	0.1	0.1	—	—	—	—	—
1, 2-Dibromo-3-chloropropane; see OSHA std 29 CFR 1910.1044	96–12–8										

Substance	CAS No.	ppm	mg/m³	Skin	ppm	mg/m³	ppm	mg/m³	ppm	mg/m³	Skin
2-N-Dibutylaminoethanol	102-81-8	—	—	—	2	14	—	—	—	—	—
Dibutyl phosphate	107-66-4	1	5	—	1	5	2	10	—	—	—
Dibutyl phthalate	84-74-2	—	5	—	—	5	—	—	—	—	—
Dichloroacetylene	7572-29-4	—	—	—	—	—	—	—	0.1	0.4	—
o-Dichlorobenzene	95-50-1	(C)50	(C)300	—	—	—	—	—	50	300	—
p-Dichlorobenzene	106-46-7	75	450	—	75	450	110	675	—	—	—
3,3'-Dichlorobenzidine; see OSHA std 29 CFR 1910.1007	91-94-1	—	—	X	—	—	—	—	—	—	—
Dichlorodifluoromethane	75-71-8	1000	4950	—	1000	4950	—	—	—	—	—
1,3-Dichloro-5,5-dimethyl hydantoin	118-52-5	—	0.2	—	—	0.2	—	0.4	—	—	—
1,1-Dichloroethane	75-34-3	100	400	—	100	400	—	—	—	—	—
1,2-Dichloroethylene	540-59-0	200	790	—	200	790	—	—	—	—	—
Dichloroethyl ether	111-44-4	(C)15	(C)90	X	5	30	10	60	—	—	X
Dichloromethane; see Methylene chloride											
Dichloromonofluoromethane	75-43-4	1000	4200	—	10	40	—	—	—	—	—
1,1-Dichloro-1-nitroethane	594-72-9	(C)10	(C)60	—	2	10	—	—	—	—	—
1,2-Dichloropropane; see Propylene dichloride											
1,3-Dichloropropene	542-75-6	—	—	—	1	5	—	—	—	—	X
2,2-Dichloropropionic acid	75-99-0	—	—	—	1	6	—	—	—	—	—
Dichlorotetrafluoroethane	76-14-2	1000	7000	—	1000	7000	—	—	—	—	X
Dicrotophos	141-66-2	—	—	—	—	0.25	—	—	—	—	—
Dicyclopentadiene	77-73-6	—	—	—	5	30	—	—	—	—	—
Dicyclopentadienyl iron	102-54-5										
Total dust		15	—	—	—	10	—	—	—	—	—
Respirable fraction		5	—	—	—	5	—	—	—	—	—
Dieldrin	60-57-1	—	0.25	X	—	0.25	—	—	—	—	X
Diethanolamine	111-42-2	—	—	—	3	15	—	—	—	—	—
Diethylamine	109-89-7	25	75	—	10	30	25	75	—	—	—
2-Diethylaminoethanol	100-37-8	10	50	X	10	50	—	—	—	—	X
Diethylene triamine	111-40-0	—	—	—	1	4	—	—	—	—	X
Diethyl ether; see Ethyl ether											

APPENDIX A.1 GENERAL TABLE OF PELs

Limits for Air Contaminants

Substance	CAS No.†	Transitional Limits PEL*			Final Rule Limits** TWA		STEL[c]		Ceiling		Skin Designation
		ppm[a]	mg/m³[b]	Skin Designation	ppm[a]	mg/m³[b]	ppm[a]	mg/m³[b]	ppm[a]	mg/m³[b]	
Diethyl ketone	96–22–0	—	—	—	200	705	—	—	—	—	—
Diethyl phthalate	84–66–2	—	—	—	—	5	—	—	—	—	—
Difluorodibromomethane	75–61–6	100	860	—	100	860	—	—	—	—	—
Diglycidyl ether (DGE)	2238–07–5	(C)0.5	(C)2.8	—	0.1	0.5	—	—	—	—	—
Dihydroxybenzene; see Hydroquinone											
Diisobutyl ketone	108–83–8	50	290	—	25	150	—	—	—	—	—
Diisopropylamine	108–18–9	5	20	X	5	20	—	—	—	—	X
4-Dimethylaminoazobenzene; see OSHA std 29 CFR 1910.1015	60–11–7										
Dimethoxymethane; see Methylal											
Dimethyl acetamide	127–19–5	10	35	X	10	35	—	—	—	—	X
Dimethylamine	124–40–3	10	18	—	10	18	—	—	—	—	—
Dimethylaminobenzene; see Xylidine											
Dimethylaniline (N-Dimethyl-aniline)	121–69–7	5	25	X	5	25	10	50	—	—	X
Dimethylbenzene; see Xylene											
Dimethyl-1, 2-dibromo-2, 2-dichloroethyl phosphate	300–76–5	—	3	—	—	3	—	—	—	—	X
Dimethylformamide	68–12–2	10	30	X	10	30	—	—	—	—	X
2, 6-Dimethyl-4-heptanone; see Diisobutyl ketone											
1, 1-Dimethylhydrazine	57–14–7	0.5	1	X	0.5	1	—	—	—	—	X
Dimethylphthalate	131–11–3	—	5	—	—	5	—	—	—	—	—
Dimethyl sulfate	77–78–1	1	5	X	0.1	0.5	—	—	—	—	X

Substance	CAS No.								
Dinitolmide (3, 5-Dinitro-o-toluamide)	148-01-6	—	—	—	—	5	—	—	—
Dinitrobenzene (all isomers)		—	1	X	—	1	—	—	X
(alpha-)	528-29-0								
(meta-)	99-65-0								
(para-)	100-25-4								
Dinitro-o-cresol	534-52-1	—	0.2	X	—	0.2	—	—	X
Dinitrotoluene	2532-14-6	—	1.5	X	—	1.5	—	—	X
Dioxane (Diethylene dioxide)	123-91-1	100	360	X	25	90	—	—	X
Dioxathion (Delnav)	78-34-2	—	—	—	—	0.2	—	—	X
Diphenyl (Biphenyl)	92-52-4	0.2	1	—	0.2	1	—	—	—
Diphenylamine	122-39-4	—	—	—	—	10	—	—	—
Diphenylmethane diisocyanate; see Methylene bisphenyl isocyanate									
Dipropylene glycol methyl ether	34590-94-8	100	600	X	100	600	150	900	X
Dipropyl ketone	123-19-3	—	—	—	50	235	—	—	—
Diquat	85-00-7	—	—	—	—	0.5	—	—	—
Di-sec octyl phthalate (Di-2-ethylhexylphthalate)	117-81-7	—	5	—	—	5	—	10	—
Disulfiram	97-77-8	—	—	—	—	2	—	—	—
Disulfoton	298-04-4	—	—	—	—	0.1	—	—	X
2, 6-Di-tert-butyl-p-cresol	128-37-0	—	—	—	—	10	—	—	—
Diuron	330-54-1	—	—	—	—	10	—	—	—
Divinyl benzene	1321-74-0	—	—	—	10	50	—	—	—
Emery	112-62-9								
Total dust		—	15	—	—	10	—	—	—
Respirable fraction		—	5	—	—	5	—	—	—
Endosulfan	115-29-7	—	—	—	—	0.1	—	—	—
Endrin	72-20-8	—	0.1	X	—	0.1	—	—	X
Epichlorohydrin	106-89-8	5	19	X	2	8	—	—	X
EPN	2104-64-5	—	0.5	X	—	0.5	—	—	X
1, 2-Epoxypropane; see Propylene oxide									
2, 3-Epoxy-1-propanol; see Glycidol									

APPENDIX A.1 GENERAL TABLE OF PELs

Limits for Air Contaminants

Substance	CAS No.†	Transitional Limits PEL* ppm^a	mg/m$^{3\,b}$	Skin Designation	Final Rule Limits** TWA ppm^a	mg/m$^{3\,b}$	STEL^c ppm^a	mg/m$^{3\,b}$	Ceiling ppm^a	mg/m$^{3\,b}$	Skin Designation
Ethanethiol; see Ethyl mercaptan											
Ethanolamine	141–43–5	3	6	—	3	8	6	15	—	—	—
Ethion	563–12–2	—	—	X	—	0.4	—	—	—	—	X
2-Ethoxyethanol	110–80–5	200	740	X	200	740	—	—	—	—	X
2-Ethoxyethyl acetate (Cellosolve acetate)	111–15–9	100	540	X	100	540	—	—	—	—	X
Ethyl acetate	141–78–6	400	1400	—	400	1400	—	—	—	—	—
Ethyl acrylate	140–88–5	25	100	X	5	20	25	100	—	—	X
Ethyl alcohol (Ethanol)	64–17–5	1000	1900	—	1000	1900	—	—	—	—	—
Ethylamine	75–04–7	10	18	—	10	18	—	—	—	—	—
Ethyl amyl ketone (5-Methyl-3-heptanone)	541–85–5	25	130	—	25	130	—	—	—	—	—
Ethyl benzene	100–41–4	100	435	—	100	435	125	545	—	—	—
Ethyl bromide	74–96–4	200	890	—	200	890	250	1110	—	—	—
Ethyl butyl ketone (3-Heptanone)	106–35–4	50	230	—	50	230	—	—	—	—	—
Ethyl chloride	75–00–3	1000	2600	—	1000	2600	—	—	—	—	—
Ethyl ether	60–29–7	400	1200	—	400	1200	500	1500	—	—	—
Ethyl formate	109–94–4	100	300	—	100	300	—	—	—	—	—
Ethyl mercaptan	75–08–1	(C)10	(C)25	—	0.5	1	—	—	—	—	—
Ethyl silicate	78–10–4	100	850	—	10	85	—	—	—	—	—
Ethylene chlorohydrin	107–07–3	5	16	X	—	—	—	—	1	3	X
Ethylenediamine	107–15–3	10	25	—	10	25	—	—	—	—	—
Ethylene dibromide	106–93–4	See Table A.2			See Table A.2						—
Ethylene dichloride	107–06–2	See Table A.2			1	4	2	8	—	—	—
Ethylene glycol	107–21–1	—	—		—	—	—	—	50	125	—
Ethylene glycol dinitrate	628–96–6	(C)0.2	(C)1	X	—	0.1	—	—	—	—	X

Substance	CAS No.	ppm	mg/m³	Skin	ppm	mg/m³	ppm	mg/m³	Skin
Ethylene glycol methyl acetate; see Methyl cellosolve acetate									
Ethyleneimine; see OSHA std 29 CFR 1910.1012	151-56-4								
Ethylene oxide; see OSHA std 29 CFR 1910.1047	75-21-8								
Ethylidene chloride; see 1, 1-Dichloroethane									
Ethylidene norbornene	16219-75-3	—	—	—	—	—	5	25	—
N-Ethylmorpholine	100-74-3	20	94	X	5	23	—	—	X
Fenamiphos	22224-92-6	—	—	—	—	0.1	—	—	X
Fensulfothion (Dasanit)	115-90-2	—	—	—	—	0.1	—	—	—
Fenthion	55-38-9	—	—	—	—	0.2	—	—	X
Ferbam — Total dust	14484-64-1	—	15	—	—	10	—	—	—
Ferrovanadium dust	12604-58-9	—	1	—	—	1	—	3	—
Fluorides (as F)	Varies with compound	—	2.5	—	—	2.5	—	—	—
Fluorine	7782-41-4	0.1	0.2	—	0.1	0.2	—	—	—
Fluorotrichloromethane (Trichlorofluoromethane)	75-69-4	1000	5600	—	—	—	1000	5600	—
Fonofos	944-22-9	—	—	—	—	0.1	—	—	X
Formaldehyde; see OSHA std 29 CFR 1910.1048	50-00-0	See Table A.2 for operations or sectors excluded from 1910.1048 or for which limit(s) is (are) stayed.							
Formamide	75-12-7	—	—	—	20	30	30	45	—
Formic acid	64-18-6	5	9	—	5	9	—	—	—
Furfural	98-01-1	5	20	X	2	8	—	—	X
Furfuryl alcohol	98-00-0	50	200	—	10	40	15	60	X
Gasoline	8006-61-9	—	—	—	300	900	500	1500	—
Germanium tetrahydride	7782-65-2	—	—	—	0.2	0.6	—	—	—
Glutaraldehyde	111-30-8	—	—	—	—	—	0.2	0.8	—
Glycerin (mist) — Total dust	56-81-5	—	15	—	—	10	—	—	—
Respirable fraction		—	5	—	—	5	—	—	—

APPENDIX A.1   GENERAL TABLE OF PELs

Limits for Air Contaminants

Substance	CAS No.†	Transitional Limits PEL* ppm^a	mg/m³^b	Skin Designation	Final Rule Limits** TWA ppm^a	TWA mg/m³^b	STEL^c ppm^a	STEL mg/m³^b	Ceiling ppm^a	Ceiling mg/m³^b	Skin Designation
Glycidol	556–52–5	50	150	—	25	75	—	—	—	—	—
Glycol monoethyl ether; see 2-Ethoxyethanol											
Grain dust (oat, wheat, barley)	—	—	—	—	—	10	—	—	—	—	—
Graphite, natural respirable dust	7782–42–5	See Table A.3		—	—	2.5	—	—	—	—	—
Graphite, synthetic	—										
Total dust			15	—	—	10	—	—	—	—	—
Respirable fraction			5	—	—	5	—	—	—	—	—
Guthion²; see Azinphos methyl											
Gypsum	13397–24–5										
Total dust			15	—	—	15	—	—	—	—	—
Respirable fraction			5	—	—	5	—	—	—	—	—
Hafnium	7440–58–6	—	0.5	—	—	0.5	—	—	—	—	—
Heptachlor	76–44–8	—	0.5	X	—	0.5	—	—	—	—	X
Heptane (n-Heptane)	142–82–5	500	2000	—	400	1600	500	2000	—	—	—
Hexachlorobutadiene	87–68–3	—	—	—	0.02	0.24	—	—	—	—	—
Hexachlorocyclopentadiene	77–47–4	—	—	—	0.01	0.1	—	—	—	—	—
Hexachloroethane	67–72–1	1	10	X	1	10	—	—	—	—	X
Hexachloronaphthalene	1335–87–1	—	0.2	X	—	0.2	—	—	—	—	X
Hexafluoroacetone	684–16–2	—	—	—	0.1	0.7	—	—	—	—	X
n-Hexane	110–54–3	500	1800	—	50	180	—	—	—	—	—
Hexane Isomers	Varies with compound	—	—	—	500	1800	1000	3600	—	—	—
2-Hexanone (Methyl n-butyl ketone)	591–78–6	100	410	—	5	20	—	—	—	—	—

Substance	CAS No.	OSHA PEL ppm	OSHA PEL mg/m³	Skin	ACGIH TLV ppm	ACGIH TLV mg/m³	STEL ppm	STEL mg/m³	Ceiling ppm	Ceiling mg/m³	Skin
Hexone (Methyl isobutyl ketone)	108-10-1	100	410	—	50	205	75	300	—	—	—
sec-Hexyl acetate	108-84-9	50	300	—	50	300	—	—	—	—	—
Hexylene glycol	107-41-5	—	—	—	—	—	—	—	25	125	—
Hydrazine	302-01-2	1	1.3	X	0.1	0.1	—	—	—	—	X
Hydrogenated terphenyls	61788-32-7	—	—	—	0.5	5	—	—	—	—	—
Hydrogen bromide	10035-10-6	3	10	—	—	—	3	10	—	—	—
Hydrogen chloride	7647-01-0	(C)5	(C)7	—	—	—	5	7	—	—	—
Hydrogen cyanide	74-90-8	10	11	X	—	—	—	—	4.7	5	X
Hydrogen fluoride (as F)	7664-39-3	See Table A.2	—	—	3	—	—	—	4.7	6	—
Hydrogen peroxide	7722-84-1	1	1.4	—	1	1.4	—	—	—	—	—
Hydrogen selenide (as Se)	7783-07-5	0.05	0.2	—	0.05	0.2	—	—	—	—	—
Hydrogen sulfide	7783-06-4	See Table A.2	—	—	10	14	15	21	—	—	—
Hydroquinone	123-31-9	—	2	—	—	2	—	—	—	—	—
2-Hydroxypropyl acrylate	999-61-1	0.5	—	X	0.5	3	—	—	—	—	X
Indene	95-13-6	10	—	—	10	45	—	—	—	—	—
Indium and compounds (as In)	7440-74-6	—	0.1	—	—	0.1	—	—	—	—	—
Iodine	7553-56-2	(C)0.1	(C)1	—	—	—	—	—	0.1	1	—
Iodoform	75-47-8	0.6	—	—	0.6	10	—	—	—	—	—
Iron oxide dust and fume (as Fe) Total particulate	1309-37-1	—	10	—	—	10	—	—	—	—	—
Iron pentacarbonyl (as Fe)	13463-40-6	0.1	—	—	0.1	0.8	0.2	1.6	—	—	—
Iron salts (soluble) (as Fe)	Varies with compound	—	—	—	—	1	—	—	—	—	—
Isoamyl acetate	123-92-2	100	525	—	100	525	—	—	—	—	—
Isoamyl alcohol (primary and secondary)	123-51-3	100	360	—	100	360	125	450	—	—	—
Isobutyl acetate	110-19-0	150	700	—	150	700	—	—	—	—	—
Isobutyl alcohol	78-83-1	100	300	—	50	150	—	—	—	—	—
Isooctyl alcohol	26952-21-6	—	—	—	50	270	—	—	—	—	X

APPENDIX A.1 GENERAL TABLE OF PELs

Limits for Air Contaminants

Substance	CAS No.†	Transitional Limits PEL* ppm^a	mg/m³ b	Skin Designation	Final Rule Limits** TWA ppm^a	mg/m³ b	STEL^c ppm^a	mg/m³ b	Ceiling ppm^a	mg/m³ b	Skin Designation
Isophorone	78–59–1	25	140	—	4	23	—	—	—	—	X
Isophorone diisocyanate	4098–71–9	—	—	—	0.005	—	0.02	—	—	—	—
2-Isopropoxyethanol	109–59–1	—	—	—	25	105	—	—	—	—	—
Isopropyl acetate	108–21–4	250	950	—	250	950	310	1185	—	—	—
Isopropyl alcohol	67–63–0	400	980	—	400	980	500	1225	—	—	—
Isopropylamine	75–31–0	5	12	—	5	12	10	24	—	—	—
N-Isopropylaniline	768–52–5	—	—	—	2	10	—	—	—	—	X
Isopropyl ether	108–20–3	500	2100	—	500	2100	—	—	—	—	—
Isopropyl glycidyl either (IGE)	4016–14–2	50	240	—	50	240	75	360	—	—	—
Kaolin											
Total dust		—	15	—	—	10	—	—	—	—	—
Respirable fraction		—	5	—	—	5	—	—	—	—	—
Ketene	463–51–4	0.5	0.9	—	0.5	0.9	1.5	3	—	—	—
Lead inorganic (as Pb); see OSHA std 29 CFR 1910.1025	7439–92–1										
Limestone	1317–65–3										
Total dust		—	15	—	—	15	—	—	—	—	—
Respirable fraction		—	5	—	—	5	—	—	—	—	—
Lindane	58–89–9	—	0.5	X	—	0.5	—	—	—	—	X
Lithium hydride	7580–67–8	—	0.025	—	—	0.025	—	—	—	—	—
LPG (Liquefied petroleum gas)	68476–85–7	1000	1800	—	1000	1800	—	—	—	—	—
Magnesite	546–93–0										
Total dust		—	15	—	—	15	—	—	—	—	—
Respirable fraction		—	5	—	—	5	—	—	—	—	—
Magnesium oxide fume	1309–48–4										
Total particulate		—	15	—	—	10	—	—	—	—	—

	CAS No.	ppm	mg/m³	Skin	ppm	mg/m³	ppm	mg/m³	mg/m³	Skin
Malathion	121-75-5									
Total dust		—	15	X	—	10	—	—	—	X
Maleic anhydride	108-31-6	0.25	1	—	0.25	1	—	—	—	—
Manganese compounds (as Mn)	7439-96-5	—	(C)5	—	—	—	—	—	5	—
Manganese fume (as Mn)	7439-96-5	—	(C)5	—	—	1	—	3	—	—
Manganese cyclopentadienyl tricarbonyl (as Mn)	12079-65-1	—	—	—	—	0.1	—	—	—	X
Manganese tetroxide (as Mn)	1317-35-7	—	—	—	—	1	—	—	—	—
Marble	1317-65-3									
Total dust		—	15	—	—	15	—	—	—	—
Respirable fraction		—	5	—	—	5	—	—	—	—
Mercury (aryl and inorganic) (as Hg)	7439-97-6	—	See Table A.2	—	—	—	—	—	0.1	X
Mercury (organo) Alkyl compounds (as Hg)	7439-97-6	—	See Table A.2	—	—	0.01	—	0.03	—	X
Mercury (vapor) (as Hg)	7439-97-6	—	See Table A.2	—	—	0.05	—	—	—	X
Mesityl oxide	141-79-7	25	100	—	15	60	25	100	—	—
Methacrylic acid	79-41-4	—	—	—	20	70	—	—	—	X
Methanethiol; see Methyl mercaptan		—	—	—	—	—	—	—	—	—
Methomyl (Lannate)	16752-77-5	—	—	—	—	2.5	—	—	—	—
Methoxychlor	72-43-5									
Total dust		—	15	—	—	10	—	—	—	—
2-Methoxyethanol; see Methyl cellosolve		—	—	—	—	—	—	—	—	—
4-Methoxyphenol	150-76-5	—	—	—	—	5	—	—	—	—
Methyl acetate	79-20-9	200	610	—	200	610	250	760	—	—
Methyl acetylene (Propyne)	74-99-7	1000	1650	—	1000	1650	—	—	—	—
Methyl acetylene propadiene mixture (MAPP)	—	1000	1800	—	1000	1800	1250	2250	—	—
Methyl acrylate	96-33-3	10	35	X	10	35	—	—	—	X

APPENDIX A.1 GENERAL TABLE OF PELs

Limits for Air Contaminants

Substance	CAS No.†	Transitional Limits PEL*			Final Rule Limits** TWA		STEL^c		Ceiling		Skin Designation
		ppm^a	mg/m^3^b	Skin Designation	ppm^a	mg/m^3^b	ppm^a	mg/m^3^b	ppm^a	mg/m^3^b	
Methylacrylonitrile	126-98-7	—	—	—	1	3	—	—	—	—	X
Methylal (Dimethoxymethane)	109-87-5	1000	3100	—	1000	3100	—	—	—	—	—
Methyl alcohol	67-56-1	200	260	—	200	260	250	325	—	—	X
Methylamine	74-89-5	10	12	—	10	12	—	—	—	—	—
Methyl amyl alcohol; see Methyl isobutyl carbinol											
Methyl n-amyl ketone	110-43-0	100	465	—	100	465	—	—	—	—	—
Methyl bromide	74-83-9	(C)20	(C)80	X	5	20	—	—	—	—	X
Methyl butyl ketone; see 2-Hexanone											
Methyl cellosolve (2-Methoxyethanol)	109-06-4	25	80	X	25	80	—	—	—	—	X
Methyl cellosolve acetate (2-Methoxyethyl acetate)	110-49-6	25	120	X	25	120	—	—	—	—	X
Methyl chloride	74-87-3	See Table A.2			50	105	100	210	—	—	—
Methyl chloroform (1, 1, 1-Trichloroethane)	71-55-6	350	1900	—	350	1900	450	2450	—	—	—
Methyl 2-cyanoacrylate	137-05-3	—	—	—	2	8	4	16	—	—	—
Methylcyclohexane	108-87-2	500	2000	—	400	1600	—	—	—	—	—
Methylcyclohexanol	25639-42-3	100	470	—	50	235	—	—	—	—	—
o-Methylcyclohexanone	583-60-8	100	460	X	50	230	75	345	—	—	X
Methylcyclopentadienyl manganese tricarbonyl (as Mn)	12108-13-3	—	—	—	—	0.2	—	—	—	—	X
Methyl demeton	8022-00-2	—	—	—	—	0.5	—	—	—	—	X

4, 4′-Methylene bis (2-chloroaniline) (MBOCA)	101–14–4	—	—	—	0.02	0.22	—	—	—	—	—
Methylene bis (4-cyclohexylisocyanate)	5124–30–1	—	—	—	—	—	—	—	0.01	0.11	
Methylene chloride	75–09–2	See Table A.2				See Table A.2					
Methyl ethyl ketone (MEK); see 2-Butanone											
Methyl ethyl ketone peroxide (MEKP)	1338–23–4	—	—	—	—	—	—	—	0.7	5	
Methyl formate	107–31–3	100	250	—	100	250	150	375	—	—	—
Methyl hydrazine (Monomethyl hydrazine)	60–34–4	(C)0.2	(C)0.35	X	—	—	—	—	0.2	0.35	X
Methyl iodide	74–88–4	5	28	X	2	10	—	—	—	—	X
Methyl isoamyl ketone	110–12–3	—	—	—	50	240	—	—	—	—	—
Methyl isobutyl carbinol	108–11–2	25	100	X	25	100	40	165	—	—	X
Methyl isobutyl ketone; see Hexone											
Methyl isocyanate	624–83–9	0.02	0.05	X	0.02	0.05	—	—	—	—	X
Methyl isopropyl ketone	563–80–4	—	—	—	200	705	—	—	—	—	—
Methyl mercaptan	74–93–1	(C)10	(C)20	—	0.5	1	—	—	—	—	—
Methyl methacrylate	80–62–6	100	410		100	410	—	—	—	—	—
Methyl parathion	298–00–0	—	—	—	—	0.2	—	—	—	—	X
Methyl propyl ketone; see 2-Pentanone											
Methyl silicate	681–84–5	—	—	—	1	6	—	—	—	—	—
alpha-Methyl styrene	98–83–9	(C)100	(C)480	—	50	240	100	485	—	—	—
Methylene bisphenyl isocyanate (MDI)	101–68–8	(C)0.02	(C)0.2	—	—	—	—	—	0.02	0.2	—
Metribuzin	21087–64–9	—	—	—	—	5	—	—	—	—	—
Mica; see Silicates											
Molybdenum (as Mo)	7439–98–7										
Soluble compounds		—	5	—	—	5	—	—	—	—	—
Insoluble compounds											
Total dust		—	15	—	—	10	—	—	—	—	—

APPENDIX A.1 GENERAL TABLE OF PELs

Limits for Air Contaminants

Substance	CAS No.†	Transitional Limits PEL*			Final Rule Limits** TWA		STEL^c		Ceiling		
		ppm^a	mg/m³^b	Skin Designation	ppm^a	mg/m³^b	ppm^a	mg/m³^b	ppm^a	mg/m³^b	Skin Designation
Monocrotophos (Azodrin®)	6923-22-4	—	—	—	—	0.25	—	—	—	—	—
Monomethyl aniline	100-61-8	2	9	X	0.5	2	—	—	—	—	X
Morpholine	110-91-8	20	70	X	20	70	30	105	—	—	X
Naphtha (Coal tar)	8030-30-6	100	400	—	100	400	—	—	—	—	—
Naphthalene	91-20-3	10	50	—	10	50	15	75	—	—	—
alpha-Naphthylamine; see OSHA std 29 CFR 1910.1004	134-32-7										
beta-Naphthylamine; see OSHA std 29 CFR 1910.1009	91-59-8										
Nickel carbonyl (as Ni)	13463-39-3	0.001	0.007	—	0.001	0.007	—	—	—	—	—
Nickel, metal, and insoluble compounds (as Ni)	7440-02-0	—	1	—	—	1	—	—	—	—	—
Nickel, soluble compounds (as Ni)	7440-02-0	—	1	—	—	0.1	—	—	—	—	—
Nicotine	54-11-5	—	0.5	X	—	0.5	—	—	—	—	X
Nitric acid	7697-37-2	2	5	—	2	5	4	10	—	—	—
Nitric oxide	10102-43-9	25	30	—	25	30	—	—	—	—	—
p-Nitroaniline	100-01-6	1	6	X	1	3	—	—	—	—	X
Nitrobenzene	98-95-3	1	5	X	1	5	—	—	—	—	X
p-Nitrochlorobenzene	100-00-5	—	1	X	—	1	—	—	—	—	X
4-Nitrodiphenyl; see OSHA std 29 CFR 1910.1003	92-93-3										
Nitroethane	79-24-3	100	310	—	100	310	—	—	—	—	—
Nitrogen dioxide	10102-44-0	(C)5	(C)9	—	—	—	1	1.8	—	—	—
Nitrogen trifluoride	7783-54-2	10	29	—	10	29	—	—	—	—	—

Substance	CAS No.	OSHA PEL ppm	OSHA PEL mg/m³	Skin	ACGIH TLV ppm	ACGIH TLV mg/m³	STEL ppm	STEL mg/m³	Skin
Nitroglycerin	55–63–0	(C)0.2	(C)2	X	—	—	—	0.1	X
Nitromethane	75–52–5	100	250	—	100	250	—	—	—
1-Nitropropane	108–03–2	25	90	—	25	90	—	—	—
2-Nitropropane	79–46–9	25	90	—	10	35	—	—	—
N-Nitrosodimethylamine; see OSHA std 29 CFR 1910.1016	62–79–9			X					X
Nitrotoluene									
o-isomer	88–72–2;	5	30	X	2	11	—	—	X
m-isomer	99–08–1;								
p-isomer	99–99–0								
Nitrotrichloromethane; see Chloropicrin									
Nonane	111–84–2	—	—	—	200	1050	—	—	—
Octachloronaphthalene	2234–13–1		0.1	X	—	0.1	—	0.3	X
Octane	111–65–9	500	2350	—	300	1450	375	1800	—
Oil mist, mineral	8012–95–1		5	—	—	5	—	—	—
Osmium tetroxide (as Os)	20816–12–0	—	0.002	—	0.0002	0.002	0.0006	0.006	—
Oxalic acid	144–62–7		1	—	—	1	—	2	—
Oxygen difluoride	7783–41–7	0.05	0.1	—	(C)0.05	(C)0.1	—	—	—
Ozone	10028–15–6	0.1	0.2	—	0.1	0.2	0.3	0.6	—
Paraffin wax fume	8002–74–2	—	—	—	—	2	—	—	—
Paraquat, respirable dust	1910–42–5; 2074–50–2; 4685–14–7		0.5	—	—	0.1	—	—	—
Parathion	56–38–2		0.1	X	—	0.1	—	—	X
Particulates not otherwise regulated									
Total dust			15	—	—	15	—	—	—
Respirable fraction			5	—	—	5	—	—	—
Pentaborane	19624–22–7	0.005	0.01	—	0.005	0.01	0.015	0.03	—
Pentachloronaphthalene	1321–64–8		0.5	X	—	0.5	—	—	X
Pentachlorophenol	87–86–5		0.5	X	—	0.5	—	—	X
Pentaerythritol									
Total dust	115–77–5		15	—	—	10	—	—	—
Respirable fraction			5	—	—	5	—	—	—
Pentane	109–66–0	1000	2950	—	600	1800	750	2250	—
2-Pentanone (Methyl propyl ketone)	107–87–9	200	700	—	200	700	250	875	—

APPENDIX A.1 GENERAL TABLE OF PELs

Limits for Air Contaminants

| Substance | CAS No.† | Transitional Limits PEL* | | | Final Rule Limits** TWA | | STEL^c | | Ceiling | | Skin Designation |
		ppm^a	mg/m^3^b	Skin Designation	ppm^a	mg/m^3^b	ppm^a	mg/m^3^b	ppm^a	mg/m^3^b	
Perchloroethylene (Tetrachloroethylene)	127–18–4	See Table A.2		—	25	170	—	—	—	—	—
Perchloromethyl mercaptan	594–42–3	0.1	0.8	—	0.1	0.8	—	—	—	—	—
Perchloryl fluoride	7616–94–6	3	13.5	—	3	14	6	28	—	—	—
Perlite	—										
Total dust		—	15	—	—	15	—	—	—	—	—
Respirable fraction		—	5	—	—	5	—	—	—	—	—
Petroleum distillates (Naphtha)	8002–05–9	500	2000	—	400	1600	—	—	—	—	—
Phenol	108–95–2	5	19	X	5	19	—	—	—	—	X
Phenothiazine	92–84–2	—	—	—	—	5	—	—	—	—	X
p-Phenylene diamine	106–50–3	—	0.1	X	—	0.1	—	—	—	—	X
Phenyl ether, vapor	101–84–8	1	7	—	1	7	—	—	—	—	—
Phenyl ether-biphenyl mixture, vapor	—	1	7	—	1	7	—	—	—	—	—
Phenylethylene; see Styrene											
Phenyl glycidyl ether (PGE)	122–60–1	10	60	—	1	6	—	—	—	—	—
Phenylhydrazine	100–63–0	5	22	X	5	20	10	45	—	—	X
Phenyl mercaptan	108–98–5	—	—	—	0.5	2	—	—	—	—	—
Phenylphosphine	638–21–1	—	—	—	—	—	—	—	0.05	0.25	X
Phorate	298–02–2	—	—	—	—	0.05	—	0.2	—	—	X
Phosdrin (Mevinphos^R)	7786–34–7	—	0.1	X	0.01	0.1	0.03	0.3	—	—	X
Phosgene (Carbonyl chloride)	75–44–5	0.1	0.4	—	0.1	0.4	—	—	—	—	—
Phosphine	7803–51–2	0.3	0.4	—	0.3	0.4	1	1	—	—	—

Substance	CAS No.	ppm	mg/m³	Skin		ppm	mg/m³	ppm	mg/m³	Skin
Phosphoric acid	7664-38-2	—	1	—		—	1	—	3	—
Phosphorus (yellow)	7723-14-0	—	0.1	—		—	0.1	—	—	—
Phosphorus oxychloride	10025-87-3	—	—	—		0.1	0.6	—	—	—
Phosphorus pentachloride	10026-13-8	—	1	—		—	1	—	—	—
Phosphorus pentasulfide	1314-80-3	—	1	—		—	1	—	3	—
Phosphorus trichloride	7719-12-2	0.5	3	—		0.2	1.5	0.5	3	—
Phthalic anhydride	85-44-9	2	12	—		1	6	—	—	—
m-Phthalodinitrile	626-17-5	—	—	—		—	5	—	—	—
Picloram	1918-02-1									
Total dust		—	15	—		—	10	—	—	—
Respirable fraction		—	5	—		—	5	—	—	—
Picric acid	88-89-1	—	0.1	X		—	0.1	—	—	X
Piperazine dihydrochloride	142-64-3	—	—	—		—	5	—	—	—
Pindone (2-Pivalyl-1, 3-indandione)	83-26-1	—	0.1	—		—	0.1	—	—	—
Plaster of Paris	26499-54-0									
Total dust		—	15	—		—	15	—	—	—
Respirable fraction		—	5	—		—	5	—	—	—
Platinum (as Pt)	7440-06-4									
Metal		—	—	—		—	1	—	—	—
Soluble salts		—	0.002	—		—	0.002	—	—	—
Portland cement	65997-15-1									
Total dust		See Table A.3				—	10	—	—	—
Respirable fraction		See Table A.3				—	5	—	—	—
Potassium hydroxide	1310-58-3	—	—	—		—	—	—	2	—
Propane	74-98-6	1000	1800	—		1000	1800	—	—	—
Propargyl alcohol	107-19-7	—	—	—		1	2	—	—	X
beta-Propriolactone; see OSHA std 29 CFR 1910.1013	57-57-8	—	—	—		—	—	—	—	—
Propionic acid	79-09-4	—	—	—		10	30	—	—	—
Propoxur (Baygon)	114-26-1	—	—	—		—	0.5	—	—	—
n-Propyl acetate	109-60-4	200	840	—		200	840	250	1050	—
n-Propyl alcohol	71-23-8	200	500	—		200	500	250	625	—
n-Propyl nitrate	627-13-4	25	110	—		25	105	40	170	—
Propylene dichloride	78-87-5	75	350	—		75	350	110	510	—
Propylene glycol dinitrate	6423-43-4	—	—	—		0.05	0.3	—	—	—

APPENDIX A.1 GENERAL TABLE OF PELs

Limits for Air Contaminants

Substance	CAS No.†	Transitional Limits PEL* ppm^a	PEL mg/m^3^b	PEL Skin Designation	Final Rule Limits** TWA ppm^a	TWA mg/m^3^b	STEL^c ppm^a	STEL mg/m^3^b	Ceiling ppm^a	Ceiling mg/m^3^b	Skin Designation
Propylene glycol monomethyl ether	107-98-2	—	—	—	100	360	150	540	—	—	—
Propylene imine	75-55-8	2	5	X	2	5	—	—	—	—	X
Propylene oxide	75-56-9	100	240	—	20	50	—	—	—	—	—
Propyne; see Methyl acetylene											
Pyrethrum	8003-34-7	—	5	—	—	5	—	—	—	—	—
Pyridine	110-86-1	5	15	—	5	15	—	—	—	—	—
Quinone	106-51-4	0.1	0.4	—	0.1	0.4	—	—	—	—	—
Resorcinol	108-46-3	—	—	—	10	45	20	90	—	—	—
Rhodium (as Rh), metal fume and insoluble compounds	7440-16-6	—	0.1	—	—	0.1	—	—	—	—	—
Rhodium (as Rh), soluble compounds	7440-16-6	—	0.001	—	—	0.001	—	—	—	—	—
Ronnel	299-84-3	—	15	—	—	10	—	—	—	—	—
Rosin core solder pyrolysis products, as formaldehyde	—	—	—	—	—	0.1	—	—	—	—	—
Rotenone	83-79-4	—	5	—	—	5	—	—	—	—	—
Rouge	—	—	—	—	—	—	—	—	—	—	—
Total dust		—	15		—	10	—	—	—	—	—
Respirable fraction		—	5		—	5	—	—	—	—	—
Selenium compounds (as Se)	7782-49-2	—	0.2	—	—	0.2	—	—	—	—	—
Selenium hexafluoride (as Se)	7783-79-1	0.05	0.4	—	0.05	0.4	—	—	—	—	—
Silica, amorphous precipitated and gel	—	See Table A.3			—	6	—	—	—	—	—

Substance	CAS No.	Transitional Limits ppm	Transitional Limits mg/m³	Final Rule Limits ppm	Final Rule Limits mg/m³
Silica, amorphous, diatomaceous earth, containing less than 1% crystalline silica	61790-53-2	—	See Table A.3	—	6
Silica, crystalline cristobalite (as quartz), respirable dust	14464-46-1	—	See Table A.3	—	0.5
Silica, crystalline quartz (as quartz), respirable dust	14808-60-1	—	See Table A.3	—	0.1
Silica, crystalline tripoli (as quartz), respirable dust	1317-95-9	—	See Table A.3	—	0.1
Silica, crystalline tridymite (as quartz), respirable dust	15468-32-3	—	See Table A.3	—	0.05
Silica, fused respirable dust	60676-86-0	—	See Table A.3	—	0.1
Silicates (less than 1% crystalline silica)					
Mica (respirable dust)	12001-26-2	—	See Table A.3	—	3
Soapstone, total dust	—	—	See Table A.3	—	6
Soapstone, respirable dust	—	—	See Table A.3	—	3
Talc (containing asbestos): use asbestos limit	—		See Table A.3		See OSHA std 29 CFR 1910.1001
Talc (containing no asbestos), respirable dust	14807-96-6	—	See Table A.3	—	2
Tremolite	—		See Table A.3		See OSHA std 29 CFR 1910.1101
Silicon	7440-21-3				
Total dust		—	15	—	10
Respirable fraction		—	5	—	5
Silicon carbide	409-21-2				
Total dust		—	15	—	10
Respirable fraction		—	5	—	5
Silicon tetrahydride	7803-62-5	5	—	—	7
Silver, metal and soluble compounds (as Ag)	7440-22-4	—	0.01	—	0.01
Soapstone; see Silicates					

APPENDIX A.1 GENERAL TABLE OF PELs

Limits for Air Contaminants

Substance	CAS No.†	Transitional Limits PEL* ppm^a	mg/m^3^b	Skin Designation	Final Rule Limits** TWA ppm^a	mg/m^3^b	STEL^c ppm^a	mg/m^3^b	Ceiling ppm^a	mg/m^3^b	Skin Designation
Sodium azide	26628-22-8										
(as HN₂)		—	—	—	—	—	—	—	0.1	—	X
(as NaN₃)		—	—	—	—	—	—	—	—	0.3	X
Sodium bisulfite	7631-90-5	—	—	—	—	5	—	—	—	—	—
Sodium fluoroacetate	62-74-8	—	0.05	X	—	0.05	—	0.15	—	—	X
Sodium hydroxide	1310-73-2	—	2	—	—	—	—	—	—	2	—
Sodium metabisulfite	7681-57-4	—	—	—	—	5	—	—	—	—	—
Starch	9005-25-8										
Total dust		—	15	—	—	15	—	—	—	—	—
Respirable fraction		—	5	—	—	5	—	—	—	—	—
Stibine	7803-52-3	0.1	0.5	—	0.1	0.5	—	—	—	—	—
Stoddard solvent	8052-41-3	500	2900	—	100	525	—	—	—	—	—
Strychnine	57-24-9	—	0.15	—	—	0.15	—	—	—	—	—
Styrene	100-42-5	See Table A.2		—	50	215	100	425	—^g	—	—
Subtilisins (Proteolytic enzymes)	9014-0-1	—	—	—	—	—	—	0.00006 (60 min)	—	—	—
Sucrose	57-50-1										
Total dust		—	15	—	—	15	—	—	—	—	—
Respirable fraction		—	5	—	—	5	—	—	—	—	—
Sulfur dioxide	7446-09-5	5	13	—	2	5	5	10	—	—	—
Sulfur hexafluoride	2551-62-4	1000	6000	—	1000	6000	—	—	—	—	—
Sulfuric acid	7664-93-9	—	1	—	—	1	—	—	—	—	—
Sulfur monochloride	10025-67-9	1	6	—	—	—	—	—	1	6	—
Sulfur pentafluoride	5714-22-1	0.025	0.25	—	—	—	—	—	0.01	0.1	—
Sulfur tetrafluoride	7783-60-0	—	—	—	—	—	—	—	0.1	0.4	—
Sulfuryl fluoride	2699-79-8	5	20	—	5	20	10	40	—	—	—
Sulprofos	35400-43-2	—	—	—	—	1	—	—	—	—	—
Systox^R, see Demeton											
2, 4, 5-T	93-76-5	—	10	—	—	10	—	—	—	—	—

Substance	CAS No.	Transitional Limits			Final Rule Limits				
		ppm	mg/m³	Skin	ppm	mg/m³	STEL ppm	STEL mg/m³	Skin
Talc; see Silicates									
Tantalum, metal and oxide dust	7440-25-7	—	5	—	—	5	—	—	—
TEDP (Sulfotep)	3689-24-5	—	0.2	X	—	0.2	—	—	X
Tellurium and compounds (as Te)	13494-80-9	—	0.1	—	—	0.1	—	—	—
Tellurium hexafluoride (as Te)	7783-80-4	0.02	0.2	—	0.02	0.2	—	—	—
Temephos Total dust	3383-96-8	—	15	—	—	10	—	—	—
Respirable fraction		—	5	—	—	5	—	—	—
TEPP	107-49-3	—	0.05	X	—	0.05	—	—	X
Terphenyls	26140-60-3	(C)1	(C)9	—	—	—	0.5	5	—
1,1,1-Tetrachloro-2,2-difluoroethane	76-11-9	500	4170	—	500	4170	—	—	—
1,1,2-Tetrachloro-1,2-difluoroethane	76-12-0	500	4170	—	500	4170	—	—	—
1,1,2,2-Tetrachloroethane	79-34-5	5	35	X	1	7	—	—	X
Tetrachoroethylene; see Perchloroethylene									
Tetrachloromethane; see Carbon tetrachloride									
Tetrachloronaphthalene	1335-88-2	—	2	X	—	2	—	—	X
Tetraethyl lead (as Pb)	78-00-2	—	0.075	X	—	0.075	—	—	X
Tetrahydrofuran	109-99-9	200	590	—	200	590	250	735	—
Tetramethyl lead (as Pb)	75-74-1	—	0.075	X	—	0.075	—	—	X
Tetramethyl succinonitrile	3333-52-6	0.5	3	X	0.5	3	—	—	X
Tetranitromethane	509-14-8	1	8	—	1	8	—	—	—
Tetrasodium pyrophosphate	7722-88-5	—	—	—	—	5	—	—	—
Tetryl (2,4,6-Trinitrophenyl-methyl-nitramine)	479-45-8	—	1.5	X	—	1.5	—	—	X
Thallium, soluble compounds (as Tl)	7440-28-0	—	0.1	X	—	0.1	—	—	X

APPENDIX A.1 GENERAL TABLE OF PELs

Limits for Air Contaminants

Substance	CAS No.†	Transitional Limits PEL* ppm[a]	Transitional Limits PEL* mg/m³[b]	Transitional Limits PEL* Skin Designation	Final Rule Limits** TWA ppm[a]	Final Rule Limits** TWA mg/m³[b]	Final Rule Limits** STEL[c] ppm[a]	Final Rule Limits** STEL[c] mg/m³[b]	Final Rule Limits** Ceiling ppm[a]	Final Rule Limits** Ceiling mg/m³[b]	Final Rule Limits** Skin Designation
4,4'-Thiobis (6-tert, Butyl-m-cresol)	96-69-5										
Total dust		—	15	—	—	10	—	—	—	—	—
Respirable fraction		—	5	—	—	5	—	—	—	—	—
Thioglycolic acid	68-11-1	—	—	—	1	4	—	—	—	—	X
Thionyl chloride	7719-09-7	—	—	—	—	—	—	—	1	5	—
Thiram	137-26-8	—	5	—	—	5	—	—	—	—	—
Tin, inorganic compounds (except oxides) (as Sn)	7440-31-5	—	2	—	—	2	—	—	—	—	—
Tin, organic compounds (as Sn)	7440-31-5	—	0.1	—	—	0.1	—	—	—	—	X
Tin oxide (as Sn)	21651-19-4	—	—	—	—	2	—	—	—	—	—
Titanium dioxide	13463-67-7										
Total dust		—	15	—	—	10	—	—	—	—	—
Toluene	108-88-3	See Table A.2		—	100	375	150	560	—	—	—
Toluene-2, 4-diisocyanate (TDI)	584-84-9	(C)0.02	(C)0.14	—	0.005	0.04	0.02	0.15	—	—	—
m-Toluidine	108-44-1	—	—	—	2	9	—	—	—	—	X
o-Toluidine	95-53-4	5	22	X	5	22	—	—	—	—	X
p-Toluidine	106-49-0	—	—	—	2	9	—	—	—	—	X
Toxaphene; see Chlorinated camphene											
Tremolite; see Silicates											
Tributyl phosphate	126-73-8	—	5	—	0.2	2.5	—	—	—	—	—
Trichloroacetic acid	76-03-9	—	—	—	1	7	—	—	—	—	—
1, 2, 4-Trichlorobenzene	120-82-1	—	—	—	—	—	—	—	5	40	—
1, 1, 1-Trichloroethane; see Methyl chloroform											
1, 1, 2-Trichloroethane	79-00-5	10	45	X	10	45	—	—	—	—	X

Substance	CAS No.										
Trichloroethylene	79-01-6	—	—	—	50	270	200	1080	—	—	—
Trichloromethane; see Chloroform	See Table A.2										
Trichloronaphthalene	1321-65-9	—	5	X	—	5	—	—	—	—	X
1, 2, 3-Trichloropropane	96-18-4	50	300	—	10	60	—	—	—	—	—
1, 1, 2-Trichloro-1, 2, 2-trifluoroethane	76-13-1	1000	7600	—	1000	7600	1250	9500	—	—	—
Triethylamine	121-44-8	25	100	—	10	40	15	60	—	—	—
Trifluorobromomethane	75-63-8	1000	6100	—	1000	6100	—	—	—	—	—
Trimellitic anhydride	552-30-7	—	—	—	0.005	0.04	—	—	—	—	—
Trimethylamine	75-50-3	—	—	—	10	24	15	36	—	—	—
Trimethyl benzene	25551-13-7	—	—	—	25	125	—	—	—	—	—
Trimethyl phosphite	121-45-9	—	—	—	2	10	—	—	—	—	—
2, 4, 6-Trinitrophenyl; see Picric acid											
2, 4, 6-Trinitrophenylmethyl nitramine; see Tetryl											
2, 4, 6-Trinitrotoluene (TNT)	118-96-7	—	1.5	X	—	0.5	—	—	—	—	X
Triorthocresyl phosphate	78-30-8	—	0.1	—	—	0.1	—	—	—	—	X
Triphenyl amine	603-34-9	—	—	—	—	5	—	—	—	—	—
Triphenyl phosphate	115-86-6	—	3	—	—	3	—	—	—	—	—
Tungsten (as W)	7440-33-7										
Insoluble compounds		—	—	—	—	5	—	10	—	—	—
Soluble compounds		—	—	—	—	1	—	3	—	—	—
Turpentine	8006-64-2	100	560	—	100	560	—	—	—	—	—
Uranium (as U)	7440-61-1										
Soluble compounds		—	0.05	—	—	0.05	—	—	—	—	—
Insoluble compounds		—	0.25	—	—	0.2	—	0.6	—	—	—
n-Valeraldehyde	110-62-3	50	—	—	50	175	—	—	—	—	—
Vanadium	1314-62-1										
Respirable dust (as V_2O_5)		—	(C)0.5	—	—	0.05	—	—	—	—	—
Fume (as V_2O_5)		—	(C)0.1	—	—	0.05	—	—	—	—	—
Vegetable oil mist	—										
Total dust		—	15	—	—	15	—	—	—	—	—
Respirable fraction		—	5	—	—	5	—	—	—	—	—

APPENDIX A.1 GENERAL TABLE OF PELs

Limits for Air Contaminants

Substance	CAS No.†	Transitional Limits PEL* ppm^a	mg/m³^b	Skin Designation	Final Rule Limits** TWA ppm^a	mg/m³^b	STEL^c ppm^a	mg/m³^b	Ceiling ppm^a	mg/m³^b	Skin Designation
Vinyl acetate	108-05-4	—	—	—	10	30	20	60	—	—	—
Vinyl benzene; see Styrene											—
Vinyl bromide	593-60-2	—	—		5	20	—	—	—	—	—
Vinyl chloride see OSHA std 29 CFR 1910.1017	75-01-4										
Vinylcyanide; see Acrylonitrile											
Vinyl cyclohexene dioxide	106-87-6	—	—		10	60	—	—	—	—	X
Vinylidene chloride (1, 1-Dichloroethylene)	75-35-4	—	—		1	4	—	—	—	—	—
Vinyl toluene	25013-15-4	100	480		100	480	—	—	—	—	—
VM & P Naphtha	8032-32-4	—	—		300	1350	400	1800	—	—	—
Warfarin	81-81-2	—	0.1		—	0.1	—	—	—	—	—
Welding fumes (total particulate)***	—	—	—		—	5	—	—	—	—	—
Wood dust, all soft and hard woods, except Western red cedar		—	—		—	5	—	10	—	—	—
Wood dust, Western red cedar		—	—		—	2.5	—	—	—	—	—
Xylenes (o-, m-, p- isomers)	1330-20-7	100	435		100	435	150	655	—	—	—
m-Xylene alpha, alphadiamine	1477-55-0	—	—		—	—	—	—	—	0.1	X
Xylidine	1300-73-8	5	25	X	2	10	—	—	—	—	X
Yttrium	7440-65-5	—	1		—	1	—	—	—	—	—

Substance	CAS No.							
Zinc chloride fume	7646-85-7	—	1	—	1	—	2	—
Zinc chromate (as CrO₃)	Varies with Compound	See Table A.2	—	—	—	—	—	0.1
Zinc oxide fume	1314-13-2	—	5	—	5	—	10	—
Zinc oxide	1314-13-2							
Total dust		—	15	—	10	—	—	—
Respirable fraction		—	5	—	5	—	—	—
Zinc stearate	557-05-1							
Total dust		—	15	—	10	—	—	—
Respirable fraction		—	5	—	5	—	—	—
Zirconium compounds (as Zr)	7440-6	—	5	—	5	—	10	—

* The transitional PELs are 8-hour TWAs unless otherwise noted; a (C) designation denotes a ceiling limit.

** Unless otherwise noted, employers in general industry (i.e., those covered by 29 *CFR* 1910) may use any combination of controls to achieve these limits until December 31, 1992, as set forth in OSHA std 29 CFR 1910.1000(f).

*** As determined from breathing-zone air samples.

a Parts of vapor or gas per million parts of contaminated air by volume at 25°C and 760 torr.

b Approximate milligrams of substance per cubic meter of air.

c Duration is for 15 minutes, unless otherwise noted.

d The final benzene standard in 1910.1028 applies to all occupational exposures to benzene except some subsegments of industry where exposures are consistently under the action level (i.e., distribution and sale of fuels, sealed containers and pipelines, coke production, oil and gas drilling and production, natural gas processing, and the percentage exclusion for liquid mixtures); for the excepted subsegments, the benzene limits in Table A.2 apply.

e Exposures under 10,000 ppm to be cited de minimus.

f The CAS number is for information only. Enforcement is based on the substance name. For an entry covering more than one metal compound measured as the metal, the CAS number for the metal is given—not CAS numbers for the individual compounds.

g Compliance with the subtilisins PEL is assessed by sampling with a high volume sampler (600–800 liters per minute) for at least 60 minutes.

Source: Code of Federal Regulations 29 CFR 1910.1000.

APPENDIX A.2

PELs and STELs for Some Specialized Materials

Material	8-hour time-weighted average	Acceptable ceiling concentration	Acceptable maximum peak above the acceptable ceiling concentration for an 8-hour shift	
			Concentration	Maximum duration
Benzene (Z37.40–1969)[1]	10 ppm	25 ppm	50 ppm	10 minutes
Beryllium and Beryllium compounds (Z37.29–1970)	2 $\mu g/m^3$	5 $\mu g/m^3$	25 $\mu g/m^3$	30 minutes
Cadmium fume (Z37.5–1970)	0.1 mg/m^3	0.3 mg/m^3		
Cadmium dust (Z37.5–1970)	0.2 mg/m^3	0.6 mg/m^3		
Carbon disulfide (Z37.3–1968)	20 ppm	30 ppm	100 ppm	30 minutes
Carbon Tetrachloride (Z37.17–1967)	10 ppm	25 ppm	200 ppm	5 minutes in any 4 hours
Chromic acid and chromates (Z37.7–1971)		1 mg/10 m^3		
Ethylene dibromide (Z37.31–1970)	20 ppm	30 ppm	50 ppm	5 minutes
Ethylene dichloride (Z37.21–1969)	50 ppm	100 ppm	200 ppm	5 minutes in any 3 hours
Formaldehyde (Z37.16–1967)[2]	3 ppm	5 ppm	10 ppm	30 minutes
Hydrogen fluoride (Z37.28–1969)	3 ppm			
Hydrogen sulfide (Z37.2–1966)		20 ppm	50 ppm	10 minutes once only if no other measurable exposure occurs
Fluoride as dust (Z37.38–1969)	2.5 mg/m^3			
Mercury (Z37.8–1971)		1 mg/10 m^3		
Methyl chloride (Z37.18–1969)	100 ppm	200 ppm	300 ppm	5 minutes in any 3 hours
Methylene chloride (Z37.23–1969)	500 ppm	1,000 ppm	2,000 ppm	5 minutes in any 2 hours
Organo (alkyl) mercury (Z37.30–1969)	0.01 mg/m^3	0.04 mg/m^3		
Styrene (Z37.15–1969)	100 ppm	200 ppm	600 ppm	5 minutes in any 3 hours
Tetrachloroethylene (Z37.22–1967)	100 ppm	200 ppm	300 ppm	5 minutes in any 3 hours
Toluene (Z37.12–1967)	200 ppm	300 ppm	500 ppm	10 minutes
Trichloroethylene (Z37.19–1967)	100 ppm	200 ppm	300 ppm	5 minutes in any 2 hours

[1]This standard applies to the industry segments exempt from the 1 ppm 8-hour TWA and 5 ppm STEL of the benzene standard at OSHA Standard 29 CFR 1910.1028. This standard also applies to any industry for which OSHA Standard 29 CFR 1910.1028 is stayed or otherwise not in effect.

[2]This standard applies to any industry for which OSHA Standard 29 CFR 1910.1048 is stayed or otherwise not in effect.

APPENDIX A.3

PELs for Mineral Dusts

Substance	mppcf[e]	mg/m^3
Silica:		
Crystalline		
Quartz (Respirable)	$\dfrac{250^f}{\%SiO_2 + 5}$	$\dfrac{10 \text{ mg/m}^{3m}}{\%SiO_2 + 2}$
Quartz (Total dust)		$\dfrac{30 \text{ mg/m}^3}{\%SiO_2 + 2}$
Cristobalite: Use ½ the value calculated from the count or mass formulae for quartz		
Tridymite: Use ½ the value calculated from the formulae for quartz		
Amorphous, Including natural diatomaceous earth	20	$\dfrac{80 \text{ mg/m}^3}{\%SiO_2}$
Silicates (less than 1% crystalline silica):		
Mica	20	
Soapstone	20	
Talc (not containing asbestos)	20^n	
Talc (containing asbestos). Use asbestos limit		
Tremolite (see 29 CFR 1910.1101)		
Portland cement	50	
Graphite (natural)	15	
Coal dust (respirable fraction less than 5% SiO$_2$)		2.4 mg/m^3
For more than 5% SiO$_2$		$\dfrac{10 \text{ mg/m}^3}{\%SiO_2 + 2}$
Inert or nuisance dust:		
Respirable fraction	15	5 mg/m^3
Total dust	50	15 mg/m^3

Note. Conversion factors—mpcf × 35.3 = million particles per cubic meter = particles per cc

[e]Millions of particles per cubic foot of air, based on impinger samples counted by light-field techniques.

[f]The percentage of crystalline silica in the formula is the amount determined from airborne samples, except in those instances in which other methods have been shown to be applicable.

[m]Both concentration and percent quartz for the application of this limit are to be determined from the fraction passing a size-selector with the following characteristics:

Aerodynamic diameter (unit density sphere)	Percent passing selector
2	90
2.5	75
3.5	50
5.0	25
10	0

[n]Containing less than 1% quartz, use quartz limit.

The measurements under this note refer to the use of an AEC (now NRC) instrument. The respirable fraction of coal dust is determined with an MRD: the figure corresponding to that of 2.4 mg/m^3 in the table for coal dust is 4.5 mg/m^3.

Appendix B

Medical Treatment

- Treatment of INFECTION
- Application of ANTISEPTICS during second or subsequent visit of medical personnel
- Treatment of SECOND OR THIRD DEGREE BURN(S)
- Application of SUTURES (stitches)
- Application of BUTTERFLY ADHESIVE DRESSING(S) or STERI STRIP(S) in lieu of sutures
- Removal of FOREIGN BODIES EMBEDDED IN EYE
- Removal of FOREIGN BODIES from wound if procedure is COMPLICATED because of depth of embedment, size, or location
- Use of PRESCRIPTION MEDICATIONS (except a single dose administered on first visit for minor injury or discomfort)
- Use of hot or cold SOAKING THERAPY during second or subsequent visit to medical personnel
- Application of hot or cold COMPRESS(ES) during second or subsequent visit to medical personnel
- CUTTING AWAY DEAD SKIN (surgical debridement)
- Application of HEAT THERAPY during second or subsequent visit to medical personnel
- Use of WHIRLPOOL BATH THERAPY during second or subsequent visit to medical personnel
- POSITIVE X-RAY DIAGNOSIS (fractures, broken bones, etc.)
- ADMISSION TO A HOSPITAL or equivalent medical facility for treatment

Source: Ref. 74.

Appendix C

First-Aid Treatment

- Application of ANTISEPTICS during first visit to medical personnel
- Treatment of FIRST DEGREE BURN(S)
- Application of BANDAGE(S) during first visit to medical personnel
- Use of ELASTIC BANDAGE(S) during first visit to medical personnel
- Removal of FOREIGN BODIES NOT EMBEDDED IN EYE if only irrigation is required
- Removal of FOREIGN BODIES from wound if procedure is UNCOMPLICATED, and is, for example, by tweezers or other simple technique
- Use of NONPRESCRIPTION MEDICATION AND administration of single dose of PRESCRIPTION MEDICATION on first visit for minor injury or discomfort
- SOAKING THERAPY on initial visit to medical personnel or removal of bandages by SOAKING
- Application of hot or cold COMPRESS(ES) during first visit to medical personnel
- Application of OINTMENTS to abrasions to prevent drying or cracking
- Application of HEAT THERAPY during first visit to medical personnel
- Use of WHIRLPOOL BATH THERAPY during first visit to medical personnel
- NEGATIVE X-RAY DIAGNOSIS
- OBSERVATION of injury during visit to medical personnel

The following procedure, by itself, is not considered medical treatment:

- Administration of TETANUS SHOT(S) or BOOSTER(S). However, these shots are often given in conjunction with the more serious injuries; consequently, injuries requiring tetanus shots may be recordable for other reasons.

Source: Ref. 74.

Appendix D

Classification of Medical Treatment

7a. Occupational Skin Diseases or Disorders

- *Examples:* Contact dermatitis, eczema, or rash caused by primary irritants and sensitizers or poisonous plants; oil acne; chrome ulcers; chemical burns or inflammations; etc.

7b. Dust Diseases of the Lungs (Pneumoconioses)

- *Examples:* Silicosis, asbestosis and other asbestos-related diseases, coal worker's pneumoconiosis, byssinosis, siderosis, and pneumoconioses.

7c. Respiratory Conditions Due to Toxic Agents

- *Examples:* Pneumonitis, pharyngitis, rhinitis or acute congestion due to chemicals, dusts, gases, or fumes; farmer's lung; etc.

7d. Poisoning (Systemic Effect of Toxic Materials)

- *Examples:* Poisoning by lead, mercury, cadmium, arsenic, or other metals; poisoning by carbon monoxide, hydrogen sulfide, or other gases; poisoning by benzol, carbon tetrachloride, or other organic solvents; poisoning by insecticide sprays such as parathion or lead arsenate; poisoning by other chemicals such as formaldehyde, plastics, or resins; etc.

7e. Disorders Due to Physical Agents (Other than Toxic Materials)

- *Examples:* Heatstroke, sunstroke, heat exhaustion, and other effects of environmental heat; freezing, frostbite, and effects of exposure to low temperatures; caisson disease; effects of ionizing radiation (isotopes, X-rays, radium); effects of non-ionizing radiation (welding flash, ultra-violet rays, microwaves, sunburn); etc.

7f. Disorders Associated with Repeated Trauma

- *Examples:* Noise-induced hearing loss; synovitis, tenosynovitis, and bursitis; Raynaud's phenomena; and other conditions due to repeated motion, vibration, or pressure.

7g. All Other Occupational Illnesses

- *Examples:* Anthrax, brucellosis, infectious hepatitis, malignant and benign tumors, food poisoning, histoplasmosis, coccidioidomycosis, etc.

Appendix E

EPA List of Extremely Hazardous Substances [Requiring Reporting under Section 313, Title III, Superfund Amendment and Reauthorization Act of 1986 [SARA]]

Name	Chemical Abstract Service (CAS) No.	Reportable Quantity	Threshold Planning Quantities
Acetaldehyde	75-07-0	1000	
Acetamide	60-35-5		
Acetone	67-64-1	5000	
Acetonitrile	75-05-8	5000	
2-Acetylaminofluorene	53-96-3	1	
Acrolein	107-02-8	1	500
Acrylamide	79-06-1	5000	1000/10000*
Acrylic acid	79-10-7	5000	
Acrylonitrile	107-13-1	100	10000
Aldrin	309-00-2	1	500/10000*
Allyl chloride	107-05-1	1000	
Alpha-naphthylamine	134-32-7	1	
Aluminum oxide	1344-28-1		
2-Aminoanthraquinone	117-79-3		
4-Aminoazobenzene	60-09-3		
4-Aminobiphenyl	92-67-1		

Name	Chemical Abstract Service (CAS) No.	Reportable Quantity	Threshold Planning Quantities
1-Amino-4-methoxyanthraquinone	82-28-0		
Ammonium nitrate (solution)	6484-52-2		
Ammonium sulfate (solution)	7783-20-2		
Aniline	65-53-3	5000	1000
Anthracene	120-12-7	5000	
Antimony compounds			
Antimony	7440-36-0	5000	
Arsenic compounds			
Arsenic	7440-38-2	1	
Asbestos	1332-21-4	1	
Barium compounds			
Barium	7440-39-3		
Benzal chloride	98-87-3	5000	500
Benzamide	55-21-0		
Benzene	71-43-2	1000	
Benzidine	92-87-5	1	
Benzotrichloride	98-07-7	1	100
Benzoyl peroxide	94-36-0		
Benzoyl chloride	98-88-4	1000	
Benzyl chloride	100-44-7	100	500
Beryllium compounds			
Beryllium	7440-41-7	1	
Beta-naphthylamine	91-59-8	1	
Beta-propiolactone	57-57-8	1	500
Biphenyl	92-52-4		
Bis (2-chloro-1-methylethyl) ether	108-60-1	1000	
Bis (2-chloroethyl) ether	111-44-4	1	10000
Bis (2-ethylhexyl) adipate	103-23-1		
Bis (chloromethyl) ether	542-88-1	1	100
Bromoform	75-25-2	100	
Bromomethane	74-83-9	1000	1000
Butadiene	106-99-0	1	10000
Butyl acrylate	141-32-2		
Butyl benzyl phthalate	85-68-7	100	
1,2-Butylene oxide	106-88-7		
Butyraldehyde	123-72-8		
C.I. Acid blue 9, diammonium salt	2650-18-2		
C.I. Acid blue 9, disodium salt	3844-45-9		
C.I. Acid green 3	4680-78-8		
C.I. Basic green 4	569-64-2		
C.I. Basic red 1	989-38-8		
C.I. Direct black 38	1937-37-7		

Name	Chemical Abstract Service (CAS) No.	Reportable Quantity	Threshold Planning Quantities
C.I. Direct brown 95	16071-86-6		
C.I. Disperse yellow 3	2832-40-8		
C.I. Food red 15	81-88-9		
C.I. Food red 5	3761-53-3		
C.I. Solvent yellow 3	97-56-3		
C.I. Solvent yellow 34 (auramine)	492-80-8	1000	
C.I. Solvent yellow 14	842-07-9		
C.I. Solvent orange 7	3118-97-6		
C.I. Vat yellow 4	128-66-5		
Cadmium compounds			
Cadmium	7440-43-9	1	
Calcium cyanamide	156-62-7		
Captan	133-06-2	10	
Carbaryl	63-25-2	100	
Carbon disulfide	75-15-0	100	10000
Carbon tetrachloride	56-23-5	5000	
Carbonyl sulfide	463-58-1		
Catechol	120-80-9		
Chloramben	133-90-4		
Chlordane	57-74-9	1	1000
Chlorinated phenols			
Chlorinated fluorocarbon (313–freon 113 only)	76-13-1		
Chlorine	7782-50-5	10	100
Chlorine dioxide	10049-04-4		
Chloroacetic acid	79-11-8	1	100/10000*
2-Chloroacetophenone	532-27-4		
Chlorobenzene	108-90-7	100	
Chlorobenzilate	510-15-6	1	
Chloroethane	75-00-3	100	
Chloroform	67-66-3	5000	10000
Chloromethane	74-87-3	1	
Chloromethyl methyl ether	107-30-2	1	100
Chloroprene	126-99-8		
Chlorothalinol	1897-45-6		
Chromium compounds			
Chromium	7440-47-3	1	
Cobalt compounds			
Cobalt	7440-48-4	1	10000
Copper compounds			
Copper	7440-50-8	5000	
Creosol (mixed isomers)	1319-77-3	1000	
Cumene	98-82-8	5000	

Name	Chemical Abstract Service (CAS) No.	Reportable Quantity	Threshold Planning Quantities
Cumene hydroperoxide	80-15-9	10	
Cupferron	135-20-6		
Cyanide compounds (Cn—only)			
Cyclohexane	110-82-7	1000	
2,4-D	94-75-7	100	
Decabromodiphenyl oxide	1163-19-5		
Diallate	2303-16-4	1	
Di (2-ethylhexyl) phthalate	117-81-7	1	
2,4-Diaminoanisole	615-05-4		
2,4-Diaminoanisole sulfate	39156-41-7		
4,4-Diaminodiphenyl ether	101-80-4		
Diaminotoluene	25367-45-8	1	
2,4-Diaminotoluene	95-80-7	1	
Diazamethane	334-88-3		
Dibenzofuran	132-64-9		
1,2-Dibromo-3-chloropropane	96-12-8	1	
1,2-Dibromoethane	106-93-4	1000	
Dibutyl phthalate	84-74-2	10	
Dichloromethane (methylene chloride)	75-09-2	1000	
Dichlorobromomethane	75-27-4	5000	
Dichlorobenzene (mixed)	25321-22-6	100	
1,2-Dichlorobenzene	95-50-1	100	
1,3-Dichlorobenzene	541-73-1	100	
1,4-Dichlorobenzene	106-46-7	100	
3,3'-Dichlorobenzidine	91-94-1	1	
1,2-Dichloroethane	107-06-2	5000	
1,2-Dichloroethylene	540-59-0		
2,4-Dichlorophenol	120-83-1	100	
1,2-Dichloropropane	78-87-5	1000	
1,3-Dichloropropylene	542-75-6	100	
Dichlorvos	62-73-7	10	1000
Dicofol	115-32-2	10	
Diepoxybutane	1464-53-5	1	500
Diethanolamine	111-42-2		
1,4-Diethylene dioxide	123-91-1	1	
Diethyl phthalate	84-66-2	1000	
Diethyl sulfate	64-67-5		
3,3'-Dimethoxybenzidine	119-90-4	1	
4-Dimethylaminoazobenzene	60-11-7	1	
3,3'-Dimethylbenzidine	119-93-7	1	
Dimethylcarbamoyl chloride	79-44-7	1	
1,1-Dimethylhydrazine	57-14-7	1	1000
2,4-Dimethylphenol	105-67-9	100	

Name	Chemical Abstract Service (CAS) No.	Reportable Quantity	Threshold Planning Quantities
Dimethyl phthalate	131-11-3	5000	10000
Dimethyl sulfate	77-78-1	1	500
4,6-Dinitro-o-cresol	534-52-1	10	10/10000*
2,4-Dinitrophenol	51-28-5	10	
2-4 Dinitrotoluene	121-14-2	1000	
2,6-Dinitrotoluene	606-20-2	1000	
Dioctyl phthalate	117-84-0	5000	10000
1,4-Dioxane	123-31-9	1	500/10000*
1,2-Diphenylhydrazine	122-66-7	1	
Direct blue 6	2602-46-2		
Epichlorohydrin	106-89-8	1000	1000
2-Ethoxyethanol	110-80-5	1	
Ethyl acrylate	140-88-5	1000	
Ethyl chloroformate	541-41-3	1000	
Ethylbenzene	100-41-4	1000	
Ethylene, liquid	74-85-1		
Ethylene oxide	75-21-8	1	1000
Ethylenethiourea	96-45-7	1	
Ethylene glycol	107-21-1		
Ethyleneimine (aziridine)	151-56-4	1	500
Fluometuron	2164-17-2		
Formaldehyde	50-00-0	1000	500
Glycol ethers			
Heptachlor	76-44-8	1	
Hexachloroethane	67-72-1	1	
Hexachlorocyclopentadiene	77-47-4	1	100
Hexachloro-1,3-butadiene	87-68-3		
Hexachlorobenzene	118-74-1	1	
Hexachloronaphthalene	1335-87-1	1	10000
Hexamethylphosphoramide	680-31-9		
Hydrazine	302-01-2	1	1000
Hydrazine sulfate	10034-93-2		
Hydrogen chloride	7647-01-0	1	500
Hydrogen cyanide	74-90-8	10	100
Hydrogen fluoride	7664-39-3	100	100
Isobutyraldehyde	78-84-2		
Isopropyl alcohol	67-63-0		
(313—Manufacture only by strong acid process)			
4,4'-Isopropylidenediphenol	80-05-7		
Lead compounds			
Lead	7439-92-1	1	
Lindane	58-89-9	1	1000/10000*
M-Cresol	108-39-4	1000	
M-Xylene	108-38-3	1000	

Name	Chemical Abstract Service (CAS) No.	Reportable Quantity	Threshold Planning Quantities
Maleic anhydride	108-31-6	5000	
Maneb	12427-38-2		
Manganese compounds			
Manganese	7439-96-5		
Melamine	108-78-1		
Mercury compounds			
Mercury	7439-97-6	1	
Methanol	67-56-1	5000	
Methoxychlor	72-43-5	1	
2-Methoxyethanol	109-86-4		
Methyl acrylate	96-33-3		
Methyl ethyl ketone	78-93-3	5000	
Methyl iodide	74-88-4	1	
Methyl isobutyl ketone	108-10-1	5000	
Methyl isocyanate	624-83-9	1	500
Methyl methacrylate	80-62-6	1000	
Methyl tert-butyl ether	1634-04-4		
4,4'-Methylene bis- (2-chloroaniline)	101-14-4	1	
4,4' Methylene bis (N,N-dimethyl) benzeneamine	101-61-1		
Methylene bis (phenylisocyanate)	101-68-8		
Methylene bromide	74-95-3	1000	
4,4'-Methylene dianiline	101-77-9		
Methylhydrazine	60-34-4	10	500
Michler's ketone	90-94-8		
Molybdenum trioxide	1313-27-5		
Mustard gas	505-60-2	1	500
N'-Nitrosodimethylamine	62-75-9	1	1000
N,N-Dimethylaniline	121-69-7		
N-Butyl alcohol	71-36-3	5000	
N-Nitrosodiethylamine	55-18-5	1	
N-Nitrosodi-n-butylamine	924-16-3	1	
N-Nitrosodi-n-propylamine	621-64-7	1	
N-Nitrosodiphenylamine	86-30-6	100	
N-Nitroso-n-ethylurea	759-73-9	1	
N-Nitroso-n-methylurea	684-93-5	1	
N-Nitrosomethylvinylamine	4549-40-0	1	
N-Nitrosomorpholine	59-89-2		
N-Nitrosonornicotine	16543-55-8		
N-Nitrosopiperidine	100-75-4	1	
Naphthalene	91-20-3	100	
Nickel compounds			

Name	Chemical Abstract Service (CAS) No.	Reportable Quantity	Threshold Planning Quantities
Nickel	7440-02-0	1	10000
Nitric acid	7697-37-2	1000	1000
Nitrilotriacetic acid	139-13-9		
Nitrobenzene	98-95-3	1000	1000
4-Nitrobiphenyl	92-93-3		
Nitrofen	1836-75-5		
Nitrogen mustard	51-75-2	1	10
Nitroglycerin	55-63-0	10	
5-Nitro-o-anisidine	99-59-2		
2-Nitrophenol	88-75-5	100	
4-Nitrophenol	100-02-7	100	
2-Nitropropane	79-46-9	1	
O-Anisidine	90-04-0		
O-Anisidine hydrochloride	134-29-2		
O-Cresol	95-48-7	1000	1000/10000*
O-Toluidine	95-53-4	1	
O-Toluidine hydrochloride	636-21-5	1	
O-Xylene	95-47-6	1000	
Octachloronaphthalene	2234-13-1		
Osmium tetroxide	20816-12-0	1000	10000
P-Ansidine	104-94-9		
P-Cresidine	120-71-8		
P-Cresol	106-44-5	1000	
P-Nitrosodiphenylamine	156-10-5		
P-Phenylenediamine	106-50-3		
P-Xylene	106-42-3	1000	
Parathion	56-3802	1	100
Pentachlorophenol	87-86-5	10	10000
Peracetic acid	79-21-0	1	500
Perchloroethylene (see 1,1,2,2-tetrachloroethene)			
Phenol	108-95-2	1000	500/10000*
2-Phenylphenol	90-43-7		
Phosgene	75-44-5	10	10
Phosphoric acid	7664-38-2	5000	
Phosphorus, white or yellow	7723-14-0	1	100
Phthalic anhydride	85-44-9		
Picric acid	88-89-1		
Polybrominated biphenyls (PBBs)			
Polychlorinated biphenyls (PCBs)	1336-36-3	10	
Propane sultone	1120-71-4	1	
Propionaldehyde	123-38-6		
Propoxur	114-26-1		

Name	Chemical Abstract Service (CAS) No.	Reportable Quantity	Threshold Planning Quantities
Propyleneimine	75-55-8	1	10000
Propylene oxide	75-56-9	100	10000
Propylene (propene)	115-07-1		
Pyridine	110-86-1	1000	
Quinoline	91-22-5	5000	
Quinone	106-51-4	10	
Quintozene	82-68-8	1	
Saccharin and salts (313—manufacture only)	81-07-2	1	
Safrole	94-59-7	1	
Sec-butyl alcohol	78-92-2		
Selenium compounds			
Selenium	7782-49-2	100	
Silver compounds			
Silver	7440-22-4	1000	
Sodium hydroxide (solution)	1310-73-2	1000	
Sodium sulfate (solution)	7757-82-6		
Styrene oxide	96-09-3		
Styrene (monomer)	100-42-5	1000	
Sulfuric acid	7664-93-9	1000	1000
Terephthalic acid	100-21-0		
Tert-butyl alcohol	75-65-0		
1,1,2,2-Tetrachloroethane	79-34-5	1	
1,1,2,2-Tetrachloroethene (perchloroethylene)	127-18-4	1	
Tetrachlorvinphos	961-11-5		
Thallium compounds			
Thallium	7440-28-0	1000	
Thioacetamide	62-55-5	1	
4,4'-Thiodianiline	139-65-1		
Thiourea	62-56-6	1	
Thorium dioxide	1314-20-1		
Titanium tetrachloride	7550-45-0	1	100
Toluene 2,6-diisocyanate	91-08-7	100	100
Toluene	108-88-3	1000	
Toluene 2,4-diisocyanate	584-84-9	100	500
Toxaphene	8001-35-2	1	500/10000*
Triaziquone	68-76-8		
1,2,4-Trichlorobenzene	120-82-1	100	
1,1,1-Trichloroethane	71-55-6	1000	
1,1,2-Trichloroethane	79-00-5	1	
Trichlorofon	52-68-6	100	
Trichloroethylene	79-01-6	1000	
2,4,5-Trichlorophenol	95-95-4	10	

Name	Chemical Abstract Service (CAS) No.	Reportable Quantity	Threshold Planning Quantities
2,4,6-Trichlorophenol	88-06-2	10	
Trifluralin	1582-09-8		
1,2,4-Trimethyl benzene	95-63-6	1	10000
Tris (2,3-dibromopropyl) phosphate	126-72-7	1	
Urethane (ethyl carbamate)	51-79-6	1	
Vanadium (fume or dust)	7440-62-2		
Vinyl acetate monomer	108-05-4	5000	1000
Vinyl bromide	593-60-2		
Vinyl chloride	75-01-4	10	
Vinylidene chloride	75-35-4	5000	
Xylene (mixed isomers)	1330-20-7	1000	
2,6-Xylidine	87-62-7		
Zinc compounds			
Zinc (fume or dust)	7440-66-6	1000	
Zineb	12122-67-7		

*If the threshold planning quantity is shown as two numbers (00/00000), the first number pertains if
1. The solid is a powder with a particle size less than 100 microns (1 micron = a millionth part of a meter), or
2. It is handled in solution or molten form, or
3. It has a National Fire Protection Rating of 2, 3, or 4 for reactivity.

Otherwise, the second number pertains.

Appendix F

Standard Industrial Classification [SIC] Code

PRINCIPAL MANUFACTURING CATEGORIES

SIC CODE	DESCRIPTION
20	Food and Kindred Products
21	Tobacco Manufacturers
22	Textile Mill Products
23	Apparel and Other Finished Products made from Fabrics and Other Similar Materials
24	Lumber and Wood Products Except Furniture
25	Furniture and Fixtures
26	Paper and Allied Products
27	Printing, Publishing, and Allied Industries
28	Chemicals and Allied Products
29	Petroleum Refining and Related Industries
30	Rubber and Miscellaneous Plastics Products
31	Leather and Leather Products
32	Stone, Clay, Glass and Concrete Products
33	Primary Metal Industries
34	Fabricated Metal Products, except Machinery and Transportation Equipment
35	Machinery, except Electrical
36	Electrical and Electronic Machinery, Equipment and Supplies
37	Transportation Equipment
38	Measuring, Analyzing and Controlling Instruments; Photographic, Medical and Optical Goods; Watches and Clocks
39	Miscellaneous Manufacturing Industries

Appendix G

States Having Federally Approved[1] State Plans for Occupational Safety and Health Standards and Enforcement

Alaska

Arizona — New York

California — North Carolina

Connecticut — Oregon

Hawaii — South Carolina

Indiana — Tennessee

Iowa — Utah

Kentucky — Vermont

Maryland — Virginia

Michigan — Washington

Minnesota — Wyoming

Nevada — also Puerto Rico and the

New Mexico — Virgin Islands

[1]Approved by authority of Section 18.b of Public Law 91-596. The list is current as of June 1989; The most recent change was in 1984.

Glossary

ACGIH American Conference of Governmental Industrial Hygienists

AIHA American Industrial Hygiene Association

AL action level

amp ampere

ANSI American National Standards Institute

anti-two-block device a mechanism for preventing a crane hook block from being drawn up to the point at which it contacts the boom point

API American Petroleum Institute

ASME American Society of Mechanical Engineers

asphyxiants substances which prevent oxygen from reaching the body cells

ASSE American Society of Safety Engineers

ASTM American Society for Testing and Materials

BCSP Board of Certified Safety Professionals of America

BLEVE boiling liquid expanding vapor explosion

block a mechanical assembly containing one or more freely rotating pulleys

block and tackle an assembly consisting of (usually two) blocks reeved together to achieve mechanical advantage

boom a long pivoting structure or arm on a crane

C ceiling value; a maximum acceptable exposure concentration

carcinogens substances which are known to cause, or are suspected to cause, cancer

CAS Chemical Abstracts number, a reference list for chemical substances

CERCLA Comprehensive Environmental Response, Compensation, and Liability Act

CIH Certified Industrial Hygienist

combustible having a flashpoint higher than 100 °F

CPSC Consumer Product Safety Commission

CSP Certified Safety Professional

dB decibel(s)

dBA decibel, by the A-weighted scale

egregious violation glaring or flagrant safety violation which invokes high penalties from OSHA

EPA Environmental Protection Agency

ergonomics study of human capability in relation to the work environment

experience rating insurance rating based on a company's claim history

fail-safe principles principles of engineering design which consider the consequences of component failure within the system

fault tree analysis logic diagram used to analyze probabilities associated with various causes and their adverse effects

FCAW flux core arc welding

flammable having a flashpoint lower than 100 °F

FMEA failure modes and effects analysis

GFCI ground fault circuit interrupter

GMAW gas metal arc welding (also known as MIG)

GTAW gas tungsten arc welding (also known as TIG)

Hz Hertz (cycles per second)

IBP initial boiling point

ICHD industrial chemical hazards database

IDL immediately dangerous to life

IDLH immediately dangerous to life or health

incidence rate rate of job-related injuries and illnesses, including fatalities, lost-workdays-cases, number of lost workdays, specific hazard incidence, lost-workday-injury rate

irritants substances which inflame surfaces of parts of the body by their corrosive action

LBW laser beam welding

LEL lower explosive limit (for flammable vapors); the percent concentration in air below which the mixture is too lean to ignite

load block   the pulley assembly to which the load is attached

LPG   liquefied petroleum gas

LWDI   lost workday injuries rate

ma   milliampere

MAC   maximum acceptable ceiling

mechanical advantage   a favorable ratio of output force to the required input force
for a mechanism

MIG   metal inert gas (welding); also known as GMAW

MSHA   Mine Safety and Health Administration

MSDS   material safety data sheet

mutagens   substances which affect chromosomes and are thus a hazard to the species

NEC   National Electrical Code®

NFPA   National Fire Protection Association

NIOSH   National Institute for Occupational Safety and Health

NSC   National Safety Council

OSHA   Occupational Safety and Health Administration

parts of rope   mechanical advantage provided by block and tackle; number of lines
supporting the load block

PE   Registered Professional Engineer

PEL   permissible exposure limit

RCRA   Resource Conservation and Recovery Act

reeving   the threading of ropes through pulleys

ROPS   rollover protective structures

RSEW   resistance seam welding

RSW   resistance spot welding

SARA   Superfund Amendments and Reauthorization Act

SAW   submerged arc welding

SCBA   self-contained breathing apparatus

SIC   standard industrial classification

SLM   sound level meter

SMAW   shielded metal arc welding (also known as "stick electrode" welding)

STEL   short-term exposure limit

teratogens   substances which have harmful effects on fetuses

two-block   to draw a block and tackle assembly up so tightly that the two blocks
make physical contact

TAG   a popular test method for determining flashpoint

TIG tungsten inert gas (welding); also known as GTAW

TLV threshold limit value

TOSCA Toxic Substances Control Act

TW thermit welding

TWA time weighted average

UEL upper explosive limit (for flammable vapors); the percent concentration in air above which the mixture is too rich to ignite

Workers Compensation statutory compensation levels to be paid by the employer for various injuries that may be incurred by the worker

Index

A

A-weighted scale (decibel reading), 155
Aboveground tanks, 179
Accident cause analysis, 27
Accidents:
  general categories, 31
  uninsured costs, 31, 34
Accident statistic summaries, 8
Acetone, 318, 319
Acetylene:
  as depressant, 122
  for welding, 314, 318, 319, 320
  welding service piping, 324
Acid:
  asphyxiant, 122
  dip tank hazards, 187
  eyewash stations, 211
  irritants, 120
  presence of slings, 250
  toxic substances, 132
Action Level (AL):
  for noise, 155, 156, 157, 161
  for toxic substances, 134
Acute effects, 119
Acute exposure to toxic substances, 138
Acute hazards, 4
Adjustable barriers for machine guarding, 272
Administrative controls:
  hazard avoidance, 46
  for noise, 163
Aerial baskets, 114

Aerial lifts in construction, 378
AIDS, 37
Ainlay, John A., 178
Airborne contaminants size comparison (figure), 127
Air cleaning device (figure), 144
Air contaminants, 125
  hazardous operations (table) 136, 137
  industries (table) 136, 137
Air Force fault tree analysis, 53
Air-line respirator, 200
Air-purifying respiratory protection, 197
Air sampling, 136
Aisles, 104
  blockage hazards, 226
  marking, 106
  width, 106
Alarms:
  filter, 145
  fire, 215
Alcoholic beverages, 35–37, 86
Alpha particles, 167
Alternating currents (AC) circuits, 337
Aluminosis, 120
Aluminum (cutting), 305
American Board of Industrial Hygiene, 7
American Conference of Government Industrial
  Hygienists (ACGIH), 7, 128
American Industrial Hygiene Association
  (AIHA), 7
American National Standards Institute (ANSI),
  8, 15, 71, 195, 247

American Petroleum Institute (API), 179
American Society of Mechanical Engineers
    (ASME), 8
American Society of Safety Engineers (ASSE), 7
American Society of Testing and Materials
    (ASTM), 8, 176
Ammonia, 94, 121
Analysis:
   fault tree, 53, 54
   in hazard avoidance, 52
Analytical approach to hazard avoidance, 52
Anchor clamp for ropes and sheaves, 241
Anchoring machines, 266
Angiosarcoma, 61, 124
Angle of repose (trench), 382
Anhydrous ammonia storage, 189
Animal studies (toxicology), 60
Anti-kickback protection (table saws), 303, 305
Anti-two-block devices, 236
Anti-repeat mechanism for power presses, 281
Appeal of OSHA citations, 78
Arcing, electrical:
   hazards, 349
   source of ignition, 180
   welding, 325
Argon, 122, 332
Arkansas missile silo fire, 327
Arsenic in lunchrooms, 117
Articles, 86–87
Artificial intelligence, 96
Asbestos, 60, 126
Asbestosis, 61, 120
Asphyxiants, 122, 332
Asthma, 206
Atmosphere-supplying respiratory protection,
   197
Atmospheric pressure of make-up air, 145
Atmospheric sampling, 331
Atomic energy, 168
Atomizing hazards, 126
Audiologist, 167
Audiometric testing:
   in hearing conservation program, 167
   recordkeeping, 167
Automatic rail clamps for cranes, 235
Automatic sprinkler systems, 186, 224
Awareness barrier for machine guards, 274

B

Bag houses for air filters, 150
Banana oil, 206
Band saws, 305
Barlow Decision, 74, 82
Barrier creams for skin, 209
Barriers (sound), 163
Baseline:
   examinations, 120

hearing acuity, 167
Battery acid in refueling areas, 229
Beard:
   discrimination suit, 206
   fire brigade, 216
   HAZMAT teams, 92
   with respiratory protection, 201, 206
Bell Laboratories fault tree analysis, 53
Belts and pulleys:
   guarding, 306
   decision diagram (figure), 307
Benzene, 122
Blacklisting employees, 79
Black lung disease, 61
Black powder explosives, 188
Bleed-off valve, 189
BLEVE (boiling liquid expanding vapor explo-
   sion), 189
Blood:
   affected by toxic substances, 121
   hemoglobin, 123
Board of Certified Safety Professionals of
   America, 7
Boiler inspection fines, 78
Boilers, 115, 116
Boiling point, 176
Bonding containers, 181
Brain, 121
Brakes:
   monitor for power press, 293, 296
   crane, 239
Breaks for employees, 164
Breathing:
   air compressor, 206
   toxic substances, 125
Bridge plate, 111
Bromine, 121
Brown lung disease, 61
Building codes, 100
Building and facilities (general), 100
Burnability ranges (figure), 179
Butane:
   flammable liquid, 176
   hazardous material, 188
Byssinosis, 120

C

Cadmium:
   hazards in welding, 331
   as systemic poison, 121
Calibration (sound level meters), 160
Caliper for wire rope, 247
Cancer, 123
Carbon dioxide:
   as air contaminant, 125, 136
   as asphyxiant, 123
   fixed extinguishing system, 222

sprinkler system, 186
in welding, 332
Carbon disulfide, 121
Carbon monoxide:
    as air contaminant, 125, 136, 145
    as asphyxiant, 123
    from industrial trucks, 229
    in respiration, 207
    respiratory protection canisters, 205
Carbon tetrachloride:
    health hazards, 142
    extinguisher systems, 222
Carboxyhemoglobin, 123
Carcinogens, 123, 124
Cardox method of filling tank trucks, 183
Carpal tunnel syndrome, 83, 168
Catch platforms, 102
Categories (accident), 31
Caustics:
    eyewash stations, 211
    soda (cleaning welding equipment), 324
    as toxic substance, 132
Ceiling levels, 133
Cement industry, 147
Central nervous system:
    affected by electricity, 303
    affected by systemic poisons, 96
    affected by toxic substances, 95
Centrifugal collectors (figure), 147
Ceramics industry, 147
Chemical:
    asphyxiant, 97
    dry particulates, 150
    engineering, 138
    industry, 150
    processes, 141
    reactions, 126
    skin hazard source, 208
Chemical Abstracts Number (CAS), 97
Chlorinated hydrocarbons, 121, 142, 331
Chlorine, 121, 125
Chlorobromomethane, 222
Chlorothane (1, 1, 1-trichloroethylene), 324
Chrome holes, 121, 209, 331
Chromic acid, 121, 209, 331
Chromium:
    plating, 209
    trioxide (in welding), 331
Chromosomes (mutagens), 124
Chronic exposures, 119, 138
Chronic hazards, 3
Circuit tester, 355
Citations, 76
Class:
    explosives, 176, 188
    fires, 218
    flammable liquids, 175–177
    hazardous locations, 349

industrial trucks, 227–228
    magazines, 188
    wiring, 186
Classification scale, 62
Cleveland Open Cup Test, 176
CLOUT®, 97–98
Closed circuit breathing apparatus, 202
CO₂ (see Carbon dioxide)
Coal dust:
    as air contaminant, 126
    as irritant, 121
Coal miners' black lung disease, 61
Combustible and flammable liquids classification
    (figure), 177
Combustible residues in spray booth fires, 186
Committees (safety and health), 28
Community Right-to-Know Act, 92
Complaint (employee, inspection), 75
Comprehensive Environmental Response, Com-
    pensation and Liability Act (CERCLA),
    91, 95
Compressed air:
    for cleaning, 309
    in power presses, 279
    tools, 50
Computer, 96, 107, 168
Concrete work hazards, 386
Conduction (electrical), 342
Confidentiality, 87
Confined spaces, 123
Congress, endorsement of OSHA concept, 82
Construction:
    aerial lifts, 378
    blasting, 388
    concrete work, 385
    cranes and hoists, 373–378
    demolition, 387
    electrical, 368–370
    electric utilities, 389–390
    fall protection, 364
    fire protection, 366
    floors and stairways, 373
    heavy equipment, 378–381
    ladders and scaffolds, 370–372
    lighting, 362
    material and personnel hoists, 376
    materials handling and storage, 363
    personal protective equipment, 363
    steel erection hazards, 349–350
    tool hazards, 366–368
    trenching and excavation, 381–385
    vertical standards, 72
Consultation, 9, 81
Consumer Product Safety Commission (CPSC),
    6, 69, 86, 123, 306
Contaminants, 125
    airborne (figure), 127
    in welding, 329
Continuity tester, 357

Controls for power presses, 290–291
Conveyors:
  belt type, 252
  hook orientation, 252
  in-running nip points, 253
  overhead type, 252
  remote control, 226
  screw type, 253
Cosmetics, 86
Cost/benefit analysis, 61
Cost of compliance, 63
Costs:
  hidden costs of accidents, 29–31
  uninsured costs of accidents, 31, 33
Cotton:
  as air contaminant, 126
  for hearing protection, 193
  for protective clothing, 329
Court's challenge of OSHA standards, 80
Cracking process, 188
Cranes:
  bridge cranes, 233
  cantilever gantry cranes, 233, 234
  construction, 373
  defective hooks (figure), 245
  electrical shock hazard, 237
  footwalks, 236
  gantry cranes, 233
  hoists, 40
  importance of brakes, 239
  inspections, 243
  monorails, 235
  overhead, 233, (figure), 234
  overtravel hazards, 235
  pendant control, 233
  pulpit control, 233
  trolley, 235
  types, 233
  underhung, 235
  wind hazards, 235
Cresol, 94
Cutoff saw, 301
Cutting oil, 141, 209
Cyclones, 147
Cylinder:
  LPG, 189
  valves, 189

D

Database, 96, 97
dB (*see* Decibels)
dBA (*see* Decibels)
"Deadman control" for saws, 305
Deaths:
  accidental work (figure), 16
  fire (table), 214
Decibels:
  measure of audible pressures, 152

scale for combining (figure), 154
Decision diagram:
  automatic extinguishing systems for dip tanks
    (figure), 187
  for classifying hazardous locations (figure),
    351
  for hazardous substance mixtures, 87,
    (diagram), 90
  for wiring (figure), 185
Defibrillation, 336
Degreasers, 121
Deluge-type shower, 211
De minimus violations, 63, 76
Demolition, 387
Demotion of employee, 79
Department of Health and Human Services, 74
Department of Health, Education and Welfare,
    74
Department of Transportation:
  approved vessels, 115
  explosive blasting, 388
Depressants, 122
Design principles for ventilation, 142
Detecting air contaminants, 136
Di-2-ethylhexyl phthalate (DEHP), 206
Die casting, 141
Die enclosures, 271
Diesel industrial trucks, 227, 230
Digestive system, 121
Dikes, 190
Dilution ventilation, 143
Dip tanks:
  automatic extinguishing facilities (figure), 187
  hazardous materials, 186
Discrimination, 75, 78
Disinfectant, 121
Dismissal of employee, 79
Distal cause, 57
Ditches, 103
Division:
  hazardous locations, 350, 351
  industrial trucks, 227
  wiring, 186
Dockboards, 111
Doors (locked), (*see also* Exits), 112
Dosimeter:
  air contamination, 138, 141, (figure), 140
  noise exposure, 160
Double insulation for hand tools, 345
Drugs, 35–37, 86
Dry chemical fire extinguishing systems, 186,
    222
Drying area, 186
Dust, 126
  mask, 197
  plastic, 147
  removed by recirculating system, 144
Dust and fume particle size comparison (figure),
    126

Dump trucks, 381
Dynamite, 188

E

Eardrum ruptures, 206, 216
Earmuffs, 194
Ear plugs, 194
Economics of safety and health, 28
Egregious violation, 77, 82
Egress, 101
Electricity:
  alternating current, 337, 338
  arcs and sparks, 349
  circuit tester, 355
  construction, 389–90
  continuity tester, 357
  disconnects, 358
  double insulation, 345
  exposed live parts, 358
  fire hazards, 348
  flexible cords, 358
  frequent violations, 357–58
  ground fault circuit interruptor (GFCI), 344
  grounding, 341–44, 357
  hazardous locations, 349
  industrial trucks, 227
  miswiring dangers, 346
  Ohm's Law, 337
  physiological effects, 336
  receptacle wiring tester, 355
  shock, 237
Electrocution, fault tree analysis, 54
Electromagnetic field, 284
Electrons, 167
Electroplating air contaminants, 126
Electrostatic precipitators (figure), 148
Electrostatic sprinkler system, 186
Elevators, 114, 376
Emergency:
  action plan, 215
  evacuation, 215
  eyewash stations, 211
  rescue equipment, 207
  showers, 211
  standards, 134
Emphysema, 120, 206, 331
Employee:
  complaint inspections, 75
  discrimination, 78
  dismissal, 79
  fitness, 216
End-of-service-life indicator, 205
Energy:
  losses, 145
  savings in ventilation system, 144
Enforcement, 43, 74, 81
Engineering, 46, 50, 142, 161, 372

Entry of toxic substances (figure), 125
Environmental controls, 101
Environmental Protection Agency (EPA), 91,
  92, 94, 95, 99, 138, 144, 183
Epidemiological studies, 60
Epilepsy:
  in fire brigade, 216
  with respiratory protection, 206
Ergonomics, 83
Ethanol, 122
Ethyl alcohol, 122
Evacuation, 215
Excavation and trench, 80, 381
Exhaust, 143
Exits, 112, 226
Expandable foam, 205
Experience rating, 15
Expert systems, 96
Explosion-proof equipment, 49, 352, 353
Explosives, 188
Exterior maintenance, 113
Extinguishing systems, 221
Eye protection:
  in construction, 363
  safety glasses, 195, 196
  for welding, 328
Eyewash stations, 211

F

Fabric filters for exhaust air (figure), 150
Factory Mutual Engineering Corporation:
  approval of industrial trucks, 229
  crane testing, 244
  electrical equipment approval, 352
  welding apparatus, 322
Fail-safe practices, 47
Failure Modes and Effects Analysis (FEMA),
  52
Failure-to-abate penalty, 77
Falls, 102, 105, 364
False sense of security, 50
Fan, 142, 143, 267
Fatal accidents inspections, 75
Fault tree analysis, 53
Federal agencies, 81
Federal regulations, impact, 69
Fetus, 124
Fibrillation, 336
Fibrosis, 120
Filter:
  alarms, 145
  exhaust air, 147 (figure), 150
  recirculating system, 144
  sampling devices, 141
Filtration of exhaust air, 144
Final rule limits, 130, 131
Fines collected by OSHA, 77, 78

Fire:
brigade, 216
classes and extinguishing media (table), 218
detection systems, 216
electrical, 349
industrial trucks, 227
ingredients, 178
LPG, 190
prevention, 215
protection, 213, 366
refueling areas, 229
welding, 327
Fire extinguishers:
fire classes, 217
inspection, testing, mounting, 218
training and education, 219
Firefighter training, 216
Firepoint, 176
First aid, 192, 210
First-line supervisors, 35
Fit testing respirators, 206
Fixed barriers for machine guards, 271, 272
Fixed extinguishing systems, 221
Fixed ladders, 109
Flammable and combustible liquids class, 176,
    177
Flashback (welding), 323, 324
Flashpoint, 176
Flesh burn, 189
Floors:
construction, 373
openings, 101
structure, 107
uneven, 106
water on, 105
Floors and aisles, 104
Floor-load-marking plates, 107
Flow process chart, air contaminants, 136–37
Fluorides in welding, 332
Fluorine:
as irritant, 120
in welding, 332
Flux (welding), 315
Flux-cored arc welding (FCAW), 315
Fly ash, 148
Flywheel, 294
Food, 86
storage, 125
Foot switch, 284
Footwalks, 236
Forging, 141
Forklift trucks, 190, 227, 229
Foundries, 150
Fourth Amendment, 74
Freon, 142
Full-face mask, 198, 200
Full-revolution presses, 279
Fume and dust particles (size comparison)
    (figure), 126

Fumes, 126, 143

G

Galvanized metals, 331
Gamma rays, 167
Gases as contaminants, 125
Gas mask, 198, 201
Gas metal arc welding (GMAW), 314, 315
Gasoline:
as air contaminant, 126
burnability, 178
electric trucks, 227
flammable liquid, 177
ignition temperature, 178
octane rating, 180
for powering trucks, 230
Gas tungsten arc welding (GTAW), 314, 315
Gates (guards) for power presses, 281
Gauges for wiring, 349
Gears lubrication, 163
General Duty Clause, 37, 70, 76, 85
General Fail-Safe Principle, 47, 286, 297
Good-faith effort, 76
Grain mills, 215
Grease hazards, 323
Grinding, air contaminants, 126
Grinding machines:
hazards, 298–299
Ground fault circuit interruptor (GFCI), 344,
    369
Grounding, 341–344
Guarding by location or distance, 262
Guard opening size, 270
Guardrails, 52, 101, 386
Guards for power presses, 268, 281

H

Hair, 125
Half mask, 198, 199
Halogen, 121
Hand feeding power presses, 269, 284
Hand-held circular saws, 305
Handrail, 108
Harassment of employee, 79
Hardhats, 207, 208, 363
Hazard:
acute, 4
chronic, 4
classification scale, 62 (figure), 65
communication, 86
levels, 63, (table), 64
priority, 63
Hazard avoidance, 42, 52
Hazardous materials, 175
Hazardous operations, 136
HAZMAT teams, 92
Headache ball, 374, 377

Head protection, 207
Health and environmental control, 119
Health professionals vs. safety professionals, 119
Hearing acuity, 167
Hearing protection:
  construction, 363
  personal protective equipment, 165
  types, 193
Heart, 216, 336
Heat exchangers, 145
Heating and air conditioning design principles, 142
Helicopters, 376
Helium, 122, 332
Helmet:
  for hearing protection, 194
  law, 44
Hemp and linen standpipe/hose systems, 220
Hertz, 151
Hidden costs of accidents, 29, 30
High-hazard industry inspection, 75
High-voltage power lines, 114
Hitchhikers on industrial trucks, 231
Hoist chains, 244
Hoists in construction, 376
Holdouts (restraint devices), 289
Hooks for conveyors, 253
Horizontal standards, 71
Hose mask, 201
Hospital, 210
Hot machinery and processes, 215
Housekeeping:
  fire prevention, 215
  hazards, 226
Hydraulic construction tools, 367
Hydrogen cyanide, 123
Hydrogen sulfide, 125, 137
Hz (see Hertz)

I

Ignition:
  sources, 180
  temperature, 178
Immediately Dangerous to Life (IDL), 196
Immediately Dangerous to Life and Health (IDLH), 196, 207
Imminent danger, 62, 64, 65, 75
Incidence rates, recordkeeping, 17
Indexes, traditional recordkeeping, 16
Industrial fires (dangerous industries), 215
Industrial hygienists, 4, 5, 119
Industrial trucks:
  design class summary (table), 228
  operations, 229
  types, 227
Inert gases:
  asphyxiants, 122
  welding, 314, 315

Ingestion, 125
Inhalation, 125
Initial Boiling Point (IBP), 176
Injunction, 75
In-running nip points:
  conveyors, 252, 253
  general machinery machanical hazards, 260
  welding, 327
Inspection:
  breathing apparatus, 207
  cranes, 243, 244
  employee complaint, 75
  extinguishing systems, 222
  fatalities, 75
  high-hazard industries, 75
  imminent danger, 75
  industrial trucks, 233
  pullbacks, 287
Instrumentation, 138
Interlocked barriers for machine guards, 272
Interlocks, 264
Inventory records of flammable liquids, 183
Ionizing radiation, 168
Iron oxide, 331
Irritants, 120
Isoamyl acetate, 206

J

Jackhammer, 366
Jack safety, 310
Jig guards, 275

K

Kansas City 1981 Hyatt Regency Hotel accident, 47
Kentucky supper club fire, 214
Kickback hazards, 304–305
Kidney impairment, 331

L

Labeling, 86
Laboratory:
  analysis of toxic substances, 138
  analyzing air samples, 140, 141
Ladder cages, 110
Ladders, 108–111
  construction, 370–371
Laser welding, 314
Latency (carcinogens), 124
Lay (wire rope), 246
Lead:
  air contaminant, 126
  paint, 205
  poison, 121
  welding, 332

Leaks:
   air contaminants, 137
   flammable liquids, 190
   noise in enclosures, 163
   ventilation system, 143
Leak testing (respirators), 206
Leather protective clothing, 328
Leukemia, 122
Levels of hazards, 63
Lifeline, 102
Lifting, 229, 254
Lighting, 113, 362
Light screen for power presses, 283
Liquefied Petroleum Gas (LPG):
   fire, 189
   hazardous materials, 189
   power for industrial trucks, 227, 230
   storage tanks, 189
Liver, 124
Loading docks, 102
Lobl, L., hand speed constant, 293
Local exhaust, 143
Logarithmic relation of decibels, 152
Log/summary, recordkeeping (figure), 22
Los Angeles, beard discrimination suit, 206
Loss control representative, 15
Loss incident causation model, 57
Lost-workday injuries rate (LWDI), 18
Lower exposure limits (LEL), 349
Lubrication of gears, 163
Lumens, 113
Lungs:
   disorders (preemployment), 120
   fibrosis, 60
   hazards from silica, 142
   irritants, 120, 121

M

Machine guarding:
   adjustable barriers, 272
   awareness barriers, 274
   blade guards, 264, 266
   die enclosures, 271
   fixed barriers, 271
   guard opening size, 270
   hand feeding tools, 269
   interlocked barriers, 272
   interlocks, 264
   jig guards, 275
   mechanical hazards, 260
   tagouts and lockouts, 262
   trip bars, 264
Machine-tool cutting oils, 126
Magnesium, 121
Maintenance, 104, 113
Make-up air, 143
Mandatory standards, 43, 44
Manganese, 121

Manholes, 202
Manlifts, 101, 115
Manometer, 144
MAPP gas, 314, 320
Material and personnel hoists, 376
Materials handling and storage:
   construction, 363
   housekeeping, 226
Material Safety Data Sheets (MSDS), 85, 87
Maximum acceptable ceiling (MAC or C), 133
Means of egress, 101
Measurement of toxic substances, 136
Medical:
   exams, 92, 93, 120, 167, 210, 211
   treatment, 17
Mercury:
   systemic poison, 121
   vapor, 204
   welding, 332
Metal cutting industry, 150
Metal fumes, 126, 331
Methane:
   asphyxiant, 122
   hazardous material, 188
   testing for presence, 138
Methanol, 204
Methyl alcohol (methanol), 136
Microwave radiation, 168
Mine Safety and Health Administration
      (MSHA), 69
Mining industry, 148, 215
Mists (air contaminants), 126, 127
Miswiring hazards, 346
Mixtures, 87, 90, 132
Molded ear caps, 194
Molecular weight, 134
Motorized hand trucks, 227
Motors (explosion-proof), 49
Mouthpiece respirator, 199
Mucous membrane, 125
Mutagens, 124

N

Nailers, 367
National consensus, 8, 71, 128
National Electrical Code:
   ground fault circuit interruptors, 369
   grounding, 341
   industrial trucks class, 227
   sources of ignition, 180
   wiring, 184, 350
National Emphasis Program (NEP), 75
National Fire Protection Association (NFPA), 8,
      71, 179
   accidental death statistics, 214
   automatic sprinkler systems, 186
   bonding requirements, 182

National Institute of Occupational Safety and
    Health, 9, 74, 91, 123, 134
  air contaminants measurement, 138
  respiratory protection, 206
  safe lifting with back, 255
National Safety Council, 7
  accident incident rates, 18
  accident incident rates comparison (figure), 20
  categories of hidden accident costs, 29, 30
  eye injury film, 196
  first system of recordkeeping, 15
  general categories of accidents, 31
  materials handling accidents, 225
Natural gas:
  as asphyxiant, 122
  hazardous material, 189
  welding, 314, 320
New York, Triangle Shirtwaist Company fire,
    215
Nitric oxide, 332
Nitrogen, 122, 125, 136, 318, 332
Nitroglycerin, 188
"No hands in the dies," 278
Noise:
  acute exposure, 151
  chronic exposure, 151
  control by plant layout, 163
  engineering controls, 161
  measurement, 159
  metal gears, 163
  nylon gears, 163
  OSHA standards, 155
  PELs (table), 158
  threshold limit values, 151
  time-weighted average, 151
  work practices control, 165
Nomex, 208, 328
Nonionizing radiation, 168
Nonserious category of hazards, 65
Nonserious violations, 62
Nose irritants, 121
Notification, 95
Nuisance tripping (of GFCIs), 344, 369
Nurse (first aid), 210
Nylon:
  gears, 163
  mesh fan guards, 266, 267

O

Octane rating, 180
Octave band analyzer, 161
Odorant for propane, 188
Ohio, fire extinguisher training, 219
Ohm's Law, 337
Oil:
  folliculitis, 209
  welding hazard, 323
Olfactory system, 136

Open circuit breathing apparatus, 202
Open pits, 103
Open-surface tanks, 208
Organic vapor respirators, 204
Overcurrent protection, 349
Overhead transmission lines, 237
Overheated bearings, 215
Overtravel of cranes, 236
Oxides of nitrogen, 121
Oxyacetylene torch welding, 314
Oxygen:
  cylinders, 320, 321
  deficiency, 123, 138
  welding, 314, 322

P

Paint:
  drying area, 186
  lead hazard, 142
  organic vapors, 204
  spray booth, 184
  systemic poison, 121
Palletized loads, 233
Parapet, 102
Particulates, 126
Part-revolution presses, 279
Penalty:
  for failure to abate, 78
  summary (table), 77
Pendant control, 237, 238
Pensky-Martens Closed Tester, 176
Perchloroethylene, 142
Performance standards, 72
Permissible Exposure Limits (PELs)
  noise, 155–59 (table), 157
  respiratory protection, 196
  toxic substances, 128
Personal hygiene, 210
Personal protection, 192
Personal protective equipment:
  construction, 363
  hearing, 165
  role in hazard avoidance, 46
Personnel manager, 5
Pesticide, 86, 126
Phenolic vapors, 250
Phosgene, 121, 124, 331
Photoelectric cells, 283
Physical exams, 120
Pigments in paints, 142
Pitch, 152
Plant layout for noise control, 163
Plastic dusts, 147
Platform lift truck selection, 227
Platforms, powered, 101
Plating:
  dip tanks hazards, 186–87
  irritants, 120

Plugging (cranes), 239
Pneumatic tools, 366
Pneumoconioses, 120
Point-of-operation safeguarding (power presses),
    280
Poison:
   human exposure, 128
   routes of entry, 125
   study of, 60
   systemic, 121
Polyvinyl chloride (PVC), 124
Posters, 45
Posting citation, 76
Potassium dichromate, 209
Powder-actuated tools, 367
Powered platforms, lifts, 101
Power hacksaws, 306
Power presses:
   brake monitor, 293, 296
   controls vs. trips, 292
   design, 279
   gates, 282
   hazards, 276
   holdouts, 289
   points-of-operation safeguarding, 280
   presence-sensing devices, 283
   pullbacks, 287
   safeguarding summary (figure), 297–298
   safety distances, 292
   sweeps, 289
   two-hand controls, 290
   types of guards, 281
Pregnancy, 124
Product Safety and Liability Act, 6
Professional certification, 6
Professional societies, 7
Proximal cause, 57
Punch presses (see Power presses)
Presence-sensing device, 283–286
Preventive maintenance:
   in hazard avoidance, 53
   to reduce noise, 163, 164
Priority (hazard), 63
Probability theory, 56, 57, 62
Promulgation of standards, 70
Propane:
   alternative to Freon, 142
   flammable liquid, 176
   hazardous material, 189, 190
   welding, 314, 320
Protective clothing:
   fire brigade, 217
   skin diseases, 209
   welding, 328
Protons, 167
Proximity warning devices, 114
Psychological approach to hazard avoidance, 45
Pullbacks (pullouts), 287
Pulmonary irritants in welding, 330

   irritants, 330
   problems, 206
Purchasing, 5
Purification devices, 146

Q

Quarter mask, 198

R

Radial saws, 301
Radiation, 167
Radioactive dusts, 126
Radio frequencies, 168
Radio-frequency sensor, 284
Railings, 101, 103, 108
Ranking workplace hazards, 62, 65
R:BASE®, 97
Reagan, Ronald, inspection priority, 75
Receptacle wiring tester, 355
Recirculating system, 144
Recordkeeping, 16, 17, 91, 92
   audiometric testing, 167
   changes, 15, 80
   fire extinguishers, 218
   Log and Summary of Occupational Injuries
      and Illnesses (figure), 19, 22
   Supplementary Record of Occupational In-
      juries and Illnesses, 21 (figure), 24
Redundancy (General Fail-Safe Principle), 48
Reeving (figure), 240
Refineries, 215
Refueling with LPG, 190
Registered Professional Engineer, 371
Repeat violations, 77
Resistance spot welding (RSW), 316, 326
Resource Conservation and Recovery Act
      (RCRA), 91
Respiratory protection:
   air line respirator, 200
   classifications, 197
   dust masks, 197
   fire brigade, 216
   fitness, 92
   full-face mask, 198
   gas mask, 198
   half mask, 198
   hose mask, 201
   mouthpiece, respirator, 199
   protection plan, 203
   quarter mask, 198
   self-contained breathing apparatus, 202
Restricted work activity, 18
Restrooms, 210
Reversed polarity, 306, 347
Right-to-know, 26
Ring test, 301
Risk in hazard avoidance, 62

Robots:
  guardrails, 52
  role in hazard avoidance, 51
Rollover protection structures (ROPS)
  construction, 378, 379
  industrial trucks, 233
Roofing industry, 205
Rooftops, 102
Ropes, 239
Routes of toxic substance entry, 125
Rubber industry:
  machine guarding, 264
  use of wet scrubbers, 148
Runovers, 378, 380

S

Safeguarding point of operation, 267
Safety and health:
  committees, 28
  economics, 28
Safety belts, 114, 364
Safety distances for power presses, 292
Safety factor:
  hazard avoidance, 47
  ropes and sheaves, 239
Safety glasses:
  industrial, 195
  management support, 45
  street, 195
Safety platform (industrial trucks), 231
Safety professionals vs. health professionals, 119
Safety shoes, 208
Sampling air contaminants, 138
Sanitation, 116, 125
SARA (*see* Superfund Amendments and Reau-
    thorization Act)
Saws:
  band saws, 305
  cutoff, 301
  hand-held circular saws, 305
  kickback hazards, 304
  power hacksaws, 306
  radial, 301
  table, 303
  types, 301–306
Scaffold:
  construction, 364, 371
  design safety factor, 47
Scale of hazard classification, 62
Scope of standards, 72
Scrubbers (air), 148
Seam welding (RSEW), 316
Secrets, 74, 87
Self-contained breathing apparatus, 202
Serious violations, 62, 65
Severity rate (accidents), 19
Shaft couplings, 308
Sheaves, 239

Shielded metal arc welding (SMAW), 314, 332
Shipping and receiving docks, 208
Short-term exposure limit (STEL) (toxic sub-
    stances), 133
Siderosis, 120, 331
Signs:
  exit, 112
  explosives, 388
  hazard avoidance, 45
  personal hygiene, 210
  smoking, 184
Silica, 120, 126, 142
Size of airborne contaminants (figure), 127
Skin hazards, 124, 125, 141, 209
Slag, 315, 323
Slings:
  comparisons of requirements, 251
  rated capacity, 248
  tensile forces (figure), 249
Slow response (noise-level metering), 159
Smell (air contaminants), 136
Smoke (welding), 329
Smoke alarms, 215
Smoking in refueling areas, 229
Sodium chromate, 209
Soldering alternatives, 141
Solvents:
  air contaminants, 126, 141
  health hazards, 142
  irritants, 120, 121
  skin hazard protection, 209, 210
Sorbent (sampling device), 141
Sound-absorbing barrier, 163
Sound Level Meter (SLM), 160
Sound waves, 151 (figure), 152
Special emphasis programs, 75, 382
Specification standards, 72
Splash loading, 182 (figure), 183
Spray finishing:
  hazardous materials, 184
  spray booths, 184
Sprinkler systems, 221
Staff functions, 13
Stainless steel, 331
Stairways, 108, 373
Standard Industrial Classification Code (SIC),
    19, 93
Standards:
  national consensus, 72
  institutes, 8
  performances, 72
  specification, 72
  structure of, 71
Standards completion project for toxic sub-
    stances, 129, 134, 135
Standpipe and hose systems, 220
Stannosis, 120
Staplers, 366
State(s), 95

States' role in OSHA, 80
Static electricity:
  during flammable liquid dispensing, 182
  source of ignition, 182
Statutory compensation, 14
Statutory limit for citation, 76
Steel:
  erection, 386, 387
  industry, 148
Stem fires (welding), 319
Stenching agent for natural gas, 136
Stick electrode welding, 314
Strand, wire rope, 246
Structural design of floors, 107
Structure (standards), 71
Submerged arc welding (SAW), 315
Summary of power press safeguarding, 297, 298
Superfund Amendments and Reauthorization
      Act (SARA), 91–93
Swedish wool, 194
Sweeps for power presses, 289
Switch loading, 182
Supreme Court, US, 74, 78
Synergistic effect, 132
Systemic poison, 121, 330

T

Table saws, 303
Tag Closed Tester method, 176
Tagouts and locks for machine guarding, 262
Tank roofs, 181
Target Industries Program (TIP), 75
Tempe, Arizona, trenching accident, 381
Ten percent (10%) point, 176
Tensile forces for slings, 249
Teratogens, 124
Termination of employee, 79
Tetraethyl lead, 121
Textile workers, brown lung disease, 61
Theory (probability), 56, 57
Thermit welding (TW), 316
Threshold limit value (TLV):
  noise, 151
  toxic substances, 128
Threshold quantities, 94
Threshold shift in hearing, 166
Time-weighted average:
  noise, 151, 160, 161
  toxic substances, 131
Tobacco, 86
Toeboards and handrails:
  general requirements, 101, 103
Tools, hazards in construction, 366
Top management, 2, 3, 12, 45
Top-stop monitor, 296
Toxicology, 60
Toxic substances, 9, 120, 125
Toxic Substances Control Act (TOSCA), 69

Trachea irritants, 121
Tractors (selection), 227
Trade:
  associations, 8
  secrets, 74, 87
Traditional indexes, 16
Training (staff function), 35
Transitional limits, 130, 131
Trenching and excavation, construction, 381
Trench shoring requirements (table), 383
Triangle Shirtwaist Company fire, 215
Trichloroethylene:
  health hazards, 142, 210
  welding, 331
Trip bars:
  machine guarding, 264
  power presses, 292
Trips and falls, 105
Trisodium phosphate, 324
Truck selection, 227
Tungsten inert gas (TIG) welding, 315
Two-blocking, 236, 374
Two-hand controls and trips, 290, 292
Type A and B gate guard, 282, 286
Type T platform, 114

U

U-bolts:
  ropes and sheaves, 242
  secure wire rope clips (figure), 242
Underwriters' Lab:
  crane testing, 243
  electric equipment approval, 352
  industrial truck approval, 229
  welding apparatus, 322
Uninsured costs of accidents, 31, 34
US Atomic Energy Commission, 205
US Constitution, Fourth Amendment, 74
US Supreme Court:
  appeal of OSHA citation, 78
  ruling on inspecting workplaces, 74
Upper exposure levels (UEL) for explosive
      vapors, 349
Utility lines, 374

V

Vacuum cleaner, 150
Vapors, 190
Vapor density, 179
Vapors, 125
Vapor-tight equipment, 352
Variances from standards, 78
Vats, 103
Vehicle-mounted work platforms, 101
Ventilation, 141
Vertical standards, 71, 72
Vessel entry, 202

Vinyl chloride:
  carcinogen, 124
  explosion hazard, 124
  lunchrooms, 117
  workers, 61
Violations:
  repeat, 77
  willful, 77
Volatility, 176
Voluntary compliance, 10

W

Washrooms, 125, 210
Wastes, 86
Wattage, 337
Welding:
  acetylene hazards, 318
  arc welding hazards, 324
  atmospheric contaminants, 329–332
  eye protection, 328
  fires and explosions, 327
  fumes, 126
  gases as asphyxiants, 122, 329–332
  ignition by sparks, 180
  oxygen cylinders, 320
  permits, 327
  process terminology, 313
  protective clothing, 328
  radiation, 121
  resistance welding hazards, 326
  respiratory protection, 207
  service piping, 324
  skin hazard source, 209
  torches and apparatus, 322
  toxic air contaminants, 141

Wet scrubber (figure), 149
Willful violations, 77
Williams-Steiger Occupational Safety and Health
    Act, 69
Willow Island, West Virginia, concrete wall col-
    lapse, 385
Wind hazards for cranes, 235
Window cleaning, 113
Wire rope:
  gauging, 247 (figure), 247
  wear, 244
Wiring:
  class, 186
  division, 186
  electric, 344
  spray areas, 184
Woodworking hazards, 147
Wood products, 86
Wool, 328
Workers' Compensation, 13, 29, 86
Work practice controls, 46
  hazard avoidance, 46
  noise, 163–165
Worst case principle, 47, 49

X

X-rays, 167

Y

Young workers, 46

Z

Zinc oxide, 331